T0324206

Undergraduate Texts in Mathematics

Editors

S. Axler
F. W. Gehring
K. A. Ribet

Springer
New York
Berlin
Heidelberg
Hong Kong
London
Milan
Paris
Tokyo

Undergraduate Texts in Mathematics

(continued after index)

Joseph H. Silverman John Tate

Rational Points on Elliptic Curves

With 34 Illustrations

Springer

Joseph H. Silverman
Department of Mathematics
Brown University
Providence, RI 02912
USA

John Tate
Department of Mathematics
University of Texas at Austin
Austin, TX 78712
USA

Mathematics Subject Classification (2000): 11G05, 11D25

Library of Congress Cataloging-in-Publication Data
Silverman, Joseph H., 1955–
     Rational points on elliptic curves / Joseph H. Silverman, John
Tate.
          p.   cm. — (Undergraduate texts in mathematics)
     Includes bibliographical references and index.
     ISBN 0-387-97825-9 — ISBN 3-540-97825-9
     1. Curves, Elliptic.   2. Diophantine analysis.   I. Tate, John
Torrence, 1925–   II. Title.   III. Series.
     QA567.2.E44S55   1992
     516.3´52—dc20   92-4669

Printed on acid-free paper.

Printed in the United States of America.   (SBA)

9 8 7 6   SPIN 11011361

Springer-Verlag is a part of *Springer Science+Business Media*

springeronline.com

Preface

In 1961 the second author delivered a series of lectures at Haverford College on the subject of "Rational Points on Cubic Curves." These lectures, intended for junior and senior mathematics majors, were recorded, transcribed, and printed in mimeograph form. Since that time they have been widely distributed as photocopies of ever decreasing legibility, and portions have appeared in various textbooks (Husemöller [1], Chahal [1]), but they have never appeared in their entirety. In view of the recent interest in the theory of elliptic curves for subjects ranging from cryptography (Lenstra [1], Koblitz [2]) to physics (Luck-Moussa-Waldschmidt [1]), as well as the tremendous purely mathematical activity in this area, it seems a propitious time to publish an expanded version of those original notes suitable for presentation to an advanced undergraduate audience.

We have attempted to maintain much of the informality of the original Haverford lectures. Our main goal in doing this has been to write a textbook in a technically difficult field which is "readable" by the average undergraduate mathematics major. We hope we have succeeded in this goal. The most obvious drawback to such an approach is that we have not been entirely rigorous in all of our proofs. In particular, much of the foundational material on elliptic curves presented in Chapter I is meant to explain and convince, rather than to rigorously prove. Of course, the necessary algebraic geometry can mostly be developed in one moderately long chapter, as we have done in Appendix A. But the emphasis of this book is on the number theoretic aspects of elliptic curves; and we feel that an informal approach to the underlying geometry is permissible, because it allows us more rapid access to the number theory. For those who wish to delve more deeply into the geometry, there are several good books on the theory of algebraic curves suitable for an undergraduate course, such as Reid [1], Walker [1] and Brieskorn-Knörrer [1]. In the later chapters we have generally provided all of the details for the proofs of the main theorems.

The original Haverford lectures make up Chapters I, II, III, and the first two sections of Chapter IV. In a few places we have added a small amount of explanatory material, references have been updated to include some discoveries made since 1961, and a large number of exercises have been added. But those who have seen the original mimeographed notes will recognize that the changes have been kept to a minimum. In particular, the emphasis is still on proving (special cases of) the fundamental theorems in the subject: (1) the Nagell-Lutz theorem, which gives a precise procedure for finding all of the rational points of finite order on an elliptic curve; (2) Mordell's theorem, which says that the group of rational points on an elliptic curve is finitely generated; (3) a special case of Hasse's theorem, due to Gauss, which describes the number of points on an elliptic curve defined over a finite field.

In the last section of Chapter IV we have described Lenstra's elliptic curve algorithm for factoring large integers. This is one of the recent applications of elliptic curves to the "real world," to wit the attempt to break certain widely used public key ciphers. We have restricted ourselves to describing the factorization algorithm itself, since there have been many popular descriptions of the corresponding ciphers. (See, for example, Koblitz [2].)

Chapters V and VI are new. Chapter V deals with integer points on elliptic curves. Section 2 of Chapter V is loosely based on an IAP undergraduate lecture given by the first author at MIT in 1983. The remaining sections of Chapter V contain a proof of a special case of Siegel's theorem, which asserts that an elliptic curve has only finitely many integral points. The proof, based on Thue's method of Diophantine approximation, is elementary, but intricate. However, in view of Vojta's [1] and Faltings' [1] recent spectacular applications of Diophantine approximation techniques, it seems appropriate to introduce this subject at an undergraduate level. Chapter VI gives an introduction to the theory of complex multiplication. Elliptic curves with complex multiplication arise in many different contexts in number theory and in other areas of mathematics. The goal of Chapter VI is to explain how points of finite order on elliptic curves with complex multiplication can be used to generate extension fields with abelian Galois groups, much as roots of unity generate abelian extensions of the rational numbers. For Chapter VI only, we have assumed that the reader is familiar with the rudiments of field theory and Galois theory.

Finally, we have included an appendix giving an introduction to projective geometry, with an especial emphasis on curves in the projective plane. The first three sections of Appendix A provide the background needed for reading the rest of the book. In Section 4 of Appendix A we give an elementary proof of Bezout's theorem, and in Section 5 we provide a rigorous discussion of the reduction modulo p map and explain why it induces a homomorphism on the rational points of an elliptic curve.

The contents of this book should form a leisurely semester course, with some time left over for additional topics in either algebraic geometry or number theory. The first author has also used this material as a supplementary special topic at the end of an undergraduate course in modern algebra, covering Chapters I, II, and IV (excluding IV §3) in about four weeks of classes. We note that the last five chapters are essentially independent of one another (except IV §3 depends on the Nagell-Lutz theorem, proven in Chapter II). This gives the instructor maximum freedom in choosing topics if time is short. It also allows students to read portions of the book on their own (e.g., as a suitable project for a reading course or an honors thesis.) We have included many exercises, ranging from easy calculations to published theorems. An exercise marked with a (*) is likely to be somewhat challenging. An exercise marked with (**) is either extremely difficult to solve with the material we cover or actually a currently unsolved problem.

It has been said that "it is possible to write endlessly on elliptic curves."[†] We heartily agree with this sentiment, but have attempted to resist succumbing to its blandishments. This is especially evident in our frequent decision to prove special cases of general theorems, even when only a few more pages would be required to prove a more general result. Our goal throughout has been to illuminate the coherence and the beauty of the arithmetic theory of elliptic curves; we happily leave the task of being encyclopedic to the authors of more advanced monographs.

Computer Packages

The first author has written two computer packages to perform basic computations on elliptic curves. The first is a stand-alone application which runs on any variety of Macintosh. The second is a collection of *Mathematica* routines with extensive documentation included in the form of Notebooks in Macintosh *Mathematica* format. Instructors are welcome to freely copy and distribute both of these programs. They may be obtained via anonymous ftp at

$$\text{gauss.math.brown.edu} \qquad (128.148.194.40)$$

in the directory dist/EllipticCurve.

Acknowledgments

The authors would like to thank Rob Gross, Emma Previato, Michael Rosen, Seth Padowitz, Chris Towse, Paul van Mulbregt, Eileen O'Sullivan, and the students of Math 153 (especially Jeff Achter and Jeff Humphrey) for reading and providing corrections to the original draft. They would also like to thank Davide Cervone for producing beautiful illustrations from their original jagged diagrams.

[†] From the introduction to *Elliptic Curves: Diophantine Analysis*, Serge Lang, Springer-Verlag, New York, 1978. Professor Lang follows his assertion with the statement that "This is not a threat," indicating that he, too, has avoided the temptation to write a book of indefinite length.

The first author owes a tremendous debt of gratitude to Susan for her patience and understanding, to Debby for her fluorescent attire brightening up the days, to Danny for his unfailing good humor, and to Jonathan for taking timely naps during critical stages in the preparation of this manuscript.

The second author would like to thank Louis Solomon for the invitation to deliver the Philips Lectures at Haverford College in the Spring of 1961.

Joseph H. Silverman
John Tate
March 27, 1992

Acknowledgments for the Second Printing

The authors would like to thank the following people for sending us suggestions and corrections, many of which have been incorporated into this second printing: G. Allison, D. Appleby, K. Bender, G. Bender, P. Berman, J. Blumenstein, D. Freeman, L. Goldberg, A. Guth, A. Granville, J. Kraft, M. Mossinghoff, R. Pries, K. Ribet, H. Rose, J.-P. Serre, M. Szydlo, J. Tobey, C.R. Videla, J. Wendel.

Joseph H. Silverman
John Tate
June 13, 1994

Contents

Introduction

The theory of Diophantine equations is that branch of number theory which deals with the solution of polynomial equations in either integers or rational numbers. The subject itself is named after one of the greatest of the ancient Greek algebraists, Diophantus of Alexandria,[1] who formulated and solved many such problems.

Most readers will undoubtedly be familiar with Fermat's Last Theorem.[2] This theorem says that if $n \geq 3$ is an integer, then the equation

$$X^n + Y^n = Z^n$$

has no solutions in non-zero integers X, Y, Z. Equivalently, the only solutions in rational numbers to the equation

$$x^n + y^n = 1$$

are those with either $x = 0$ or $y = 0$. Fermat's Theorem is now known to be true for all exponents $n \leq 125000$, so it is unlikely that anyone will find a counterexample by random guessing. On the other hand, there are still a lot of possible exponents left to check between 125000 and infinity!

As another example, we consider the problem of writing an integer as the difference of a square and a cube. In other words, we fix an integer $c \in \mathbb{Z}$ and look for solutions to the Diophantine equation[3]

$$y^2 - x^3 = c.$$

---

[1] Diophantus lived sometime before the 3rd century A.D. He wrote the *Arithmetica*, a treatise on algebra and number theory in 13 volumes, of which 6 volumes have survived.

[2] Fermat's Last "Theorem" is really a conjecture, because it is still unsolved after more than 350 years. Fermat stated his "Theorem" as a marginal note in his copy of Diophantus' *Arithmetica*; unfortunately, the margin was too small for him to write down his proof!

[3] This equation is often called Bachet's equation, after the 17th century mathematician who originally discovered the duplication formula. It is also sometimes called Mordell's equation, in honor of the 20th century mathematician L.J. Mordell, who made a fundamental contribution to the solution of this and many similar Diophantine equations. We will be proving a special case of Mordell's theorem in Chapter III.

Suppose we are interested in solutions in rational numbers $x, y \in \mathbb{Q}$. An amazing property of this equation is the existence of a *duplication formula*, discovered by Bachet in 1621. If (x, y) is a solution with x and y rational, then it is easy to check that

$$\left(\frac{x^4 - 8cx}{4y^2}, \frac{-x^6 - 20cx^3 + 8c^2}{8y^3} \right)$$

is a solution in rational numbers to the same equation. Further, it is possible to prove (although Bachet was not able to) that if the original solution has $xy \neq 0$ and if $c \neq 1, -432$, then repeating this process leads to infinitely many distinct solutions. So if an integer can be expressed as the difference of a square and a cube of non-zero rational numbers, then it can be so expressed in infinitely many ways. For example, if we start with the solution $(3, 5)$ to the equation

$$y^2 - x^3 = -2$$

and apply Bachet's duplication formula, we find a sequence of solutions that starts

$$(3, 5), \quad \left(\frac{129}{10^2}, \frac{-383}{10^3} \right), \quad \left(\frac{2340922881}{7660^2}, \frac{113259286337292}{7660^3} \right), \ldots$$

As you can see, the numbers rapidly get extremely large.

Next we'll take the same equation

$$y^2 - x^3 = c$$

and ask for solutions in integers $x, y \in \mathbb{Z}$. In the 1650's Fermat posed as a challenge to the English mathematical community the problem of showing that the equation $y^2 - x^3 = -2$ has only two solutions in integers, namely $(3, \pm 5)$. This is in marked contrast to the question of solutions in rational numbers, since we have just seen there are infinitely many of those. None of Fermat's contemporaries appears to have solved the problem, which was solved incorrectly by Euler in the 1730's, and given a correct proof 150 years later! Then in 1908, Axel Thue[4] made a tremendous breakthrough; he showed that for any non-zero integer c, the equation $y^2 - x^3 = c$ can have only a finite number of solutions in integers x, y. This is a tremendous (qualitative) generalization of Fermat's challenge problem; among the infinitely many solutions in rational numbers, there can be but finitely many integer solutions.

---

[4] Axel Thue made important contributions to the theory of Diophantine equations, especially to the problem of showing that certain equations have only finitely many solutions in integers. These theorems about integer solutions were generalized by C.L. Siegel during the 1920's and 1930's. We will prove a version of the Thue-Siegel theorem (actually a special case of Thue's original result) in Chapter V.

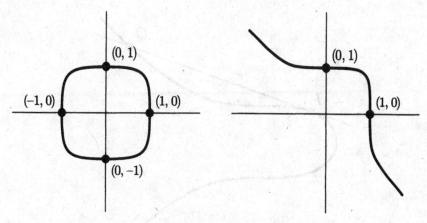

The Fermat Curves $x^4 + y^4 = 1$ and $x^5 + y^5 = 1$

Figure 0.1

The 17th century witnessed Descartes' introduction of coordinates into geometry, a revolutionary development which allowed geometric problems to be solved algebraically and algebraic problems to be studied geometrically. For example, if n is even, then the real solutions to Fermat's equation $x^n + y^n = 1$ in the xy plane form a geometric object that looks like a squashed circle. Fermat's Theorem is then equivalent to the assertion that the only points on that squashed circle having rational coordinates are the four points $(\pm 1, 0)$ and $(0, \pm 1)$. The Fermat equations with odd exponents look a bit different. We have illustrated the Fermat curves with exponents 4 and 5 in Figure 0.1.

Similarly, we can look at Bachet's equation $y^2 - x^3 = c$, which we have graphed in Figure 0.2. Recall that Bachet discovered a duplication formula which allows us to take a given rational solution and produce a new rational solution. Bachet's formula is rather complicated, and one might wonder where it comes from. The answer is, it comes from geometry! Thus, suppose we let $P = (x, y)$ be our original solution, so P is a point on the curve (as illustrated in Figure 0.2). Next we draw the tangent line to the curve at the point P, an easy exercise suitable for a first semester calculus course.[5] This tangent line will intersect the curve at one further point, which we have labeled Q. Then, if you work out the algebra to calculate the coordinates of Q, you will find Bachet's duplication formula. So Bachet's complicated algebraic formula has a simple geometric interpretation in terms of the intersection of a tangent line with a curve. This is our first intimation of the fruitful interplay that is possible among algebra, number theory, and geometry.

---

[5] Of course, Bachet had neither calculus nor analytic geometry; so he probably discovered his formula by clever algebraic manipulation.

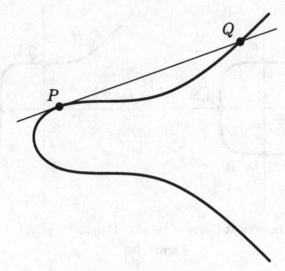

Bachet's Equation $y^2 - x^3 = c$

Figure 0.2

The simplest sort of Diophantine equation is a polynomial equation in one variable:

$$a_n x^n + a_{n-1} x^{n-1} + \cdots + a_1 x + a_0 = 0.$$

Assuming that $a_0, \ldots, a_n$ are integers, how can we find all integer and all rational solutions? Gauss' lemma provides the simple answer. If p/q is a rational solution written in lowest terms, then Gauss' lemma tells us that q divides a_n and p divides a_0. This gives us a small list of possible rational solutions, and we can substitute each of them into the equation to determine the actual solutions. So Diophantine equations in one variable are easy.

When we move to Diophantine equations in two variables, the situation changes dramatically. Suppose we take a polynomial $f(x, y)$ with integer coefficients and look at the equation

$$f(x, y) = 0.$$

For example, Fermat's and Bachet's equations are equations of this sort. Here are some natural questions we might ask:

    (a) Are there any solutions in integers?
    (b) Are there any solutions in rational numbers?
    (c) Are there infinitely many solutions in integers?
    (d) Are there infinitely many solutions in rational numbers?

In this generality, only question (c) has been fully answered, although much progress has been recently made on (d).[6]

---

[6] For polynomials $f(x_1, \ldots, x_n)$ with more than two variables, our four questions have only

The set of real solutions to an equation $f(x,y) = 0$ forms a curve in the xy plane. Such curves are often called *algebraic curves* to indicate that they are the solutions of a polynomial equation. In trying to answer questions (a)–(d), we might begin by looking at simple polynomials, such as polynomials of degree 1 (also called linear polynomials, because their graphs are straight lines.) For a linear equation

$$ax + by = c$$

with integer coefficients, it is easy to answer our questions. There are always infinitely many rational solutions, there are no integer solutions if $\gcd(a,b)$ does not divide c, and otherwise there are infinitely many integer solutions. So linear equations are even easier than equations in one variable.

Next we might turn to polynomials of degree 2 (also called quadratic polynomials). Their graphs are conic sections. It turns out that if such an equation has one rational solution, then there are infinitely many. The complete set of solutions can be described very easily using geometry. We will explain how this is done in the first section of Chapter I. We will also briefly indicate how to answer question (b) for quadratic polynomials. So although it would be untrue to say that quadratic polynomials are easy, it is fair to say that their solutions are completely understood.

This brings us to the main topic of this book, namely, the solution of degree 3 polynomial equations in rational numbers and in integers. One example of such an equation is Bachet's equation $y^2 - x^3 = c$ which we looked at earlier; some other examples which will appear during our studies are

$$y^2 = x^3 + ax^2 + bx + c \qquad \text{and} \qquad ax^3 + by^3 = c.$$

The real solutions to these equations are called *cubic curves* or *elliptic curves*. (However, they are not ellipses, since ellipses are conic sections, and conic sections are given by quadratic equations! The curious chain of events that led to elliptic curves being so named will be recounted in Chapter I, Section 3.) In contrast to linear and quadratic equations, the rational and integer solutions to cubic equations are still not completely understood; and even in those cases where the complete answers are known, the proofs involve a subtle blend of techniques from algebra, number theory, and geometry. Our main goal in this book is to introduce you to the beautiful subject of Diophantine equations by studying in depth the first case of such equations which are still imperfectly understood, namely cubic equations in two variables. To give you an idea of the sorts of results we will be studying, we briefly indicate what is known about questions (a)–(d).

---

been answered for some very special sorts of equations. Even worse, work of Davis, Matijasevič, and Robinson has shown that in general it is not possible to find a solution to question (a). That is, there does not exist an algorithm which takes as input the polynomial f and produces as output either "YES" or "NO" as an answer to question (a).

First, a cubic equation has only finitely many integer solutions[7] (Siegel, 1920's); and there is an explicit upper bound for the largest solution in terms of the coefficients of the polynomial (Baker-Coates, 1970). This provides a satisfactory answer to (a) and (c), although the actual bounds for the largest solution are generally too large to be practical. We will prove a special case of Siegel's theorem (for equations of the form $ax^3 + by^3 = c$) in Chapter V.

Second, all of the (possibly infinitely many) rational solutions to a cubic equation may be found by starting with a finite set of solutions and repeatedly applying a geometric procedure similar to Bachet's duplication formula. The fact that there exists such a finite generating set was suggested by Poincaré in 1901 and proven by L.J. Mordell in 1923. We will prove (a special case of) Mordell's theorem in Chapter III. However, we must in truth point out that Mordell's theorem does not really answer questions (b) and (d). As we will see, the proof of Mordell's theorem gives a procedure which *often* allows one to find a finite generating set for the set of rational solutions. But it is only conjectured, and not yet proven, that Mordell's method always yields a generating set. So even for special sorts of cubic equations, such as $y^2 - x^3 = c$ and $ax^3 + by^3 = c$, there is no general method (i.e., algorithm) currently known which is guaranteed to answer question (b) or (d).

We have mentioned several times the idea that the study of Diophantine equations involves an interplay among algebra, number theory, and geometry. The geometric component is clear, because the equation itself defines (in the case of two variables) a curve in the plane; and we have already seen how it may be useful to consider the intersection of that curve with various lines. The number theory is also clearly present, because we are searching for solutions in either integers or rational numbers, and what else is number theory other than a study of the relations between integers and/or rational numbers. But what of the algebra? We could point out that polynomials are essentially algebraic objects. However, algebra plays a much more important role than this.

Recall that Bachet's duplication formula can be described as follows: start with a point P on a cubic curve, draw the tangent line at P, and take the third point of intersection of the line with the curve. Similarly, if we start with two points P_1 and P_2 on the curve, we can draw the line through P_1 and P_2 and look at the third intersection point P_3. (This will work for most points, because the intersection of a line and a cubic curve will usually consist of exactly three points.) We might describe this procedure, which we illustrate in Figure 0.3, as a way to "add" two points on the curve and get a third point on the curve. Amazingly enough, we will show that with a slight modification this geometric operation takes the set of rational

---

[7] Actually, Siegel's theorem applies only to "non-singular" cubic equations. However, most cubic equations are non-singular; and in practice it is quite easy to check whether or not a given equation is non-singular.

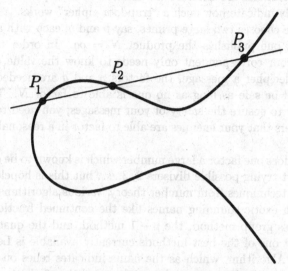

"Adding" Two Points on a Cubic Curve

Figure 0.3

solutions to a cubic equation and turns it into an abelian group! And
Mordell's theorem alluded to earlier can be rephrased by saying that this
group has a finite number of generators. So here is algebra, number theory,
and geometry all packaged together in one of the greatest theorems of this
century.

We hope that the foregoing introduction has convinced you of some of
the beauty and elegance to be found in the theory of Diophantine equations.
But the study of Diophantine equations, in particular the theory of elliptic
curves, also has its practical applications. We will study one such application in this book. Everyone is familiar with the Fundamental Theorem of
Arithmetic, which asserts that every positive integer factors uniquely into
a product of primes. It is less well known that if the integer is fairly large,
say on the order of 10^{100} or 10^{200}, it may be virtually impossible to perform
that factorization. This is true even though there are very quick ways to
check that an integer of this size is not itself a prime. In other words, if one
is presented with an integer N with (say) 150 digits, then one can easily
check that N is not prime, even though one cannot in general find any of
the prime factors of N.

This curious state of affairs has been used by Rivest, Shamir, and
Adleman to construct what is known as a public key cipher based on a
trapdoor function. These are ciphers in which one can publish, for all to
see, the method of enciphering a message; but even with the encipherment
method on hand, a would-be spy will not be able to decipher any messages.
Needless to say, such ciphers have numerous applications, ranging from
espionage to ensuring secure telecommunications between banks and other

financial institutions. To describe the relation with elliptic curves, we will need to briefly indicate how such a "trapdoor cipher" works.

First one chooses two large primes, say p and q, each with around 100 digits. Next one publishes the product $N = pq$. In order to encipher a message, your correspondent only needs to know the value of N. But in order to decipher a message, the factors p and q are needed. So your messages will be safe as long as no one is able to factor N. This means that in order to ensure the safety of your messages, you need to know the largest integers that your enemies are able to factor in a reasonable amount of time.

So how does one factor a large number which is known to be composite? One can start trying possible divisors 2, 3,..., but this is hopelessly inefficient. Using techniques from number theory, various algorithms have been devised, with exotic sounding names like the continued fraction method, the ideal class group method, the $p - 1$ method, and the quadratic sieve method. But one of the best methods currently available is Lenstra's Elliptic Curve Algorithm, which as the name indicates relies on the theory of elliptic curves. So it is essential to understand the strength of Lenstra's algorithm if one is to ensure that one's public key cipher will not be broken. We will describe how Lenstra's algorithm works in Chapter IV.

CHAPTER I

Geometry and Arithmetic

1. Rational Points on Conics

Everyone knows what a rational number is, a quotient of two integers. We call a point in the (x, y) plane a *rational point* if both its coordinates are rational numbers. We call a line a *rational line* if the equation of the line can be written with rational numbers; that is, if its equation is

$$ax + by + c = 0$$

with a, b, c, rational. Now it is pretty obvious that if you have two rational points, the line through them is a rational line. And it is neither hard to guess nor hard to prove that if you have two rational lines, then the point where they intersect is a rational point. If you have two linear equations with rational numbers as coefficients and you solve them, you get rational numbers as answers.

The general subject of these notes is rational points on curves, especially on cubic curves. But as an introduction, we will start with conics. Let

$$ax^2 + bxy + cy^2 + dx + ey + f = 0$$

be a conic. We will say that the conic is *rational* if we can write its equation with rational numbers.

Now what about the intersection of a rational line with a rational conic? Will it be true that the points of intersection are rational. By writing down some examples, it is easy to see that the answer is, in general, no. If you use analytic geometry to find the coordinates of these points, you will come out with a quadratic equation for the x coordinate of the intersection. And if the conic is rational and the line is rational, the quadratic equation you come out with will have rational coefficients. So the two points of intersection will be rational if and only if the roots of that quadratic equation are rational. In general, they might be conjugate quadratic irrationalities.

However, if one of those points is rational, then so is the other. This is true because if a quadratic equation with rational coefficients has one

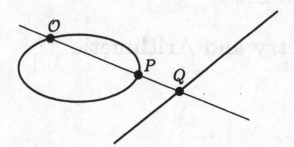

Projecting a Conic onto a Line

Figure 1.1

rational root, then the other root is rational, because the sum of the roots is the middle coefficient. This very simple idea enables one to describe the rational points on a conic completely. Given a rational conic, the first question is whether or not there are any rational points on it. (We will return to this question later.) But let us suppose that we know of one rational point $\mathcal{O}$ on our rational conic. Then we can get all of them very simply. We just draw some rational line and we project the conic onto the line from this point $\mathcal{O}$. (To project $\mathcal{O}$ itself onto the line, we use the tangent line to the conic at $\mathcal{O}$.)

A line meets a conic in two points, so for every point P on the conic we get a point Q on the line; and conversely, for every point Q on the line, by joining it to the point $\mathcal{O}$, we get a point P on the conic. (See Figure 1.1.) We get a one-to-one correspondence between the points on the conic and points on the line.[†] But now you see by the remarks we have made that if the point P on the conic has rational coordinates, then the point Q on the line will have rational coordinates. And conversely, if Q is rational, then because $\mathcal{O}$ is assumed to be rational, the line through P and Q meets the conic in two points, one of which is rational. So the other point is rational, too. Thus the rational points on the conic are in one-to-one correspondence with the rational points on the line. Of course, the rational points on the line are easily described in terms of rational values of some parameter.

Let's carry out this procedure for the circle

$$x^2 + y^2 = 1.$$

We will project from the point $(-1,0)$ onto the y axis. Let's call the point of intersection $(0,t)$. (See Figure 1.2.) If we know x and y, then we can

---

[†] More precisely, there is a one-to-one correspondence between the points of the line and all but one of the points of the conic. The missing point of the conic is the unique point $\mathcal{O}'$ on the conic such that the line connecting $\mathcal{O}$ and $\mathcal{O}'$ is parallel to the line we are projecting onto. However, if we work in the projective plane and use homogeneous coordinates, then this problem disappears and we get a perfect one-to-one correspondence. See Appendix A for details.

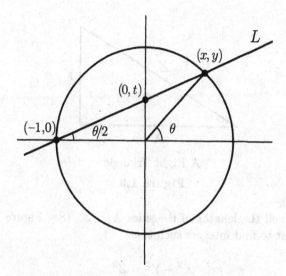

A Rational Parametrization of the Circle

Figure 1.2

get t very easily. The equation of the line L connecting $(-1,0)$ to $(0,t)$ is $y = t(1+x)$. The point (x,y) is assumed to be on the line L and also on the circle, so we get the relation

$$1 - x^2 = y^2 = t^2(1+x)^2.$$

For a fixed value of t, this is a quadratic equation whose roots are the x coordinates of the two intersections of the line L with the circle. Clearly, $x = -1$ is a root, because the point $(-1,0)$ is on both L and the circle. To find the other root, we cancel a factor of $1+x$ from both sides of the equation. This gives the linear equation $1 - x = t^2(1+x)$. Solving this for x in terms of t, and then using the relation $y = t(1+x)$ to find y, we obtain

$$x = \frac{1-t^2}{1+t^2}, \qquad y = \frac{2t}{1+t^2}.$$

This is the familiar rational parametrization of the circle . And now the assertion made above is clear from these formulas. That is, if x and y are rational numbers, then t will be a rational number. Conversely, if t is a rational number, then from these formulas it is obvious that the coordinates x and y will be rational numbers. So this is the way you get rational points on the circle: simply plug in an arbitrary rational number for t. That will give you all points except $(-1,0)$. [If you want to get $(-1,0)$, you must "substitute" infinity for t!]

These formulas can be used to solve the elementary problem of describing all right triangles with integer sides. Let us consider the problem of finding some other triangles, besides 3,4,5, which have whole number

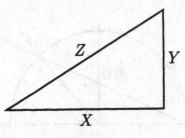

A Right Triangle

Figure 1.3

sides. Let us call the lengths of the sides X, Y, Z. (See Figure 1.3.) That means we want to find integers such that

$$X^2 + Y^2 = Z^2.$$

Now if we have such integers where $X, Y,$ and Z have a common factor, then we can take the common factor out. So we may as well assume that the three of them do not have any common factors. Right triangles whose integer sides have no common factor are called *primitive*. But then it follows that no two of the sides have a common factor, either. For example, if there is some prime dividing both Y and Z, then it would divide $X^2 = Z^2 - Y^2$, hence it would divide X, contrary to our assumption that X, Y, Z have no common factor. So if we make the trivial reduction to the case of primitive triangles, then no two of the sides have a common factor.

In particular, the point (x, y) defined by

$$x = \frac{X}{Z}, \qquad y = \frac{Y}{Z},$$

is a rational point on the circle $x^2 + y^2 = 1$. Further, the rational numbers x and y are in lowest terms.

Since X and Y have no common factor, they cannot both be even. We claim that neither can they both be odd. The point is that the square of an odd number is congruent to 1 modulo 4. If X and Y were both odd, then $X^2 + Y^2$ would be congruent to 2 modulo 4. But $X^2 + Y^2 = Z^2$, and Z^2 is congruent to either 0 or 1 modulo 4. So X and Y are not both odd. Say X is odd and Y is even.

Since (x, y) is a rational point on the circle, there is some rational number t so that x and y are given by the formulas we derived above. Write $t = m/n$ in lowest terms. Then

$$\frac{X}{Z} = x = \frac{n^2 - m^2}{n^2 + m^2}, \qquad \frac{Y}{Z} = y = \frac{2mn}{n^2 + m^2}.$$

Since X/Z and Y/Z are in lowest terms, this means that there is some integer λ so that

$$\lambda Z = n^2 + m^2, \qquad \lambda Y = 2mn, \qquad \lambda X = n^2 - m^2.$$

We want to show that $\lambda = 1$. Because λ divides both $n^2 + m^2$ and $n^2 - m^2$, it divides their sum $2n^2$ and their difference $2m^2$. But m and n have no common divisors. Hence, λ divides 2, and so $\lambda = 1$ or $\lambda = 2$. If $\lambda = 2$, then $n^2 - m^2 = \lambda X$ is divisible by 2, but not by 4, because we are assuming that X is odd. In other words, $n^2 - m^2$ is congruent to 2 modulo 4. But n^2 and m^2 are each congruent to either 0 or 1 modulo 4, so this is not possible. Hence, $\lambda = 1$.

This proves that to get all primitive triangles, you take two relatively prime integers m and n and let

$$X = n^2 - m^2, \qquad Y = 2mn, \qquad Z = n^2 + m^2,$$

be the sides of the triangle. These are the ones with X odd and Y even. The others are obtained by interchanging X and Y.

These formulas have other uses; you may have met them in calculus . In Figure 1.2, we have

$$x = \cos\theta, \quad y = \sin\theta; \qquad \text{and so} \quad t = \tan\tfrac{1}{2}\theta = \frac{\sin\theta}{1 + \cos\theta}.$$

So the formulas given above allow us to express cosine and sine rationally in terms of the tangent of the half angle:

$$x = \cos\theta = \frac{1 - t^2}{1 + t^2}, \qquad y = \sin\theta = \frac{2t}{1 + t^2}.$$

If you have some complicated identity in sine and cosine that you want to test, all you have to do is substitute these formulas, collect powers of t, and see if you get zero. (If they had told you this in high school, the whole business of trigonometric identities would have become a trivial exercise in algebra!)

Another use comes from the observation that these formulas let us express all trigonometric functions of an angle θ as rational expressions in $t = \tan(\theta/2)$. Note that

$$\theta = 2\arctan(t), \qquad d\theta = \frac{2dt}{1 + t^2}.$$

So if you have an integral which involves $\cos\theta$ and $\sin\theta$ and $d\theta$, and you make the appropriate substitutions, then you transform it into an integral in t and dt. If the integral is a rational function of $\sin\theta$ and $\cos\theta$, you obviously come out with the integral of a rational function of t. Since rational functions can be integrated in terms of elementary functions, it follows that any rational function of $\sin\theta$ and $\cos\theta$ can be integrated in terms of elementary functions.

What if we take the circle

$$x^2 + y^2 = 3.$$

and ask to find the rational points on it? That is the easiest problem of all, because the answer is that there are none. It is impossible for the sum of the squares of two rational numbers to equal 3. How can we see that it is impossible?

If there is a rational point, we can write it as

$$x = \frac{X}{Z} \quad \text{and} \quad y = \frac{Y}{Z}$$

for some integers X, Y, Z; and then

$$X^2 + Y^2 = 3Z^2.$$

If X, Y, Z have a common factor, then we can remove it; so we may assume that they have no common factor. It follows that both X and Y are not divisible by 3. This is true because if 3 divides X, then 3 divides $Y^2 = 3Z^2 - X^2$, so 3 divides Y. But then 9 divides $X^2 + Y^2 = 3Z^2$, so 3 divides Z, contradicting the fact that X, Y, Z have no common factors. Hence 3 does not divide X, and similarly for Y.

Since X and Y are not divisible by 3, we have

$$X \equiv \pm 1 \,(\text{mod } 3), \quad Y \equiv \pm 1 \,(\text{mod } 3), \quad \text{and so} \quad X^2 \equiv Y^2 \equiv 1 \,(\text{mod } 3).$$

But then
$$0 \equiv 3Z^2 = X^2 + Y^2 \equiv 1 + 1 \equiv 2 \quad (\text{mod } 3).$$

This contradiction shows that no two rational numbers have squares which add up to 3.

We have seen by the projection argument that if you have one rational point on a rational conic, then all of the rational points can be described in terms of a rational parameter t. But how do you check whether or not there is one rational point? The argument we gave for $x^2 + y^2 = 3$ gives the clue. We showed that there were no rational points by checking that a certain equation had no solutions modulo 3.

There is a general method to test, in a finite number of steps, whether or not a given rational conic has a rational point. The method consists in seeing whether a certain congruence can be satisfied. The theorem goes back to Legendre. Let us take the simple case

$$aX^2 + bY^2 = cZ^2,$$

which is to be solved in integers. Legendre's theorem states that there is an integer m, depending in a simple fashion on a, b, and c, so that the above equation has a solution in integers, not all zero, if and only if the congruence

$$aX^2 + bY^2 \equiv cZ^2 \pmod{m}$$

has a solution in integers relatively prime to m.

There is a much more elegant way to state this theorem, due to Hasse: "A homogeneous quadratic equation in several variables is solvable by integers, not all zero, if and only if it is solvable in real numbers and in p-adic numbers for each prime p." Once one has Hasse's result, then one gets Legendre's theorem in a fairly elementary way. Legendre's theorem combined with the work we did earlier provides a very satisfactory answer to the question of rational points on rational conics. So now we move on to cubics.

2. The Geometry of Cubic Curves

Now we are ready to begin our study of cubics. Let

$$ax^3 + bx^2y + cxy^2 + dy^3 + ex^2 + fxy + gy^2 + hx + iy + j = 0$$

be the equation for a general cubic. We will say that a cubic is *rational* if the coefficients of its equation are rational numbers. A famous example is

$$x^3 + y^3 = 1;$$

or, in homogeneous form,

$$X^3 + Y^3 = Z^3.$$

To find rational solutions of $x^3 + y^3 = 1$ amounts to finding integer solutions of $X^3 + Y^3 = Z^3$, the first non-trivial case of Fermat's Last "Theorem."

We cannot use the geometric principle that worked so well for conics because a line generally meets a cubic in three points. And if we have one rational point, we cannot project the cubic onto a line, because each point on the line would then correspond to two points on the curve.

But there is a geometric principle we can use. If we can find two rational points on the curve, then we can generally find a third one. Namely, draw the line connecting the two points you have found. This will be a rational line, and it meets the cubic in one more point. If we look and see what happens when we try to find the three intersections of a rational line with a rational cubic, we find that we come out with a cubic equation with rational coefficients. If two of the roots are rational, then the third must be also. We will work out some explicit examples below, but the principle is clear. So this gives some kind of composition law: Starting with two points P and Q, we draw the line through P and Q and let $P * Q$ denote the third point of intersection of the line with the cubic. (See Figure 1.4.)

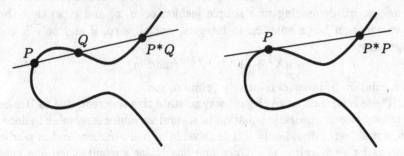

The Composition of Points on a Cubic

Figure 1.4

Even if we only have one rational point P, we can still generally get another. By drawing the tangent line to the cubic at P, we are essentially drawing the line through P and P. The tangent line meets the cubic twice at P, and the same argument will show that the third intersection point is rational. Then we can join these new points up and get more points. So if we start with a few rational points, then by drawing lines, we generally get lots of others.

One of the main theorems we want to prove in this book is the theorem of Mordell (1921) which states that if C is a non-singular rational cubic curve, then there is a *finite* set of rational points such that all other rational points can be obtained in the way we described. We will prove Mordell's theorem for a wide class of cubic curves, using only elementary number theory of the ordinary integers. The principle of the proof in the general case is exactly the same, but requires some tools from the theory of algebraic numbers.

The theorem can be reformulated to be more enlightening. To do that, we first give an elementary geometric property of cubics. We will not prove it completely, but we will make it very plausible, which should suffice. (For more details, see Appendix A.) In general, two cubic curves meet in nine points. To make that statement correct, one should first of all use the projective plane, which has extra points at infinity. Secondly, one should introduce multiplicities of intersections, counting points of tangency for example as intersections of multiplicity greater than one. And finally, one must allow complex numbers for coordinates. We will ignore these technicalities. Then a curve of degree m and a curve of degree n meet in mn points. This is Bezout's theorem, one of the basic theorems in the theory of plane curves. (See Appendix A, Section 4, for a proof of Bezout's theorem.) So two cubics meet in nine points. (See Figure 1.5.)

The theorem that we want to use is the following:

> Let C, C_1, and C_2 be three cubic curves. Suppose C goes through eight of the nine intersection points of C_1 and C_2. Then C goes through the ninth intersection point.

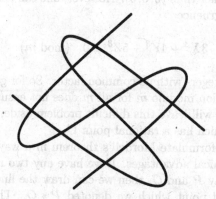

The Intersection of Two Cubics

Figure 1.5

Why is this true, at least in general? The trick is to consider the problem
of constructing a cubic curve which goes through a certain set of points. To
define a cubic curve, we have to give ten coefficients $a, b, c, d, e, f, g, h, i, j$.
If we multiply all the coefficients by a constant, then we get the same curve.
So really the set of all possible cubics is, so to speak, nine dimensional. And
if we want the cubic to go through a point whose x and y coordinates are
given, that imposes one linear condition on those coefficients. The set of
cubics which go through one given point is, so to speak, eight dimensional.
Each time you impose a condition that the cubic should go through a given
point, that imposes an extra linear condition on the coefficients. Thus, the
family of all cubics which go through the eight points of intersection of the
two given cubics C_1 and C_2 forms a one dimensional family.

Let $F_1(x, y) = 0$ and $F_2(x, y) = 0$ be the cubic equations giving C_1
and C_2. We can then find cubics going through the eight points by taking linear combinations $\lambda_1 F_1 + \lambda_2 F_2$. Because the cubics going through
the eight points form a one dimensional family, and because the set of cubics $\lambda_1 F_1 + \lambda_2 F_2$ is a one dimensional family, we see that the cubic C has
an equation $\lambda_1 F_1 + \lambda_2 F_2 = 0$ for a suitable choice of λ_1, λ_2.

Now how about the ninth point? Since that ninth point is on both C_1
and C_2, we know that $F_1(x, y)$ and $F_2(x, y)$ both vanish at that point. It
follows that $\lambda_1 F_1 + \lambda_2 F_2$ also vanishes there, and this means that C also
contains that point.

In passing, we will mention that there is no known method to determine
in a finite number of steps whether a given rational cubic has a rational
point. There is no analogue of Hasse's theorem for cubics. That question
is still open, and it is a very important question. The idea of looking
modulo m for all integers m is not sufficient. Selmer gave the example

$$3X^3 + 4Y^3 + 5Z^3 = 0.$$

This is a cubic, and Selmer shows by an ingenious argument that it has no integer solutions other than $(0, 0, 0)$. However, one can check that for every integer m, the congruence

$$3X^3 + 4Y^3 + 5Z^3 \equiv 0 \pmod{m}$$

has a solution in integers with no common factor. So for general cubics, the existence of a solution modulo m for all m does not ensure that a rational solution exists. We will leave this difficult problem aside, and assume that we have a cubic which has a rational point $\mathcal{O}$.

We want to reformulate Mordell's theorem in a way which has great aesthetic and technical advantages. If we have any two rational points on a rational cubic, say P and Q, then we can draw the line joining P to Q, obtaining the third point which we denoted $P * Q$. This has the flavor of many of the constructions you have studied in modern algebra. If we consider the set of all rational points on the cubic, we can say that set has a law of composition. Given any two points P, Q, we have defined a third point $P * Q$. We might ask about the algebraic structure of this set and this composition law; for example, is it a group? Unfortunately, it is not a group; to start with, it is fairly clear that there is no identity element.

However, by playing around with it a bit, we can make it into a group in such a way that the given rational point $\mathcal{O}$ becomes the zero element of the group. We will denote the group law by $+$ because it is going to be a commutative group. The rule is as follows:

> To add P and Q, take the third intersection point $P * Q$, join it to $\mathcal{O}$, and then take the third intersection point to be $P + Q$. Thus by definition, $P + Q = \mathcal{O} * (P * Q)$.

The group law is illustrated in Figure 1.6, and the fact that $\mathcal{O}$ acts as the zero element is shown in Figure 1.7.

It is clear that this operation is commutative, that is, $P + Q = Q + P$. We claim first that $P + \mathcal{O} = P$, so $\mathcal{O}$ acts as the zero element. Why is that? Well, if we join P to $\mathcal{O}$, then we get the point $P * \mathcal{O}$ as the third intersection. Next we must join $P * \mathcal{O}$ to $\mathcal{O}$ and take the third intersection point. That third intersection point is clearly P. So $P + \mathcal{O} = P$.

It is a little harder to get negatives, but not very hard. Draw the tangent line to the cubic at $\mathcal{O}$, and let the tangent meet the cubic at the additional point S. (We are assuming that the cubic is non-singular, so there is a tangent line at every point.) Then given a point Q, we join Q to S, and the third intersection point $Q * S$ will be $-Q$. (See Figure 1.8.) To check that this is so, let us add Q and $-Q$. To do this we take the third intersection of the line through Q and $-Q$, which is S; and then join S to $\mathcal{O}$ and take the third intersection point $S * \mathcal{O}$. But the line through S and $\mathcal{O}$, because it is tangent to the cubic at $\mathcal{O}$, meets the cubic once at S and twice at $\mathcal{O}$. (You must interpret these things properly.) So the third intersection is the second time it meets $\mathcal{O}$. Therefore, $Q + (-Q) = \mathcal{O}$.

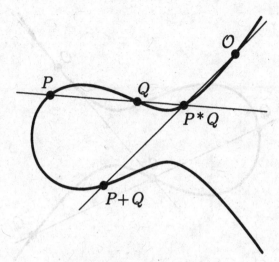

The Group Law on a Cubic

Figure 1.6

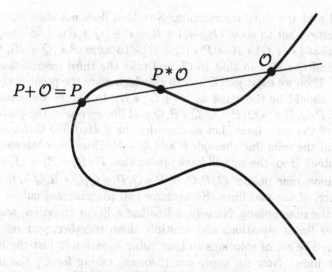

Verifying $\mathcal{O}$ Is the Zero Element

Figure 1.7

If we only knew that + was associative, then we would have a group. Let us try to prove the associative law. Let P, Q, R be three points on the curve. We want to prove that $(P+Q)+R = P+(Q+R)$. To get $P+Q$, we form $P*Q$ and take the third intersection of the line connecting it to $\mathcal{O}$. To add $P+Q$ to R, we draw the line from R through $P+Q$, and that meets the curve at $(P+Q)*R$. To get $(P+Q)+R$, we would have to join $(P+Q)*R$

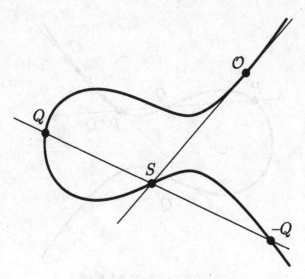

The Negative of a Point
Figure 1.8

to $\mathcal{O}$ and take the third intersection. Now that does not show up too well in the picture, but to show $(P+Q)+R = P+(Q+R)$, it will be enough to show that $(P+Q)*R = P*(Q+R)$. To form $P*(Q+R)$, we first have to find $Q*R$, join that to $\mathcal{O}$, and take the third intersection which is $Q+R$. Then we must join $Q+R$ to P, which gives the point $P*(Q+R)$; and that should be the same as $(P+Q)*R$. In Figure 1.9, each of the points $\mathcal{O}, P, Q, R, P*Q, P+Q, Q*R, Q+R$ lies on one of the dotted lines and one of the solid lines. Let us consider the dotted line through $P+Q$ and R and the solid line through P and $Q+R$. Does their intersection lie on the cubic? If so, the we will have proven that $P*(Q+R) = (P+Q)*R$.

We have nine points: $\mathcal{O}, P, Q, R, P*Q, P+Q, Q*R, Q+R$ and the intersection of the two lines. So we have two (degenerate) cubics that go through the nine points. Namely, a line has a linear equation, and if you have three linear equations and multiply them together, you get a cubic equation. The set of solutions to that cubic equation is just the union of the three lines. Now we apply our theorem, taking for C_1 the union of the three dotted lines and for C_2 the union of the three solid lines. By construction, these two cubics go through the nine points. But the original cubic curve C goes through eight of the points, and therefore it goes through the ninth. Thus, the intersection of the two lines lies on C, which proves that $(P+Q)*R = P*(Q+R)$.

We will not do any more toward proving that the operation $+$ makes the points of C into a group. Later, when we have a normal form, we will give explicit formulas for adding points. So if our use of unproven assertions bothers you, then you can spend a day or two computing with those explicit

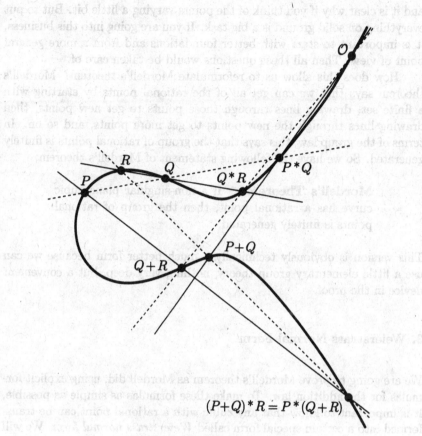

$$(P+Q)*R = P*(Q+R)$$

Verifying the Associative Law

Figure 1.9

formulas and verify directly that associativity holds.

We also want to mention that there is nothing special about our choice of $\mathcal{O}$; if we choose a different point $\mathcal{O}'$ to be the zero element of our group, then we get a group with exactly the same structure. In fact, the map

$$P \longmapsto P + (\mathcal{O}' - \mathcal{O})$$

is an isomorphism from the group "C with zero element $\mathcal{O}$" to the group "C with zero element $\mathcal{O}'$."

Maybe we should explain that we have dodged some of the subtleties. If the line through P and Q is tangent to the curve at P, then the third point of intersection must be interpreted as P. And if you think of that tangent line as the line through P and P, the third intersection point is Q. Further, if you have a point of inflection P on C, then the tangent line P meets the curve three times at P. So in this case the third point of intersection for the line through P and P is again P. [That is, if P is an inflection point, then $P * P = P$.] You just have to count intersections in the proper way, and it is clear why if you think of the points varying a little bit. But to put everything on solid ground is a big task. If you are going into this business, it is important to start with better foundations and from a more general point of view. Then all these questions would be taken care of.

How does this allow us to reformulate Mordell's theorem? Mordell's theorem says that we can get all of the rational points by starting with a finite set, drawing lines through those points to get new points, then drawing lines through the new points to get more points, and so on. In terms of the group law, this says that the group of rational points is finitely generated. So we have the following statement of Mordell's theorem.

> **Mordell's Theorem.**   If a non-singular plane cubic curve has a rational point, then the group of rational points is finitely generated.

This version is obviously technically a much better form because we can use a little elementary group theory, nothing very deep, but a convenient device in the proof.

3. Weierstrass Normal Form

We are going to prove Mordell's theorem as Mordell did, using explicit formulas for the addition law. To make these formulas as simple as possible, it is important to know that any cubic with a rational point can be transformed into a certain special form called *Weierstrass normal form*. We will not completely prove this, but we will give enough of an indication of the proof so that anyone who is familiar with projective geometry can carry out the details. (See Appendix A for an introduction to projective geometry.) Also, we will work out a specific example to illustrate the general theory. After that, we will restrict attention to cubics which are given in the Weierstrass form, which classically consists of equations that look like

$$y^2 = 4x^3 - g_2 x - g_3.$$

We will also use the slightly more general equation

$$y^2 = x^3 + ax^2 + bx + c,$$

and will call either of them Weierstrass form. What we need to show is that any cubic is, as one says, birationally equivalent to a cubic of this type. We will now explain what this means, assuming that the reader knows a (very) little bit of projective geometry.

We start with a cubic curve, which we will think of as being in the projective plane. The idea is to choose axes in the projective plane so that the equation for the curve has a simple form. We assume we are given

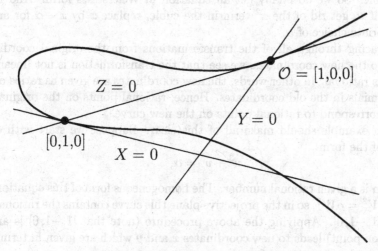

Choosing Axes to Put C into Weierstrass Form

Figure 1.10

a rational point $\mathcal{O}$ on C, so we begin by taking $Z = 0$ to be the tangent line to C at $\mathcal{O}$. This tangent line intersects C at one other point, and we take the $X = 0$ axis to be tangent to C at this new point. Finally, we choose $Y = 0$ to be any line (other than $Z = 0$) which goes through $\mathcal{O}$. (See Figure 1.10. We are assuming that $\mathcal{O}$ is not a point of inflection; otherwise we can take $X = 0$ to be any line not containing $\mathcal{O}$.)

If we choose axes in this fashion and let $x = \dfrac{X}{Z}$ and $y = \dfrac{Y}{Z}$, then we get some linear conditions on the form the equation will take in these coordinates. This is called a projective transformation. We will not work out the algebra, but will just tell you that at the end the equation for C takes the form

$$xy^2 + (ax + b)y = cx^2 + dx + e.$$

Next we multiply through by x,

$$(xy)^2 + (ax + b)xy = cx^3 + dx^2 + ex.$$

Now if we give a new name to xy, we will just call it y again, then we obtain

$$y^2 + (ax + b)y = \text{cubic in } x.$$

Replacing y by $y - \frac{1}{2}(ax + b)$, another linear transformation which amounts to completing the square on the left-hand side of the equation, we obtain

$$y^2 = \text{cubic in } x.$$

The cubic in x might not have leading coefficient 1, but we can adjust that by replacing x and y by λx and $\lambda^2 y$, where λ is the leading coefficient of the cubic. So we do finally get an equation in Weierstrass form. And if we want to get rid of the x^2 term in the cubic, replace x by $x - \alpha$ for an appropriate choice of α.

Tracing through all of the transformations from the original coordinates to the new coordinates, we see that the transformation is not linear, but it is rational. In other words, the new coordinates are given as ratios of polynomials in the old coordinates. Hence, rational points on the original curve correspond to rational points on the new curve.

An example should make all of this clear. Suppose we start with a cubic of the form

$$u^3 + v^3 = \alpha,$$

where α is a given rational number. The homogeneous form of this equation is $U^3 + V^3 = \alpha W^3$, so in the projective plane this curve contains the rational point $[1, -1, 0]$. Applying the above procedure (note that $[1, -1, 0]$ is an inflection point) leads to new coordinates x and y which are given in terms of u and v by the rational functions

$$x = \frac{12\alpha}{u + v} \quad \text{and} \quad y = 36\alpha \frac{u - v}{u + v}.$$

If you work everything out, you will see that x and y satisfy the Weierstrass equation

$$y^2 = x^3 - 432\alpha^2.$$

Further, the process can be inverted, and one finds that u and v can be expressed in terms of x and y by

$$u = \frac{36\alpha + y}{6x} \quad \text{and} \quad v = \frac{36\alpha - y}{6x}.$$

Thus if we have a rational solution to $u^3 + v^3 = \alpha$, then we get rational x and y which satisfy the equation $y^2 = x^3 - 432\alpha^2$. And conversely, if we have a rational solution of $y^2 = x^3 - 432\alpha^2$, then we get rational numbers satisfying $u^3 + v^3 = \alpha$. Of course, if $u = -v$, then the denominator in the expression for x and y is zero; but there are only a finite number of such exceptions, and they are easy to find. So the problem of finding rational points on $u^3 + v^3 = \alpha$ is the same as the problem of finding rational points on $y^2 = x^3 - 432\alpha^2$. And the general argument sketched above indicates that the same is true for any cubic. Of course, the normal form has an entirely different shape from the original equation. But there is a one-to-one correspondence between the rational points on one curve and the rational points on the other (up to a few easily catalogued exceptional points). So the problem of rational points on general cubic curves having one rational point is reduced to studying rational points on cubic curves in Weierstrass normal form.

The transformations we used to put the curve in normalized form do not map straight lines to straight lines. Since we defined the group law

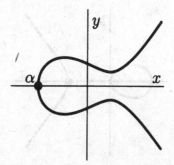

A Cubic Curve with One Real Component

Figure 1.11

on our curve using lines connecting points, it is not at all clear that our transformation preserves the structure of the group. (That is, is our transformation a homomorphism?) It is, but that is not at all obvious. The point is that our description of addition of points on the curve is not a good one, because it seems to depend on the way the curve is embedded in the plane. But in fact the addition law is an intrinsic operation which can be described on the curve and is invariant under birational transformation. This follows from basic facts about algebraic curves, but is not so easy (virtually impossible?) to prove simply by manipulating the explicit equations.

A cubic equation in normal form looks like

$$y^2 = f(x) = x^3 + ax^2 + bx + c.$$

Assuming that the (complex) roots of $f(x)$ are distinct, such a curve is called an *elliptic curve*. (More generally, any curve birationally equivalent to such a curve is called an elliptic curve.) Where does this name come from, because these curves are certainly not ellipses? The answer is that these curves arose in studying the problem of how to compute the arc length of an ellipse. If one writes down the integral which gives the arc length of an ellipse and makes an elementary substitution, the integrand will involve the square root of a cubic or quartic polynomial. So to compute the arc-length of an ellipse, one integrates a function involving $y = \sqrt{f(x)}$, and the answer is given in terms of certain functions on the "elliptic" curve $y^2 = f(x)$.

Now we take the coefficients a, b, c of $f(x)$ to be rational, so in particular they are real; hence, the polynomial $f(x)$ of degree 3 has at least one real root. In real numbers we can factor it as

$$f(x) = (x - \alpha)(x^2 + \beta x + \gamma) \qquad \text{with } \alpha, \beta, \gamma \text{ real.}$$

Of course, it might have three real roots. If it has one real root, the curve looks something like Figure 1.11, because $y = 0$ when $x = \alpha$. If $f(x)$ has three real roots, then the curve looks like Figure 1.12. In this case the real points form two components.

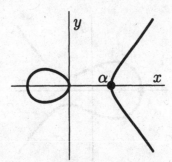

A Cubic Curve with Two Real Components

Figure 1.12

All of this is valid, provided the roots of $f(x)$ are distinct. What is the significance of that condition? We have been assuming all along that our cubic curve is non-singular. If we write the equation as $F(x,y) = y^2 - f(x) = 0$ and take partial derivatives,

$$\frac{\partial F}{\partial x} = -f'(x), \qquad \frac{\partial F}{\partial y} = 2y,$$

then by definition the curve is non-singular, provided that there is no point on the curve at which the partial derivatives vanish simultaneously. This will mean that every point on the curve has a well-defined tangent line. Now if these partial derivatives were to vanish simultaneously at a point (x_0, y_0) on the curve, then $y_0 = 0$, and hence $f(x_0) = 0$, and hence $f(x)$ and $f'(x)$ have the common root x_0. Thus x_0 is a double root of f. Conversely, if f has a double root x_0, then $(x_0, 0)$ will be a singular point on the curve.

There are two possible pictures for the singularity. Which one occurs depends on whether f has a double root or a triple root. In the case that f has a double root, a typical equation is

$$y^2 = x^2(x+1),$$

and the curve has a singularity with distinct tangent directions as illustrated in Figure 1.13.

If $f(x)$ has a triple root, then after translating x we obtain an equation

$$y^2 = x^3,$$

which is a semicubical parabola with a cusp at the origin. (See Figure 1.14.) These are examples of singular cubics in Weierstrass form, and the general case looks the same after a change of coordinates.

Why have we concentrated attention only on the non-singular cubics? It is not just to be fussy. The singular cubics and the non-singular cubics

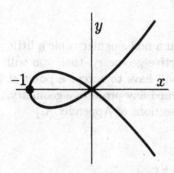

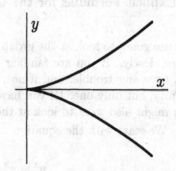

A Singular Cubic with   A Singular Cubic
Distinct Tangent Directions   with A Cusp

Figure 1.13   **Figure 1.14**

have completely different types of behavior. For instance, the singular
cubics are just as easy to treat as conics. If we project from the singular
point onto some line, we see that the line going through that singular point
meets the cubic twice at the singular point, so it meets the cubic only once
more. The projection of the cubic curve onto the line is thus one-to-one.
So just like a conic, the rational points on a singular cubic can be put in
one-to-one correspondence with the rational points on the line. In fact, it
is very easy to do that explicitly with formulas.

If we let $r = \dfrac{y}{x}$, then the equation $y^2 = x^2(x+1)$ becomes

$$r^2 = x + 1; \quad \text{and so} \quad x = r^2 - 1 \quad \text{and} \quad y = r^3 - r.$$

If we take a rational number r and define x and y in this way, then we obtain
a rational point on the cubic; and if we start with a rational point (x, y) on
the cubic, then we get the rational number r. These operations are inverses
of each other, and are defined at all rational points except for the singular
point $(0, 0)$ on the curve. So in this way we get all of the rational points on
the curve.

The curve $y^2 = x^3$ is even simpler. We just take

$$x = t^2 \quad \text{and} \quad y = t^3.$$

So the singular cubics are trivial to analyze as far as rational points go,
and Mordell's theorem does not hold for them. Actually we have not yet
explained how to get a group law for these singular curves; but if one avoids
the singularity, then one does get a group. We will see that this group is not
finitely generated when we study it in more detail at the end of Chapter III.

4. Explicit Formulas for the Group Law

We are going to look at the group of points on a non-singular cubic a little more closely. If you are familiar with projective geometry, then you will not have any trouble; and if not, then you will have to accept a point at infinity, but only one. (If you have never studied any projective geometry, you might also want to look at the first two sections of Appendix A.)

We start with the equation

$$y^2 = x^3 + ax^2 + bx + c$$

and make it homogeneous by setting $x = \dfrac{X}{Z}$ and $y = \dfrac{Y}{Z}$, yielding

$$Y^2 Z = X^3 + aX^2 Z + bX Z^2 + cZ^3.$$

What is the intersection of this cubic with the line at infinity $Z = 0$? Substituting $Z = 0$ into the equation gives $X^3 = 0$, which has the triple root $X = 0$. This means that the cubic meets the line at infinity in three points, but the three points are all the same! So the cubic has exactly one point at infinity, namely, the point at infinity where vertical lines ($x =$ constant) meet. The point at infinity is an inflection point of the cubic, and the tangent line at that point is the line at infinity, which meets it there with multiplicity three. And one easily checks that the point at infinity is a non-singular point by looking at the partial derivatives there. So for a cubic in Weierstrass form there is one point at infinity; we will call that point $\mathcal{O}$.

The point $\mathcal{O}$ is counted as a rational point, and we take it as the zero element when we make the set of points into a group. So to make the game work, we have to make the convention that the points on our cubic consist of the ordinary points in the ordinary affine xy plane together with one other point $\mathcal{O}$ that you cannot see. And now we find it is really true that every line meets the cubic in three points; namely, the line at infinity meets the cubic at the point $\mathcal{O}$ three times. A vertical line meets the cubic at two points in the xy plane and also at the point $\mathcal{O}$. And a non-vertical line meets the cubic in three points in the xy plane. Of course, we may have to allow x and y to be complex numbers.

Now we are going to discuss the group structure a little more closely. How do we add two points P and Q on a cubic equation in Weierstrass form? First we draw the line through P and Q and find the third intersection point $P * Q$. Then we draw the line through $P * Q$ and $\mathcal{O}$, which is just the vertical line through $P * Q$. A cubic curve in Weierstrass form is symmetric about the x axis, so to find $P + Q$, we just take $P * Q$ and reflect it about the x axis. This procedure is illustrated in Figure 1.15.

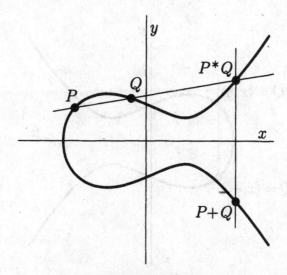

Adding Points on a Weierstrass Cubic

Figure 1.15

What is the negative of a point Q? The negative of Q is the reflected point; if $Q = (x, y)$, then $-Q = (x, -y)$. (See Figure 1.16.) To check this, suppose that we add Q to the point we claim is $-Q$. The line through Q and $-Q$ is vertical, so the third point of intersection is $\mathcal{O}$. Now connect $\mathcal{O}$ to $\mathcal{O}$ and take the third intersection. Connecting $\mathcal{O}$ to $\mathcal{O}$ gives the line at infinity, and the third intersection is again $\mathcal{O}$ because the line at infinity meets the curve with multiplicity three at $\mathcal{O}$. This shows that $Q + (-Q) = \mathcal{O}$, so $-Q$ is the negative of Q. Of course, this formula does not apply to the case $Q = \mathcal{O}$, but obviously $-\mathcal{O} = \mathcal{O}$.

Now we develop some formulas to allow us to compute $P+Q$ efficiently. Let us change notation. We set

$$P_1 = (x_1, y_1), \quad P_2 = (x_2, y_2), \quad P_1 * P_2 = (x_3, y_3), \quad P_1 + P_2 = (x_3, -y_3).$$

We assume that (x_1, y_1) and (x_2, y_2) are given, and we want to compute (x_3, y_3).

First we look at the equation of the line joining (x_1, y_1) and (x_2, y_2). This line has the equation

$$y = \lambda x + \nu, \qquad \text{where} \quad \lambda = \frac{y_2 - y_1}{x_2 - x_1} \quad \text{and} \quad \nu = y_1 - \lambda x_1 = y_2 - \lambda x_2.$$

By construction, the line intersects the cubic in the two points (x_1, y_1) and (x_2, y_2). How do we get the third point of intersection? We substitute

$$y^2 = (\lambda x + \nu)^2 = x^3 + ax^2 + bx + c.$$

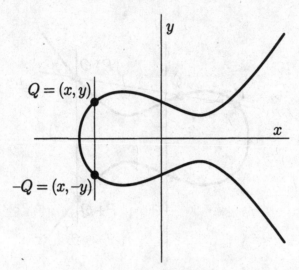

The Negative of a Point on a Weierstrass Cubic

Figure 1.16

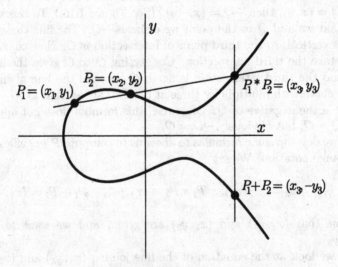

Deriving a Formula for the Addition Law

Figure 1.17

Putting everything on one side yields

$$0 = x^3 + (a - \lambda^2)x^2 + (b - 2\lambda\nu)x + (c - \nu^2).$$

This is a cubic equation in x, and its three roots x_1, x_2, x_3 give us the x

coordinates of the three intersections. Thus,

$$x^3 + (a - \lambda^2)x^2 + (b - 2\lambda\nu)x + (c - \nu^2) = (x - x_1)(x - x_2)(x - x_3).$$

Equating the coefficients of the x^2 term on either side, we find that $\lambda^2 - a = x_1 + x_2 + x_3$, and so

$$x_3 = \lambda^2 - a - x_1 - x_2, \qquad y_3 = \lambda x_3 + \nu.$$

These formulas are the most efficient way to compute the sum of two points.
Let us do an example. We look at the cubic curve

$$y^2 = x^3 + 17$$

which has the two points $P_1 = (-1, 4)$ and $P_2 = (2, 5)$. To compute $P_1 + P_2$ we first find the line through the two points, $y = \frac{1}{3}x + \frac{13}{3}$, so $\lambda = \frac{1}{3}$ and $\nu = \frac{13}{3}$. Next $x_3 = \lambda^2 - x_2 - x_1 = -\frac{8}{9}$ and $y_3 = \lambda x_3 + \nu = \frac{109}{27}$. Finally we find that $P_1 + P_2 = (x_3, -y_3) = \left(-\frac{8}{9}, -\frac{109}{27}\right)$. So computations really are not that bad.

The formulas we gave earlier involve the slope λ of the line connecting the two points. What if the two points are the same? So suppose that we have $P_0 = (x_0, y_0)$ and we want to find $P_0 + P_0 = 2P_0$. We need to find the line joining P_0 to P_0. Because $x_1 = x_2$ and $y_1 = y_2$, we cannot use our formula for λ. But the recipe we described for adding a point to itself says that the line joining P_0 to P_0 is the tangent line to the cubic at P_0. From the relation $y^2 = f(x)$ we find by implicit differentiation that

$$\lambda = \frac{dy}{dx} = \frac{f'(x)}{2y},$$

so that is what we use when we want to double a point.

Continuing with our example curve $y^2 = x^3 + 17$ and point $P_1 = (-1, 4)$, we compute $2P_1$ as follows. First, $\lambda = f'(x_1)/2y_1 = f'(-1)/8 = \frac{3}{8}$. Then, once we have a value for λ, we just substitute into the formulas as above, eventually finding that $2P_1 = \left(\frac{137}{64}, -\frac{2651}{512}\right)$.

Sometimes it is convenient to have an explicit expression for $2P$ in terms of the coordinates for P. If we substitute $\lambda = f'(x)/2y$ into the formulas given earlier, put everything over a common denominator, and replace y^2 by $f(x)$, then we find that

$$x \text{ coordinate of } 2(x, y) = \frac{x^4 - 2bx^2 - 8cx + b^2 - 4ac}{4x^3 + 4ax^2 + 4bx + 4c}.$$

This formula for $x(2P)$ is often called the *duplication formula*. It will come in very handy later for both theoretical and computational purposes. We will leave it to you verify this formula, as well as to derive a companion formula for the y coordinate of $2P$.

These are the basic formulas for the addition of points on a cubic when the cubic is in Weierstrass form. We will use the formulas extensively to prove many facts about rational points on cubic curves, including Mordell's theorem. Further, if you were not satisfied with our incomplete proof that the addition law is associative, you can just take three points at random and compute. Of course, there are an awful lot of special cases to consider, such as when one of the points is the negative of the other or two of the points coincide. But in a few days you will be able to check associativity using these formulas. So we need say nothing more about the proof of the associative law!

EXERCISES

1.1. (a) If P and Q are distinct rational points in the (x, y) plane, prove that the line connecting them is a rational line.
(b) If L_1 and L_2 are distinct rational lines in the (x, y) plane, prove that their intersection is a rational point (or empty).

1.2. Let C be the conic given by the equation

$$F(x, y) = ax^2 + bxy + cy^2 + dx + ey + f = 0,$$

and let δ be the determinant

$$\det \begin{pmatrix} 2a & b & d \\ b & 2c & e \\ d & e & 2f \end{pmatrix}.$$

(a) Show that if $\delta \neq 0$, then C has no singular points. That is, show that there are no points (x, y) where

$$F(x, y) = \frac{\partial F}{\partial x}(x, y) = \frac{\partial F}{\partial y}(x, y) = 0.$$

(b) Conversely, show that if $\delta = 0$ and $b^2 - 4ac \neq 0$, then there is a unique singular point on C.
(c) Let L be the line $y = \alpha x + \beta$ with $\alpha \neq 0$. Show that the intersection of L and C consists of either zero, one, or two points.
(d) Determine the conditions on the coefficients which ensure that the intersection $L \cap C$ consists of exactly one point. What is the geometric significance of these conditions? (Note. There will be more than one case.)

1.3. Let C be the conic given by the equation

$$x^2 - 3xy + 2y^2 - x + 1 = 0.$$

(a) Check that C is non-singular (cf. the previous exercise).
(b) Let L be the line $y = \alpha x + \beta$. Suppose that the intersection $L \cap C$ contains the point $P_0 = (x_0, y_0)$. Assuming that $L \cap C$ consists of two distinct points, find the second point of $L \cap C$ in terms of α, β, x_0, y_0.
(c) If L is a rational line and P_0 is a rational point (i.e., $\alpha, \beta, x_0, y_0 \in \mathbb{Q}$), prove that the second point of $L \cap C$ is also a rational point.

1.4. Find all primitive integral right triangles whose hypotenuse has length less than 30.

1.5. Find all of the rational points on the circle

$$x^2 + y^2 = 2$$

by projecting from the point $(1,1)$ onto an appropriate rational line. (Your formulas will be simpler if you are clever in your choice of the line.)

1.6. (a) Let a, b, c, d, e, f be non-zero real numbers. Use the substitution $t = \tan(\theta/2)$ to transform the integral

$$\int \frac{a + b\cos\theta + c\sin\theta}{d + e\cos\theta + f\sin\theta}\, d\theta$$

into the integral of a rational function of t.
(b) Evaluate the integral

$$\int \frac{a + b\cos\theta + c\sin\theta}{1 + \cos\theta + \sin\theta}\, d\theta.$$

1.7. For each of the following conics, either find a rational point or prove that there are no rational points.
(a) $x^2 + y^2 = 6$
(b) $3x^2 + 5y^2 = 4$
(c) $3x^2 + 6y^2 = 4$

1.8. Prove that for every exponent $e \geq 1$, the congruence

$$x^2 + 1 \equiv 0 \pmod{5^e}$$

has a solution $x_e \in \mathbb{Z}/5^e\mathbb{Z}$. Prove further that these solutions can be chosen to satisfy

$$x_1 \equiv 2 \pmod 5, \qquad \text{and} \quad x_{e+1} \equiv x_e \pmod{5^e} \quad \text{for all } e \geq 1.$$

(This is equivalent to showing that the equation $x^2 + 1 = 0$ has a solution in the 5-adic numbers. It is a special case of Hensel's lemma. *Hint.* Use induction on e.)

1.9. Let C_1 and C_2 be the cubics given by the following equations:

$$C_1 : x^3 + 2y^3 - x - 2y = 0, \qquad C_2 : 2x^3 - y^3 - 2x + y = 0.$$

(a) Find the nine points of intersection of C_1 and C_2.
(b) Let $\{(0,0), P_1, \ldots, P_8\}$ be the nine points from part (a). Prove that if a cubic curve goes through $P_1, \ldots, P_8$, then it must also go through the ninth point $(0,0)$.

1.10. Define a composition law on the points of a cubic C by the following rule as described in the text: Given $P, Q \in C$, then $P * Q$ is the point on C so that P, Q, and $P * Q$ are colinear.

(a) Explain why this law is commutative, $P * Q = Q * P$.

(b) Prove that there is no identity element for this composition law; that is, there is no element $P_0 \in C$ such that $P_0 * P = P$ for all $P \in C$.

(c) Prove that this composition law is not associative; that is, in general, $P * (Q * R) \neq (P * Q) * R$.

(d) Explain why $P * (P * Q) = Q$.

(e) Suppose that the line through $\mathcal{O}$ and S is tangent to C at $\mathcal{O}$. Explain why

$$\mathcal{O} * \big(Q * (Q * S)\big) = \mathcal{O}.$$

(This is an algebraic verification that the point we called $-Q$ is the additive inverse of Q.)

1.11. Let S be a set with a composition law $*$ satisfying the following two properties:

(i) $P * Q = Q * P$ for all $P, Q \in S$.

(ii) $P * (P * Q) = Q$ for all $P, Q \in S$.

Fix an element $\mathcal{O} \in S$, and define a new composition law $+$ by the rule

$$P + Q = \mathcal{O} * (P * Q).$$

(a) Prove that $+$ is commutative and has $\mathcal{O}$ as identity element (i.e., prove that $P + Q = Q + P$ and $P + \mathcal{O} = P$).

(b) Prove that for any given $P, Q \in S$, the equation $X + P = Q$ has the unique solution $X = P * (Q * \mathcal{O})$ in S. In particular, if we define $-P$ to be $P * (\mathcal{O} * \mathcal{O})$, then $-P$ is the unique solution in S of the equation $X + P = \mathcal{O}$.

(c) Prove that $+$ is associative (and so $(S, +)$ is a group) if and only if

(iii) $R * (\mathcal{O} * (P * Q)) = P * (\mathcal{O} * (Q * R))$ for all $P, Q, R \in S$.

(d) Let $\mathcal{O}' \in S$ be another point, and define a composition law $+'$ by $P +' Q = \mathcal{O}' * (P * Q)$. Suppose that both $+$ and $+'$ are associative, so we obtain two group structures $(S, +)$ and $(S, +')$ on S. Prove that the map

$$P \longmapsto \mathcal{O} * (\mathcal{O}' * P)$$

is a (group) isomorphism from $(S, +)$ to $(S, +')$.

(e) * Find a set S with a composition law $*$ satisfying (i) and (ii) such that $(S, +)$ is not a group.

1.12. The cubic curve $u^3 + v^3 = \alpha$ (with $\alpha \neq 0$) has a rational point $[1, -1, 0]$ at infinity. (That is, this is a point on the homogenized equation $U^3 + V^3 = \alpha W^3$.) Taking this rational point to be $\mathcal{O}$, we can make the points on the curve into a group.

(a) Derive a formula for the sum $P_1 + P_2$ of two points $P_1 = (u_1, v_1)$ and $P_2 = (u_2, v_2)$.

(b) Derive a duplication formula for $2P$ in terms of $P = (u, v)$.

1.13. (a) Verify that if u and v satisfy the relation $u^3 + v^3 = \alpha$, then the quantities

$$x = \frac{12\alpha}{u + v} \quad \text{and} \quad y = 36\alpha \frac{u - v}{u + v}$$

satisfy the relation $y^2 = x^3 - 432\alpha^2$.

(b) Part (a) gives a birational transformation from the curve $u^3 + v^3 = \alpha$ to the curve $y^2 = x^3 - 432\alpha^2$. Each of these cubic curves has a group law defined on it. Prove that the birational transformation described in (a) is an isomorphism of groups.

1.14. Let C be the cubic curve $u^3 + v^3 = u + v + 1$. In the projective plane, this curve has a point $[1, -1, 0]$ at infinity. Find rational functions $x = x(u, v)$ and $y = y(u, v)$ so that x and y satisfy a cubic equation in Weierstrass normal form with the given point still at infinity.

1.15. Let $g(t)$ be a quartic polynomial with distinct (complex) roots, and let α be a root of $g(t)$. Let $\beta \neq 0$ be any number.

(a) Prove that the equations

$$x = \frac{\beta}{t - \alpha}, \quad y = x^2 u = \frac{\beta^2 u}{(t - \alpha)^2}$$

give a birational transformation between the curve $u^2 = g(t)$ and the curve $y^2 = f(x)$, where $f(x)$ is the cubic polynomial

$$f(x) = g'(\alpha)\beta x^3 + \tfrac{1}{2}g''(\alpha)\beta^2 x^2 + \tfrac{1}{6}g'''(\alpha)\beta^3 x + \tfrac{1}{24}g''''(\alpha)\beta^4.$$

(b) Prove that if g has distinct (complex) roots, then f also has distinct roots, and so $u^2 = g(t)$ is an elliptic curve.

1.16. Let $0 < \beta \leq \alpha$, and let E be the ellipse

$$\frac{x^2}{\alpha^2} + \frac{y^2}{\beta^2} = 1.$$

(a) Prove that the arc length of E is given by the integral

$$4\alpha \int_0^{\pi/2} \sqrt{1 - k^2 \sin^2 \theta} \, d\theta.$$

for an appropriate choice of the constant k depending on α and β.

(b) Check your value for k in (a) by verifying that when $\alpha = \beta$, the integral yields the correct value for the arc length of a circle.

(c) Prove that the integral in (a) is also equal to

$$4\alpha \int_0^1 \sqrt{\frac{1 - k^2 t^2}{1 - t^2}}\, dt = 4\alpha \int_0^1 \frac{1 - k^2 t^2}{\sqrt{(1 - t^2)(1 - k^2 t^2)}}\, dt.$$

(d) Prove that if the ellipse E is not a circle, then the equation

$$u^2 = (1 - t^2)(1 - k^2 t^2)$$

defines an elliptic curve (cf. the previous exercise). Hence the problem of determining the arc length of an ellipse comes down to evaluating the integral

$$\int_0^1 \frac{1 - k^2 t^2}{u}\, dt \qquad \text{on the ``elliptic'' curve } u^2 = (1 - t^2)(1 - k^2 t^2).$$

1.17. Let C be a cubic curve in the projective plane given by the homogeneous equation

$$Y^2 Z = X^3 + aX^2 Z + bXZ^2 + cZ^3.$$

Verify that the point $[0, 1, 0]$ at infinity is a non-singular point of C.

1.18. The cubic curve
$$y^2 = x^3 + 17$$

has the following five rational points:

$$P_1 = (-2, 3), \ P_2 = (-1, 4), \ P_3 = (2, 5), \ P_4 = (4, 9), \ P_5 = (8, 23).$$

(a) Show that P_2, P_4, and P_5 can each be expressed as $mP_1 + nP_3$ for an appropriate choice of integers m and n.

(b) Compute the points

$$P_6 = -P_1 + 2P_3 \qquad \text{and} \qquad P_7 = 3P_1 - P_3.$$

(c) Notice that the points $P_1, P_2, P_3, P_4, P_5, P_6, P_7$ all have integer coordinates. There is exactly one more rational point on this curve which has integer coordinates and $y > 0$. Find that point. (You will probably need at least a programmable calculator or else a lot of patience.)

(d) ** Prove the assertion in (c) that there are exactly eight rational points (x, y) on this curve with $y > 0$ and x, y both integers. (This is an extremely difficult problem, and you will almost certainly not be able to do it with the tools we have developed. But it is also an extremely interesting problem which is well worth thinking about.)

1.19. Suppose that $P = (x, y)$ is a point on the cubic curve

$$y^2 = x^3 + ax^2 + bx + c.$$

(a) Verify that the x coordinate of the point $2P$ is given by the *duplication formula*

$$x(2P) = \frac{x^4 - 2bx^2 - 8cx - 4ac + b^2}{4y^2}.$$

(b) Derive a similar formula for the y coordinate of $2P$ in terms of x and y.

(c) Find a polynomial in x whose roots are the x coordinates of the points $P = (x, y)$ satisfying $3P = \mathcal{O}$. (*Hint.* The relation $3P = \mathcal{O}$ can also be written $2P = -P$.)

(d) For the particular curve $y^2 = x^3 + 1$, solve the equation in (c) to find all of the points satisfying $3P = \mathcal{O}$. Note you will have to use complex numbers.

1.20. Consider the point $P = (3, 8)$ on the cubic curve $y^2 = x^3 - 43x + 166$. Compute P, $2P$, $3P$, $4P$, and $8P$. Comparing $8P$ with P, what can you conclude?

CHAPTER II

Points of Finite Order

1. Points of Order Two and Three

An element P of any group is said to have *order m* if

$$mP = \underbrace{P + P + \cdots + P}_{m \text{ summands}} = \mathcal{O},$$

but $m'P \neq \mathcal{O}$ for all integers $1 \leq m' < m$. If such an m exists, then P has *finite order*; otherwise it has *infinite order*. We begin our study of points of finite order on cubic curves by looking at points of order two and order three. As usual, we will assume that our non-singular cubic curve is given by a Weierstrass equation

$$y^2 = f(x) = x^3 + ax^2 + bx + c,$$

and that the point at infinity $\mathcal{O}$ is taken to be the zero element for the group law.

Which points in our group satisfy $2P = \mathcal{O}$, but $P \neq \mathcal{O}$? Instead of $2P = \mathcal{O}$, it is easier to look at the equivalent condition $P = -P$. Since $-(x, y)$ is just $(x, -y)$, these are the points with $y = 0$:

$$P_1 = (\alpha_1, 0), \qquad P_2 = (\alpha_2, 0), \qquad P_3 = (\alpha_3, 0),$$

where $\alpha_1, \alpha_2, \alpha_3$ are the roots of the cubic polynomial $f(x)$. So if we allow complex coordinates, there are exactly three points of order two, because the non-singularity of the curve ensures that $f(x)$ has distinct roots. If all three roots of $f(x)$ are real, then the picture looks like Figure 2.1.

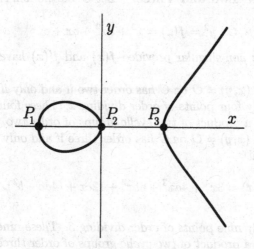

Points of Order Two

Figure 2.1

If we take all of the points satisfying $2P = \mathcal{O}$, including $P = \mathcal{O}$, then we get the set $\{\mathcal{O}, P_1, P_2, P_3\}$. It is easily seen that in any abelian group, the set of solutions to the equation $2P = \mathcal{O}$ forms a subgroup. So we have a group of order four; and since every element has order one or two, it is obvious that this group is the *Four Group*, a direct product of two groups of order two. This means that the sum of any two of the points P_1, P_2, P_3 should equal the third, which is obvious from the fact that the three points are collinear. So now we know exactly what the group of points P such that $2P = \mathcal{O}$ looks like. If we allow complex coordinates, it is the Four Group; if we allow only real coordinates, it is either the Four Group or cyclic of order two, depending on whether $f(x)$ has three or one real root; and if we restrict our attention to rational coordinates, it is either the Four Group, cyclic of order two, or trivial, depending on whether $f(x)$ has three, one, or zero rational roots.

Next we look at the points of order three. Instead of $3P = \mathcal{O}$, we write $2P = -P$, so a point of order 3 will satisfy $x(2P) = x(-P) = x(P)$. Conversely, if $P \neq \mathcal{O}$ satisfies $x(2P) = x(P)$, then $2P = \pm P$, so either $P = \mathcal{O}$ (excluded by assumption) or $3P = \mathcal{O}$. In other words, the points of order three are exactly the points satisfying $x(2P) = x(P)$.

To find the points satisfying this condition, we use the duplication formula and set the x coordinate of $2P$ equal to the x coordinate of P. If we write $P = (x, y)$, then we have shown in Chapter I that the x coordinate of $2P$ equals

$$\frac{x^4 - 2bx^2 - 8cx + b^2 - 4ac}{4x^3 + 4ax^2 + 4bx + 4c}.$$

Setting this expression equal to x, cross-multiplying, and doing a little algebra, we have completed the proof of part (c) of the following proposition.

Points of Order Two and Three. *Let C be the non-singular cubic curve*

$$C : y^2 = f(x) = x^3 + ax^2 + bx + c.$$

[Recall that C is non-singular provided $f(x)$ and $f'(x)$ have no common complex roots.]
(a) A point $P = (x, y) \neq \mathcal{O}$ on C has order two if and only if $y = 0$.
(b) C has exactly four points of order dividing 2. These four points form a group which is a product of two cyclic groups of order two.
(c) A point $P = (x, y) \neq \mathcal{O}$ on C has order three if and only if x is a root of the polynomial

$$\psi_3(x) = 3x^4 + 4ax^3 + 6bx^2 + 12cx + (4ac - b^2).$$

(d) C has exactly nine points of order dividing 3. These nine points form a group which is a product of two cyclic groups of order three.

PROOF. (a,b) We proved these above.
(c) We proved (c) above except for a little bit of algebra, which we will leave to you.

(d) Because the x coordinate of $2P$ equals $\dfrac{f'(x)^2}{4f(x)} - a - 2x$, we see that an alternative expression for $\psi_3(x)$ is

$$\psi_3(x) = 2f(x)f''(x) - f'(x)^2.$$

We claim that $\psi_3(x)$ has four distinct (complex) roots. To verify this, we need to check that $\psi_3(x)$ and $\psi_3'(x)$ have no common roots. But

$$\psi_3'(x) = 2f(x)f'''(x) = 12f(x);$$

so a common root of $\psi_3(x)$ and $\psi_3'(x)$ would be a common root of

$$2f(x)f''(x) - f'(x)^2 \qquad \text{and} \qquad 12f(x),$$

and so would be a common root of $f(x)$ and $f'(x)$. This is contrary to our assumption that C is non-singular, so we conclude that $\psi_3(x)$ indeed has four distinct complex roots.

Let $\beta_1, \beta_2, \beta_3, \beta_4$ be the four complex roots of $\psi_3(x)$; and for each β_i, let δ_i be one of the square roots $\delta_i = \sqrt{f(\beta_i)}$. Then from (c), the set

$$\{(\beta_1, \pm\delta_1), \ (\beta_2, \pm\delta_2), \ (\beta_3, \pm\delta_3), \ (\beta_4, \pm\delta_4)\}$$

is the complete set of points of order three on C. Further, we observe that no δ_i can equal zero, because otherwise the point $(\beta_i, \delta_i) = (\beta_i, 0)$ would have order two, contradicting the fact that it has order three. Therefore, this set contains exactly eight distinct points, so C contains eight points of

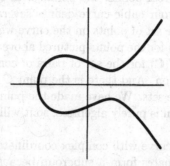

A Cubic Curve with One Real Component

Figure 2.2

order three. The only other point on C with order dividing 3 is the point of order one, namely $\mathcal{O}$, which completes the proof that C has exactly nine points of order dividing 3.

Finally, we note that there is only one (abelian) group with nine elements such that every element has order dividing 3, namely, the product of two cyclic groups of order three.   □

So we now know that if we allow complex numbers, then the points of order dividing three form a group of order nine which is the direct product of two cyclic groups of order three. It turns out that the real points of order three always form a cyclic group of order three, whereas the rational points of order three form either a cyclic group of order three or the trivial group. We will discuss this further in the next section.

There is also a geometric way to describe the points of order three; they are inflection points, the points where the tangent line to the cubic has a triple order contact. We can see that geometrically. To say that $2P = -P$ means that when we draw the tangent at the point P, then take the third intersection and connect it with $\mathcal{O}$, we get $-P$. Now that is the case only if the third point of intersection of the tangent at P is the same point P. So $2P = -P$ if and only if P is a point of inflection. Of course, this can also be shown analytically; we will leave the analytic proof as an exercise.

2. Real and Complex Points on Cubic Curves

The real points on our cubic curve

$$y^2 = f(x) = x^3 + ax^2 + bx + c$$

form either one or two components, depending on whether $f(x)$ has one or three real roots. We illustrated the case of three real roots in Figure 2.1, and, of course, the case of one real root looks like Figure 2.2.

This picture shows the real points, the points with real coordinates. Actually, the equation for our cubic curve defines several sets of points. We could write $C(\mathbb{Q})$ for the set of points on the curve whose coordinates happen to be rational, $C(\mathbb{R})$ for the points pictured above which have arbitrary real coordinates, and $C(\mathbb{C})$ for the set of pairs of complex numbers (x, y) which satisfy the equation. And there is the point $\mathcal{O}$ at infinity, which we take to be in all of these sets. We have made the points of the curve into a group. This construction is purely algebraic, so it will work in any of these three cases.

The points on the curve with complex coordinates form a group. The points with real coordinates form a subgroup because if two points have real coordinates, then so do their sum and difference. And since we are assuming the coefficients a, b, c are rational numbers, it is even true that the rational points form a subgroup of the group of real points.

So we have a big group and some subgroups:

$$\{\mathcal{O}\} \subset C(\mathbb{Q}) \subset C(\mathbb{R}) \subset C(\mathbb{C}).$$

To study the group of real points or complex points, one can use the methods of analysis.

It is obvious that the addition of real points on the curve is continuous. So the group of real points is a one-dimensional Lie group; and this set is in fact compact, although it does not look it because it has the point at infinity. There is only one such connected group. Any one dimensional compact connected Lie group is isomorphic to the group of rotations of the circle; that is, the multiplicative group of complex numbers of absolute value one. So if the group of real points on the curve is connected, then it is isomorphic to the circle group; and in any case, the component of the curve which contains the zero point is isomorphic to the circle group. And from this description we can see immediately what the real points of finite order look like.

If we think of the circle group as the multiplicative group of complex numbers of absolute value one, then the points of finite order in that group are the roots of unity. And for each integer m, the points of order dividing m form a cyclic group of order m. Explicitly, this set of complex numbers is

$$\left\{ 1, e^{2\pi i/m}, e^{4\pi i/m}, \ldots, e^{2(m-1)\pi i/m} \right\}.$$

So the points of order dividing m in $C(\mathbb{R})$ form a cyclic group of order m, at least in the case that the group is connected.

If there are two connected components, then the group $C(\mathbb{R})$ is the direct product of the circle group with a group of order two. In this case, there are two possibilities for the points of order dividing m. If m is odd, we again get a cyclic group of order m; whereas if m is even, then we find the direct product of a cyclic group of order two and a cyclic group of order m.

In particular, we see that the real points of order dividing 3 always form a cyclic group of order three. Since we have seen that there are eight points of order three, it is never possible for all of the complex points of order three to be real, and certainly they cannot all be rational. Notice that the x coordinates of the points of order three are the roots of the quartic polynomial $\psi_3(x)$ described in Section 1. This quartic has real coefficients, so it has either zero, two, or four real roots. Because each value for x gives two possible values for y, this shows that the curve has either zero, four, or eight points of order three with real x coordinate. However, our discussion shows that there must be exactly one real value of x for which the two corresponding y's are real. This can also be proven directly from the equations, a task which we leave for the exercises.

Before continuing with our discussion of rational points, we briefly digress to describe the structure of $C(\mathbb{C})$. Substituting $x - \frac{1}{3}a$ for x, we can eliminate the ax^2 term; and then replacing x and y by $4x$ and $4y$, we end up with the classical form of Weierstrass equation:

$$y^2 = 4x^3 - g_2 x - g_3.$$

As always, the cubic polynomial on the right is assumed to have distinct roots.

In the Weierstrass theory of elliptic functions, it is shown that whenever you have two complex numbers g_2, g_3 so that the polynomial $4x^3 - g_2 x - g_3$ has distinct roots (i.e., such that $g_2^3 - 27g_3^2 \neq 0$), then you can find complex numbers ω_1, ω_2 (called *periods*) in the complex u plane by evaluating certain definite integrals. These periods are $\mathbb{R}$-linearly independent, and one looks at the group formed by taking all of their $\mathbb{Z}$-linear combinations:

$$L = \mathbb{Z}\omega_1 + \mathbb{Z}\omega_2 = \{n_1\omega_1 + n_2\omega_2 : n_1, n_2 \in \mathbb{Z}\}.$$

(Such a subgroup of the complex plane is called a *lattice*.) Although there are many choices for the generators ω_1, ω_2 of L, it turns out that the coefficients g_2 and g_3 uniquely determine the group L itself. Conversely, the group L uniquely determines g_2 and g_3 via the formulas

$$g_2 = 60 \sum_{\substack{\omega \in L \\ \omega \neq 0}} \frac{1}{\omega^4}, \qquad g_3 = 140 \sum_{\substack{\omega \in L \\ \omega \neq 0}} \frac{1}{\omega^6}.$$

One uses the periods to define a function $\wp(u)$ by the series

$$\wp(u) = \frac{1}{u^2} + \sum_{\substack{\omega \in L \\ \omega \neq 0}} \left(\frac{1}{(u-\omega)^2} - \frac{1}{\omega^2} \right).$$

This meromorphic function is called the *Weierstrass $\wp$ function*. It visibly has poles at the points of L, and no other poles in the complex u plane. Less obvious is the fact that $\wp$ is doubly periodic; that is,

$$\wp(u + \omega_1) = \wp(u) \quad \text{and} \quad \wp(u + \omega_2) = \wp(u) \quad \text{for all complex numbers } u.$$

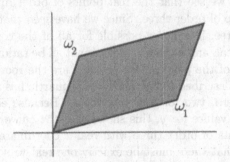

The Period Parallelogram

Figure 2.3

From this it follows that $\wp(u + \omega) = \wp(u)$ for all $u \in \mathbb{C}$ and all $\omega \in L$. Notice the similarity with trigonometric and exponential functions, which have single periods: $f(u) = \sin(u)$ has the period 2π, and $f(u) = e^u$ has the period $2\pi i$.

This doubly periodic function $\wp(u)$ satisfies the differential equation

$$(\wp')^2 = 4\wp^3 - g_2\wp - g_3, \qquad \text{where } \wp' = \frac{d\wp}{du}.$$

Thus for every complex number u we get a point

$$P(u) = \big(\wp(u), \wp'(u)\big)$$

on the given curve, in general a point with complex coordinates. So we obtain a map from the complex u plane to $C(\mathbb{C})$. (Of course, we send the points in L, which are the poles of $\wp$, to $\mathcal{O}$.)

The facts about this map are as follows. The map is onto the curve; every pair (x, y) of complex numbers satisfying $y^2 = 4x^3 - g_2x - g_3$ comes from some u. Because $\wp$ is doubly periodic, the map cannot be one-to-one. If u and v have the property that their difference $u - v$ equals $m_1\omega_1 + m_2\omega_2$ for some integers m_1, m_2 (i.e., if $u - v \in L$), then $P(u) = P(v)$. So instead we just look at values of u which lie in the *period parallelogram*, which is the parallelogram whose sides are the periods ω_1 and ω_2. (See Figure 2.3.) Then it is true that to a given point (x, y) on the curve there is exactly one u in the period parallelogram which is mapped to (x, y). (One must make suitable conventions about the boundary of the parallelogram.)

Thus, the period parallelogram is mapped one-to-one onto the complex points of the curve. The mapping $u \mapsto P(u)$ has the property

$$P(u_1 + u_2) = P(u_1) + P(u_2).$$

Note that the sum $u_1 + u_2$ is just the ordinary addition of complex numbers, whereas $P(u_1) + P(u_2)$ is the addition law on the cubic curve. This

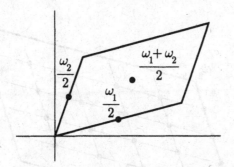

Points of Order Two on a Complex Torus

Figure 2.4

equation amounts to the famous addition formula for $\wp(u)$. It says that the functions $\wp$ and $\wp'$, evaluated at $u_1 + u_2$, can be expressed rationally in terms of their values at u_1 and u_2. The formulas are the ones we gave earlier in Chapter I, Section 4, expressing $(x_3, -y_3) = P_1 + P_2$ in terms of (x_1, y_1) and (x_2, y_2).

The mapping $u \mapsto P(u)$ is thus a homomorphism from the additive group of complex numbers onto the group of complex points on our cubic; and the kernel of that homomorphism is the lattice L considered above. The factor group of the complex u plane modulo the lattice L is isomorphic to the group of complex points on our curve. Thus the group of complex points is a torus, the direct product of two circle groups.

Using this description, we can describe what the complex points of finite order look like. Suppose we want a point of order two. This means we need a point $u \notin L$ such that $2u$ is in L. Looking modulo L, there are three such points, $\dfrac{\omega_1}{2}$, $\dfrac{\omega_2}{2}$, and $\dfrac{\omega_1 + \omega_2}{2}$, as illustrated in Figure 2.4.

Similarly, to find the points of order dividing m, we look for points u in the period parallelogram such that $mu \in L$. The case $m = 5$ is illustrated in Figure 2.5. There are 25 such points in all, and it is clear that they form the direct product of two cyclic groups of order five. In general, the complex points of order dividing m form a group of order m^2 which is the direct product of two cyclic groups of order m. So over the complex numbers and over the real numbers, we have a very good description of the points of finite order on our cubic curve.

Before returning to the rational numbers, we briefly comment on other fields. If F is any subfield of the complex numbers and if the coefficients a, b, c of the cubic equation lie in F, then we can look at the set of solutions (x, y) of the equation for which x and y both lie in F. Let $C(F)$ denote this set of "F-valued" points, together with the point $\mathcal{O}$ which we always include. Then $C(F)$ forms a subgroup of $C(\mathbb{C})$; this is clear from the formulas giving the addition law.

More generally, there is no need in all of this to start with the field of

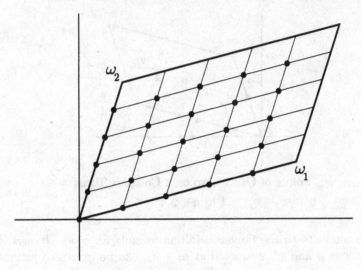

Points of Order Five on a Complex Torus
Figure 2.5

complex numbers. All of our operations, such as the addition law, are purely algebraic. If, for instance, we take F to be the field of integers modulo p, and let a, b, c be in that finite field, then we can look for solutions of the equation in the finite field. Of course, there are only a finite number of solutions, since there are only finitely many possible values for x and y. But again those solutions, together with the point at infinity, form a group; just use the formulas giving the addition law. You can't visualize it, but the formulas work perfectly well for any field.[†]

Because in this case the group is finite, we see that every point has finite order; and one can ask about points of various orders. It turns out that the points of order p form either a cyclic group of order p or a trivial group; but if q is some prime different from p, then the points of order q form either a trivial group, a cyclic group of order q, or else the direct product of two cyclic groups of order q.

[†] However, we should note that the formulas do not work for a field of characteristic 2. The problem occurs when we try to go from a general cubic equation to an equation of the form $y^2 = f(x)$; this transformation requires dividing by 2 and completing a square (cf. Chapter I, Section 3). So to work with cubic equations in characteristic 2, one has to consider more general equations of the form $y^2 + a_1 xy + a_3 y = x^3 + a_2 x^2 + a_4 x + a_6$.

3. The Discriminant

After our digression into real and complex analysis, we return to the field of rational numbers. As always, we take our curve in its normal form

$$y^2 = f(x) = x^3 + ax^2 + bx + c,$$

where a, b, c are rational numbers. If we let $X = d^2 x$ and $Y = d^3 y$, then our equation becomes $Y^2 = X^3 + d^2 a X^2 + d^4 b X + d^6 c$. By choosing a large integer d, we can clear any denominators in a, b, c. So from now on we will assume that our cubic curve is given by an equation having integer coefficients.

Our goal in this chapter is to prove a theorem, first proved by Nagell and Lutz, which will tell us how to find all of the rational points of finite order. Their theorem says that a rational point of finite order (x, y) must have integer coordinates, and either $y = 0$ (for points of order two) or else $y | D$, where D is the discriminant of the polynomial $f(x)$. In particular, a cubic curve has only a finite number of rational points of finite order.

The *discriminant of* $f(x)$ is the quantity

$$D = -4a^3 c + a^2 b^2 + 18abc - 4b^3 - 27c^2.$$

You may be familiar with this when $a = 0$, in which case $D = -4b^3 - 27c^2$. If we factor f over the complex numbers,

$$f(x) = (x - \alpha_1)(x - \alpha_2)(x - \alpha_3),$$

then one can check that

$$D = (\alpha_1 - \alpha_2)^2 (\alpha_1 - \alpha_3)^2 (\alpha_2 - \alpha_3)^2;$$

and so the non-vanishing of D tells us that the roots of $f(x)$ are distinct.

Thus the question of finding the rational points of finite order can be settled in a finite number of steps. You take the integer D, and consider each of the finitely many integers y with $y | D$. You take all those values of y and substitute them in the equation $y^2 = f(x)$. The polynomial $f(x)$ has integer coefficients and leading coefficient 1. If it has an integer root, that root will divide the constant term. Thus, there are a finite number of things to check, and in this way we will be sure to find all the points of finite order in a finite number of steps.

Warning. We are not asserting that a point (x, y) with integer coordinates and $y | D$ must have finite order. The Nagell-Lutz Theorem is not an "if and only if" statement.

If $f(x)$ is any polynomial with leading coefficient 1 in the ring $\mathbb{Z}[x]$ of polynomials with integer coefficients, then the discriminant of $f(x)$ will always be in the ideal of $\mathbb{Z}[x]$ generated by $f(x)$ and $f'(x)$. This follows from the general theory of discriminants, but for our particular polynomial $f(x) = x^3 + ax^2 + bx + c$, the quickest proof is just to write out an explicit formula:

$$D = \left\{(18b - 6a^2)x - (4a^3 - 15ab + 27c)\right\}f(x)$$
$$+ \left\{(2a^2 - 6b)x^2 + (2a^3 - 7ab + 9c)x + (a^2b + 3ac - 4b^2)\right\}f'(x).$$

We leave it to you to multiply this out and verify it is correct. The important thing to remember is that there are polynomials $r(x)$ and $s(x)$ with *integer* coefficients so that D can be written as

$$D = r(x)f(x) + s(x)f'(x).$$

Why do we want this formula for D? If we assume the first part of the Nagell-Lutz theorem, namely that points of finite order have integer coordinates, then we can use the formula to prove the second part, that either $y = 0$ or $y|D$. More precisely, if P has finite order, then clearly $2P$ also has finite order. So we will prove the following.

Lemma. Let $P = (x, y)$ be a point on our cubic curve such that both P and $2P$ have integer coordinates. Then either $y = 0$ or $y|D$.

PROOF. We assume that $y \neq 0$ and prove that $y|D$. Because $y \neq 0$, we know that $2P \neq \mathcal{O}$, so we may write $2P = (X, Y)$. By assumption, x, y, X, Y are all integers. The duplication formula says that

$$2x + X = \lambda^2 - a, \qquad \text{where} \quad \lambda = \frac{f'(x)}{2y}.$$

Since x, X, and a are all integers, it follows that λ is also an integer. Since $2y$ and $f'(x)$ are integers, we see that $2y|f'(x)$; and, in particular, $y|f'(x)$. But $y^2 = f(x)$, so also $y|f(x)$. Now we use the relation

$$D = r(x)f(x) + s(x)f'(x).$$

The coefficients of r and s are integers, so $r(x)$ and $s(x)$ take on integer values when evaluated at the integer x. It follows that y divides D.   □

4. Points of Finite Order Have Integer Coordinates

Now we come to the most interesting part of the Nagell-Lutz theorem, the proof that a rational point (x, y) of finite order must have integer coordinates. We will show that x and y are integers in a rather indirect way. We observe that one way to show a positive integer equals 1 is to show that it is not divisible by any primes. Thus we can break the problem up into an infinite number of subproblems: namely, we show that when the rational numbers x and y are written in lowest terms, there are no 2's in the denominators, no 3's in the denominators, no 5's, and so on.

So we let p be some prime, and try to show that p does not divide the denominator of x and does not divide the denominator of y. That leads us to consider the rational points (x, y) where p does divide the denominator of x or y.

It will be helpful to set some notation. Every non-zero rational number may be written uniquely in the form $\dfrac{m}{n}p^\nu$, where m, n are integers prime to p, $n > 0$, and the fraction m/n is in lowest terms. We define the *order* of such a rational number to be the integer ν, and write

$$\operatorname{ord}\left(\frac{m}{n}p^\nu\right) = \nu.$$

To say that p divides the denominator (respectively, the numerator) of a rational number is the same as saying that its order is negative (respectively, positive). The order of a rational number is zero if and only if p divides neither its numerator nor its denominator.

Let us look at a rational point (x, y) on our cubic curve, where p divides the denominator of x, say

$$x = \frac{m}{np^\mu} \quad \text{and} \quad y = \frac{u}{wp^\sigma},$$

where $\mu > 0$ and p does not divide m, n, u, w. We plug this point into the equation for our cubic curve. Putting things over a common denominator, we find

$$\frac{u^2}{w^2 p^{2\sigma}} = \frac{m^3 + am^2np^\mu + bmn^2p^{2\mu} + cn^3p^{3\mu}}{n^3p^{3\mu}}.$$

Now $p \nmid u^2$ and $p \nmid w^2$, so

$$\operatorname{ord}\left(\frac{u^2}{w^2 p^{2\sigma}}\right) = -2\sigma.$$

Since $\mu > 0$ and $p \nmid m$, it follows that

$$p \nmid (m^3 + am^2np^\mu + bmn^2p^{2\mu} + cn^3p^{3\mu});$$

and hence

$$\text{ord}\left(\frac{m^3 + am^2np^\mu + bmn^2p^{2\mu} + cn^3p^{3\mu}}{n^3p^{3\mu}}\right) = -3\mu.$$

Thus, $2\sigma = 3\mu$. In particular, $\sigma > 0$, and so p divides the denominator of y. Further, the relation $2\sigma = 3\mu$ means that $2|\mu$ and $3|\sigma$, so we have $\mu = 2\nu$ and $\sigma = 3\nu$ for some integer $\nu > 0$.

Similarly, if we assume that p divides the denominator of y, we find by the same calculation that the exact same result holds, namely, $\mu = 2\nu$ and $\sigma = 3\nu$ for some integer $\nu > 0$. Thus, if p appears in the denominator of either x or y, then it is in the denominator of both of them; and in this case the exact power is $p^{2\nu}$ in x and $p^{3\nu}$ in y for some positive integer $\nu > 0$.

This suggests we make the following definition. We will let $C(p^\nu)$ be the set of rational points of the cubic curve such that $p^{2\nu}$ divides the denominator of x and $p^{3\nu}$ divides the denominator of y. In other words,

$$C(p^\nu) = \left\{(x,y) \in C(\mathbb{Q}) : \text{ord}(x) \leq -2\nu \text{ and } \text{ord}(y) \leq -3\nu\right\}.$$

For instance, $C(p)$ is the set where p is in the denominator of x and y, and then there is at least a p^2 in x and a p^3 in y. Obviously, we have inclusions

$$C(\mathbb{Q}) \supset C(p) \supset C(p^2) \supset C(p^3) \supset \cdots.$$

By convention, we will also include the zero element $\mathcal{O}$ in every $C(p^\nu)$.

Recall that our objective is to show that if (x,y) is a point of finite order, then x and y are integers; and our strategy is to show that for every prime p, the denominators of x and y are not divisible by p. With our new notation, this means we want to show that a point of finite order cannot lie in $C(p)$. The first step in showing this is to prove that each of the sets $C(p^\nu)$ is a subgroup of $C(\mathbb{Q})$.

Those of you who know about p-adic numbers will see that it makes good sense to consider this descending chain of subgroups. A high power of p in the denominator means, in the p-adic sense, that the number is very big. As we go down the chain of subgroups $C(p^\nu)$, we find points (x,y) with bigger and bigger coordinates in the p-adic sense; points which are closer and closer to infinity, and hence closer and closer to the zero element of our group. The $C(p^\nu)$'s are neighborhoods of $\mathcal{O}$ in the p-adic topology. But this is all by way of motivation; we will not actually need to know anything about p-adic numbers for the proof.

First we are going to change coordinates and move the point at infinity to a finite place. We will let

$$t = \frac{x}{y} \qquad \text{and} \qquad s = \frac{1}{y}.$$

Then $y^2 = x^3 + ax^2 + bx + c$ becomes

$$s = t^3 + at^2s + bts^2 + cs^3$$

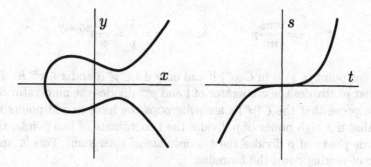

Two Views of a Cubic Curve

Figure 2.6

in the (t, s) plane. We can always get back the old coordinates, of course, because $y = 1/s$ and $x = t/s$. In the (t, s) plane we have all of the points in the old (x, y) plane except for the points where $y = 0$; and the zero element $\mathcal{O}$ on our curve is now at the origin $(0, 0)$ in the (t, s) plane.

You can visualize the situation this way. We have two views of the curve. The view in the (x, y) plane shows us everything except $\mathcal{O}$. The view in the (t, s) plane shows us $\mathcal{O}$ and everything except the points of order two. Except for $\mathcal{O}$ and the points of order two, there is a one-to-one correspondence between points on the curve in the (x, y) plane and points on the curve in the (t, s) plane. (See Figure 2.6.)

Further, a line $y = \lambda x + \nu$ in the (x, y) plane corresponds to a line in the (t, s) plane. Namely, if we divide $y = \lambda x + \nu$ by νy, we get

$$\frac{1}{\nu} = \frac{\lambda}{\nu}\frac{x}{y} + \frac{1}{y}, \qquad \text{so} \qquad s = -\frac{\lambda}{\nu}t + \frac{1}{\nu}.$$

Thus, we can add points in the (t, s) plane by the same procedure as in the (x, y) plane. We need to find the explicit formula.

It is convenient to look at a certain ring which we denote by R or R_p. This ring R will be the set of all rational numbers with no p in the denominator. Note that R is a ring, because if α and β have no p in their denominators, then the same is true of $\alpha \pm \beta$ and $\alpha\beta$.

Another way of describing R is to say that it consists of zero together with all non-zero rational numbers x such that $\text{ord}(x) \geq 0$. The ring R is a certain subring of the field of rational numbers. It is a marvelous ring in the sense that it has unique factorization; and it has only one prime, the prime p. The units of R are just the rational numbers of order zero, that is, numbers with numerator and denominator prime to p.

Let's look at the divisibility of our new coordinates s, t by powers of p, in particular for points in $C(p)$. Let (x, y) be a rational point of our curve in the (x, y) plane lying in $C(p^\nu)$, so we can write

$$x = \frac{m}{np^{2(\nu+i)}} \qquad \text{and} \qquad y = \frac{u}{wp^{3(\nu+i)}}$$

for some $i \geq 0$. Then

$$t = \frac{x}{y} = \frac{mw}{nu}p^{\nu+i} \qquad \text{and} \qquad s = \frac{1}{y} = \frac{w}{u}p^{3(\nu+i)}.$$

Thus, our point (t, s) is in $C(p^\nu)$ if and only if $t \in p^\nu R$ and $s \in p^{3\nu} R$. This says that p^ν divides the numerator of t and $p^{3\nu}$ divides the numerator of s.

To prove that the $C(p^\nu)$'s are subgroups, we have to add points and show that if a high power of p divides the t coordinate of two points, then the same power of p divides the t coordinate of their sum. This is just a matter of writing down the formulas.

Let $P_1 = (t_1, s_1)$ and $P_2 = (t_2, s_2)$ be distinct points. If $t_1 = t_2$, then $P_1 = -P_2$, so $P_1 + P_2$ is certainly in $C(p^\nu)$. Assume now that $t_1 \neq t_2$, and let $s = \alpha t + \beta$ be the line through P_1 and P_2. The slope α is given by

$$\alpha = \frac{s_2 - s_1}{t_2 - t_1}.$$

We can rewrite this as follows. The points (t_1, s_1) and (t_2, s_2) satisfy the equation

$$s = t^3 + at^2 s + bts^2 + cs^3.$$

Subtracting the equation for P_1 from the equation for P_2 and factoring gives

$$
\begin{aligned}
s_2 - s_1 &= (t_2^3 - t_1^3) + a(t_2^2 s_2 - t_1^2 s_1) + b(t_2 s_2^2 - t_1 s_1^2) + c(s_2^3 - s_1^3) \\
&= (t_2^3 - t_1^3) + a\{(t_2^2 - t_1^2)s_2 + t_1^2(s_2 - s_1)\} \\
&\qquad + b\{(t_2 - t_1)s_2^2 + t_1(s_2^2 - s_1^2)\} + c(s_2^3 - s_1^3).
\end{aligned}
$$

Some of the terms are divisible by $s_2 - s_1$, and some of the terms are divisible by $t_2 - t_1$. Factoring these quantities out, we can express their ratio in terms of what is left, finding (after some calculation)

$$\alpha = \frac{s_2 - s_1}{t_2 - t_1} = \frac{t_2^2 + t_1 t_2 + t_1^2 + a(t_2 + t_1)s_2 + bs_2^2}{1 - at_1^2 - bt_1(s_2 + s_1) - c(s_2^2 + s_1 s_2 + s_1^2)}. \qquad (*)$$

The point of all this, as we will see, was to get the 1 in the denominator of α, so the denominator of α will be a unit in R.

Similarly, if $P_1 = P_2$, then the slope of the tangent line to C at P_1 is

$$\alpha = \frac{ds}{dt}(P_1) = \frac{3t_1^2 + 2at_1 s_1 + bs_1^2}{1 - at_1^2 - 2bt_1 s_1 - 3cs_1^2}.$$

Notice that this is the same slope we get if we substitute $t_2 = t_1$ and $s_2 = s_1$ into the right-hand side of $(*)$. So we may use $(*)$ in all cases.

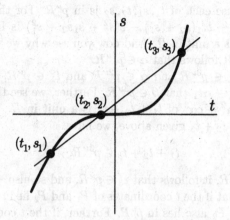

Adding Points in the (t, s) Plane

Figure 2.7

Let $P_3 = (t_3, s_3)$ be the third point of intersection of the line $s = \alpha t + \beta$ with the curve. (See Figure 2.7.) To get the equation whose roots are t_1, t_2, t_3, we substitute $\alpha t + \beta$ for s in the equation of the curve:

$$\alpha t + \beta = t^3 + at^2(\alpha t + \beta) + bt(\alpha t + \beta)^2 + c(\alpha t + \beta)^3.$$

Multiplying this out and collecting powers of t gives

$$0 = (1 + a\alpha + b\alpha^2 + c\alpha^3)t^3 + (a\beta + 2b\alpha\beta + 3c\alpha^2\beta)t^2 + \cdots.$$

This equation has roots t_1, t_2, t_3, so the right hand side equals

$$\text{constant} \cdot (t - t_1)(t - t_2)(t - t_3).$$

Comparing the coefficients of t^3 and t^2, we find that the sum of the roots is

$$t_1 + t_2 + t_3 = -\frac{a\beta + 2b\alpha\beta + 3c\alpha^2\beta}{1 + a\alpha + b\alpha^2 + c\alpha^3}.$$

That is all the formulas we will need except for the trivial one

$$\beta = s_1 - \alpha t_1$$

saying that the line goes through P_1.

We now have a formula for t_3, so how do we find $P_1 + P_2$? We draw the line through (t_3, s_3) and the zero element $(0, 0)$ and take the third intersection with the curve. It is clear at once from the equation of the curve that if (t, s) is on the curve, then so is $(-t, -s)$. So this third intersection is $(-t_3, -s_3)$.

Let's look more closely at the expression for α. The numerator of α lies in $p^{2\nu}R$, because each of t_1, s_1, t_2, s_2 is in $p^\nu R$. For the same reason, the quantity $-at_1^2 - bt_1(s_2 + s_1) - c(s_2^2 + s_1 s_2 + s_1^2)$ is in $p^{2\nu}R$, so the denominator of α is a unit in R. And now you see why we wanted the 1 in the denominator. It follows that $\alpha \in p^{2\nu}R$.

Next, since $s_1 \in p^{3\nu}R$ and $\alpha \in p^{2\nu}R$ and $t_1 \in p^\nu R$, it follows from the formula $\beta = s_1 - \alpha t_1$ that $\beta \in p^{3\nu}R$. Further, we see that the denominator $1 + a\alpha + b\alpha^2 + c\alpha^3$ of $t_1 + t_2 + t_3$ is a unit in R. Looking at the expression for $t_1 + t_2 + t_3$ given above, we have

$$t_1 + t_2 + t_3 \in p^{3\nu}R.$$

Because $t_1, t_2 \in p^\nu R$, it follows that $t_3 \in p^\nu R$, and so also $-t_3 \in p^\nu R$.

This proves that if the t coordinates of P_1 and P_2 lie in $p^\nu R$, then the t coordinate of $P_1 + P_2$ also lies in $p^\nu R$. Further, if the t coordinate of $P = (t, s)$ lies in $p^\nu R$, then it is clear that the t coordinate of $-P = (-t, -s)$ also lies in $p^\nu R$. This shows that $C(p^\nu)$ is closed under addition and taking negatives; hence it is a subgroup of $C(\mathbb{Q})$.

In fact, we have proven something a bit stronger. We have shown that if $P_1, P_2 \in C(p^\nu)$, then

$$t(P_1) + t(P_2) - t(P_1 + P_2) \in p^{3\nu}R.$$

Here we are writing $t(P)$ to denote the t coordinate of P; so if P is given in (x, y) coordinates as $\big(x(P), y(P)\big)$, then $t(P) = x(P)/y(P)$.

This last formula tells us more than the mere fact that $C(p^\nu)$ is a subgroup. A more suggestive way to write it is

$$t(P_1 + P_2) \equiv t(P_1) + t(P_2) \pmod{p^{3\nu}R}.$$

Note that the $+$ in $P_1 + P_2$ is the addition on our cubic curve, whereas the $+$ in $t(P_1) + t(P_2)$ is addition in R, which is just addition of rational numbers. So the map $P \mapsto t(P)$ is practically a homomorphism from $C(p^\nu)$ into the additive group of rational numbers. It does not quite define a homomorphism, because $t(P_1 + P_2)$ is not actually equal to $t(P_1) + t(P_2)$. However, what we do get is a homomorphism from $C(p^\nu)$ to the quotient group $p^\nu R/p^{3\nu}R$ by sending P to $t(P)$; and the kernel of this homomorphism consists of all points P with $t(P) \in p^{3\nu}R$. Thus, the kernel is just $C(p^{3\nu})$, so we finally obtain a one-to-one homomorphism

$$\frac{C(p^\nu)}{C(p^{3\nu})} \longrightarrow \frac{p^\nu R}{p^{3\nu}R},$$

$$P = (x, y) \longmapsto t(P) = \frac{x}{y}.$$

It is not hard to see that the quotient group $p^\nu R/p^{3\nu}R$ is a cyclic group of order $p^{2\nu}$. It follows that the quotient group $C(p^\nu)/C(p^{3\nu})$ is a cyclic group of order p^σ for some $0 \le \sigma \le 2\nu$.

We summarize our results so far in the following proposition.

Proposition. Let p be a prime, R the ring of rational numbers with denominator prime to p, and let $C(p^\nu)$ be the set of rational points (x, y) on our curve for which x has a denominator divisible by $p^{2\nu}$, plus the point $\mathcal{O}$.

(a) $C(p)$ consists of all rational points (x, y) for which the denominator of either x or y is divisible by p.

(b) For every $\nu \geq 1$, the set $C(p^\nu)$ is a subgroup of the group of rational points $C(\mathbb{Q})$.

(c) The map

$$\frac{C(p^\nu)}{C(p^{3\nu})} \longrightarrow \frac{p^\nu R}{p^{3\nu} R}, \qquad P = (x, y) \longmapsto t(P) = \frac{x}{y}$$

is a one-to-one homomorphism. (By convention, we send $\mathcal{O} \mapsto 0$.)

Using this proposition, it is not hard to prove our claim that points of finite order have integer coordinates.

Corollary. (a) For every prime p, the subgroup $C(p)$ contains no points of finite order (other than $\mathcal{O}$).

(b) Let $P = (x, y) \neq \mathcal{O}$ be a rational point of finite order. Then x and y are integers.

PROOF. Let the order of P be m. Since $P \neq \mathcal{O}$, we know $m \neq 1$. Take any prime p. We want to show that $P \notin C(p)$. Suppose to the contrary that $P \in C(p)$. We will derive a contradiction.

The point $P = (x, y)$ may be contained in a smaller group $C(p^\nu)$, but it cannot be contained in all of the groups $C(p^\nu)$, because the denominator of x cannot be divisible by arbitrarily high powers of p. So we can find some $\nu > 0$ so that $P \in C(p^\nu)$, but $P \notin C(p^{\nu+1})$. We separate the proof into two cases depending on whether or not m is divisible by p.

Suppose first that $p \nmid m$. Repeated application of the congruence

$$t(P_1 + P_2) \equiv t(P_1) + t(P_2) \pmod{p^{3\nu} R}$$

gives the formula

$$t(mP) \equiv mt(P) \pmod{p^{3\nu} R}.$$

Since $mP = \mathcal{O}$, we have $t(mP) = t(\mathcal{O}) = 0$. On the other hand, since m is prime to p, it is a unit in R. Therefore,

$$0 \equiv t(P) \pmod{p^{3\nu} R}.$$

This means that $P \in C(p^{3\nu})$, contradicting the fact that $P \notin C(p^{\nu+1})$.

Next we suppose that $p \mid m$. This case is handled similarly. First we write $m = pn$, and look at the point $P' = nP$. Since P has order m, it is clear that P' has order p. Further, since $P \in C(p)$ and $C(p)$ is a subgroup, we see that $P' \in C(p)$. As above, we can find some $\nu > 0$ so that $P' \in C(p^\nu)$, but $P' \notin C(p^{\nu+1})$. Then, just as before, we find

$$0 = t(\mathcal{O}) = t(pP') \equiv pt(P') \pmod{p^{3\nu}R}.$$

This means that $t(P') \equiv 0 \pmod{p^{3\nu-1}R}$. Since $3\nu - 1 \geq \nu + 1$, this contradicts the fact that $P' \notin C(p^{\nu+1})$. That completes the proof of part (a) of the corollary.

But now part (b) is easy, because if $P = (x, y)$ is a point of finite order, we now know that $P \notin C(p)$ for all primes p. This means that the denominators of x and y are divisible by no primes; hence, x and y are integers. □

5. The Nagell-Lutz Theorem and Further Developments

We have really finished the proof of the Nagell-Lutz theorem, but to wrap everything up we will state it formally and remind you of the two parts of the proof.

Nagell-Lutz Theorem. *Let*

$$y^2 = f(x) = x^3 + ax^2 + bx + c$$

be a non-singular cubic curve with integer coefficients a, b, c; and let D be the discriminant of the cubic polynomial $f(x)$,

$$D = -4a^3c + a^2b^2 + 18abc - 4b^3 - 27c^2.$$

Let $P = (x, y)$ be a rational point of finite order. Then x and y are integers; and either $y = 0$, in which case P has order two, or else y divides D.

Remark. For computational purposes, the following stronger form of the Nagell-Lutz theorem is often useful. Let $P = (x, y)$ be a rational point of finite order with $y \neq 0$. Then y^2 divides the discriminant D. We leave the proof of this stronger statement for the exercises.

PROOF. In Section 4 we showed that a point of finite order has integer coordinates. If P has order two, then we know that $y = 0$, so we are done. Otherwise, $2P \neq \mathcal{O}$. But $2P$ is also a point of finite order, so it also has integer coordinates. In Section 3 we showed that if both $P = (x, y)$ and $2P$ have integer coordinates, then y divides D, which completes the proof of the theorem. □

Warning. We want to reiterate that the Nagell-Lutz theorem is not an "if and only if" statement. It is quite possible to have points with integer coordinates and with y dividing D which are not points of finite order. The Nagell-Lutz theorem can be used to compile a list of points which includes all points of finite order; but it can never be used to prove that any particular point actually has finite order. To verify that a point P has finite order, one must find an integer $n \geq 1$ such that $nP = \mathcal{O}$.

On the other hand, the Nagell-Lutz theorem can often be used to prove that a point has *infinite* order. The idea is to compute $P, 2P, \ldots$ until one arrives at some multiple nP whose coordinates are not integers. Then one knows that nP, and a fortiori also P, cannot have finite order. This computation can be accelerated by computing instead only the x coordinates of $2P, 4P, 8P, \ldots$ using the duplication formula until some x coordinate is not an integer.

The question naturally arises as to what points of finite order can occur. We have already seen that it is easy to get points of order two by taking the cubic polynomial to have a rational root. Similarly, using our description of the points of order three, it is not hard to find cubic curves such that $C(\mathbb{Q})$ has a point of order three. On the other hand, we have indicated why it is not possible to find two independent points of order three in $C(\mathbb{Q})$, because it is not even possible in the larger group $C(\mathbb{R})$.

It is also possible to find rational points of higher order. For example, the point $P = (1,1)$ on the curve

$$y^2 = x^3 - x^2 + x$$

has order four, since one easily checks that $2P = (0,0)$, and we know that $(0,0)$ has order two. Then $3P = -P = (1,-1)$ is also a point of order four.

We can use the Nagell-Lutz theorem to check that there are no other points of finite order on this curve. Here $D = -3$, so the possible values for y are ± 1 and ± 3. We already know that $y = \pm 1$ gives points of order four, so we check $y = \pm 3$. This leads to the equation $x^3 - x^2 + x - 9 = 0$. The only possible rational roots are integers dividing 9, and one quickly checks that ± 1, ± 3, and ± 9 are not roots. So the only points of finite order are the ones we know; and the subgroup which consists of all points of finite order is a cyclic group of order four.

In fact, there are infinitely many curves with a rational point of order four. For every rational number $t \neq 0, \frac{1}{4}$, the point (t,t) on the non-singular cubic curve

$$y^2 = x^3 - (2t-1)x^2 + t^2 x$$

is a point of order four. (You should check this.)

In a similar fashion, one can write down infinitely many examples of curves with rational points of orders $5, 6, 7, 8, 9, 10$, or 12. In essence, these examples were written down during the second half of the 19th century. But no one was ever able to find even a single example of a cubic curve with a rational point of order eleven. There is a good reason for this because Billing and Mahler [1] proved in 1940 that no such curve exists!

Many people worked on the problem of determining just which orders are possible, culminating in a very beautiful and very difficult theorem of Mazur [1,2]. We will not even be able to indicate how the proof goes, but the statement of the theorem is as follows.

Mazur's Theorem. *Let C be a non-singular rational cubic curve, and suppose that $C(\mathbb{Q})$ contains a point of finite order m. Then either*

$$1 \leq m \leq 10 \quad \text{or} \quad m = 12.$$

More precisely, the set of all points of finite order in $C(\mathbb{Q})$ forms a subgroup which has one of the following two forms:

 (i) *A cyclic group of order N with $1 \leq N \leq 10$ or $N = 12$.*

 (ii) *The product of a cyclic group of order two and a cyclic group of order $2N$ with $1 \leq N \leq 4$.*

EXERCISES

2.1. Let A be an abelian group, and for every integer $m \geq 1$, let A_m be the set of elements $P \in A$ satisfying $mP = \mathcal{O}$.
  (a) Prove that A_m is a subgroup of A.
  (b) Suppose that A has order M^2, and further that for every integer m dividing M, the subgroup A_m has order m^2. Prove that A is the direct product of two cyclic groups of order M.
  (c) Find an example of a non-abelian group G and an integer $m \geq 1$ so that the set $G_m = \{g \in G : g^m = e\}$ is not a subgroup of G.

2.2. Let C be a non-singular cubic curve given by the usual Weierstrass equation
$$y^2 = f(x) = x^3 + ax^2 + bx + c.$$

  (a) Prove that
$$\frac{d^2 y}{dx^2} = \frac{2f''(x)f(x) - f'(x)^2}{4yf(x)} = \frac{\psi_3(x)}{4yf(x)}.$$

Use this to deduce that a point $P = (x, y) \in C$ is a point of order three if and only if $P \neq \mathcal{O}$ and P is a point of inflection on the curve C.
  (b) Suppose now that a, b, c are in $\mathbb{R}$. Prove that $\psi_3(x)$ has exactly two real roots, say α_1 and α_2 with $\alpha_1 < \alpha_2$. Prove that $f(\alpha_1) < 0$ and $f(\alpha_2) > 0$. Use this to deduce that the points in $C(\mathbb{R})$ of order dividing 3 form a cyclic group of order three.

2.3. Let $\omega_1, \omega_2 \in \mathbb{C}$ be two complex numbers which are $\mathbb{R}$-linearly independent, and let
$$L = \mathbb{Z}\omega_1 + \mathbb{Z}\omega_2 = \{n_1\omega_1 + n_2\omega_2 : n_1, n_2 \in \mathbb{Z}\}$$

be the lattice of $\mathbb{C}$ that they generate.

(a) Show that the series

$$\wp(u) = \frac{1}{u^2} + \sum_{\substack{\omega \in L \\ \omega \neq 0}} \left(\frac{1}{(u - \omega)^2} - \frac{1}{\omega^2} \right)$$

defining the Weierstrass $\wp$ function is absolutely and uniformly convergent on any compact subset of the complex u plane which does not contain any of the points of L. Conclude that $\wp$ is a meromorphic function with a double pole at each point of L and that $\wp$ has no other poles.

(b) Prove that $\wp$ is an even function [i.e., $\wp(-u) = \wp(u)$].

(c) Show that $\wp$ is a doubly periodic function; that is, show that

$$\wp(u + \omega) = \wp(u) \quad \text{for every } u \in \mathbb{C} \text{ and every } \omega \in L.$$

(*Hint.* From (a), you can calculate the derivative $\wp'(u)$ by differentiating each term of the series defining $\wp(u)$. First prove $\wp'(u+\omega) = \wp'(u)$, then integrate.)

2.4. Let C be the cubic curve

$$y^2 = x^3 + 1.$$

(a) For each prime $5 \le p < 30$, describe the group of points on this curve having coordinates in the finite field with p elements.

(b) For each prime in (a), let M_p be the number of points in the group. (Don't forget the point at infinity.) For the set of primes satisfying $p \equiv 2 \,(\text{mod } 3)$, can you see a pattern for the values of M_p? Make a general conjecture for the value of M_p when $p \equiv 2 \,(\text{mod } 3)$ and prove that your conjecture is correct.

(c) ** Can you find a pattern for the value of M_p for the set of primes satisfying $p \equiv 1 \,(\text{mod } 3)$? Then compute M_{31} and see if it fits your pattern. If not, make a new conjecture and compute the next few M_p's to test your conjecture.

(d) Answer the same questions as in (a) and (b) for the cubic curve $y^2 = x^3 + x$. Note that in (b) you will have to replace the condition $p \equiv 2 \,(\text{mod } 3)$ by some other congruence condition.

2.5. (a) Let $f(x) = x^2 + ax + b = (x - \alpha_1)(x - \alpha_2)$ be a quadratic polynomial with the indicated factorization. Prove that

$$(\alpha_1 - \alpha_2)^2 = a^2 - 4b.$$

(b) Let $f(x) = x^3 + ax^2 + bx + c = (x - \alpha_1)(x - \alpha_2)(x - \alpha_3)$ be a cubic polynomial with the indicated factorization. Prove that

$$(\alpha_1 - \alpha_2)^2 (\alpha_1 - \alpha_3)^2 (\alpha_2 - \alpha_3)^2$$
$$= -4a^3 c + a^2 b^2 + 18abc - 4b^3 - 27c^2.$$

(c) * Let

$$f(x) = x^n + a_1 x^{n-1} + \cdots + a_n = (x - \alpha_1)(x - \alpha_2) \cdots (x - \alpha_n)$$

be a polynomial with the indicated factorization. The *discriminant of f* is defined to be

$$\mathrm{Disc}(f) = \prod_{i=1}^{n-1} \prod_{j=i+1}^{n} (\alpha_i - \alpha_j)^2,$$

so $\mathrm{Disc}(f) = 0$ if and only if f has a double root. Prove that $\mathrm{Disc}(f)$ can be expressed as a polynomial in the coefficients $a_1, \ldots, a_n$ of f.

2.6. Let p be a prime, and for a rational number $r = \dfrac{m}{n} p^\nu$ with m, n prime to p, let $\mathrm{ord}(r) = \nu$ be as in the text. [By convention, we will set $\mathrm{ord}(0) = \infty$.]

(a) Prove that for all rational numbers r_1 and r_2,

$$\mathrm{ord}(r_1 r_2) = \mathrm{ord}(r_1) + \mathrm{ord}(r_2).$$

(b) Prove that for all rational numbers r_1 and r_2,

$$\mathrm{ord}(r_1 + r_2) \geq \min\{\mathrm{ord}(r_1), \mathrm{ord}(r_2)\}.$$

Further, if $\mathrm{ord}(r_1) \neq \mathrm{ord}(r_2)$, prove that the inequality is an equality.

(c) Define an "absolute value" on the rational numbers by the rule

$$\|r\| = \frac{1}{p^{\mathrm{ord}(r)}}.$$

(By convention, we set $\|0\| = 0$.) Prove that $\| \cdot \|$ has the following properties:

(i) $\|r\| \geq 0$; and $\|r\| = 0$ if and only if $r = 0$.

(ii) $\|r_1 r_2\| = \|r_1\| \cdot \|r_2\|$.

(iii) $\|r_1 + r_2\| \leq \max\{\|r_1\|, \|r_2\|\}$.

Notice that property (iii) is stronger than the usual triangle inequality. The absolute value $\| \cdot \|$ is called the *p-adic absolute value* on the rational numbers. It can be used to define a topology on the rational numbers, the *p-adic topology*.

2.7. Let p be a prime, and let $R = R_p$ be the set described in the text,

$$R = \{\text{non-zero rational numbers } x \text{ such that } \text{ord}(x) \geq 0\} \cup \{0\}$$
$$= \{x \in \mathbb{Q} : \|x\| \leq 1\}.$$

(Here $\| \cdot \|$ is the p-adic absolute value defined in the previous exercise. So the set R is a p-adic analogue of the interval $[-1, 1]$ on the real line or of the unit disk $\{z \in \mathbb{C} : |z| \leq 1\}$ in the complex plane.)

(a) Prove that R is a subring of the rational numbers.

(b) Prove that the ideal generated by p is a maximal ideal, and describe the quotient field R/pR.

(c) Prove that the unit group of R consists of all rational numbers a/b such that p does not divide ab.

(d) Prove that R is a unique factorization domain.

(e) Describe all of the ideals of R. Use this description to prove that the ideal generated by p is the only maximal ideal of R. (A ring such as R, which has exactly one maximal ideal, is called a *local ring*.)

2.8. Let p and R be as in the previous exercise, and let $\sigma \geq \nu$ be integers. Prove that the quotient group $p^\nu R / p^\sigma R$ is a cyclic group of order $p^{\sigma - \nu}$.

2.9. Let p be a prime, and let $S = S_p$ be the set of rational numbers whose denominator is a power of p. ($p^0 = 1$ is allowed.) Thus, S consists of all rational numbers ap^ν, where a is an integer prime to p and ν is an arbitrary integer.

(a) Prove that S is a subring of the rational numbers.

(b) Prove that the unit group in S consists of all numbers $\pm p^\nu$ with ν any integer.

(c) Let q be a prime other than p. Prove that q generates a maximal ideal of S, and describe the quotient field S/qS. Prove that every maximal ideal of S is of this form.

2.10. Let $p \geq 2$ be a prime and let C be the cubic curve

$$C : y^2 = x^3 + px.$$

Find all points of finite order in $C(\mathbb{Q})$.

2.11. As usual, let C be a non-singular cubic curve given by an equation

$$y^2 = f(x) = x^3 + ax^2 + bx + c$$

with integer coefficients. We proved in Chapter I, Section 4, that if $P = (x, y)$ is a point on C, then the x coordinate of $2P$ is given by the formula

$$x(2P) = \frac{\phi(x)}{4f(x)} = \frac{x^4 - 2bx^2 - 8cx + b^2 - 4ac}{4(x^3 + ax^2 + bx + c)},$$

where $\phi(x)$ is the indicated polynomial.

(a) Let $D = -4a^3c + a^2b^2 + 18abc - 4b^3 - 27c^2$ be the discriminant of $f(x)$. Find polynomials $F(X)$ and $\Phi(X)$ with integer coefficients so that

$$F(X)f(X) + \Phi(X)\phi(X) = D.$$

(*Hint.* $F(X)$ has degree 3 and $\Phi(X)$ has degree 2.)

(b) If $P = (x,y)$ is a rational point of finite order on C, prove that either $2P = \mathcal{O}$ or else $y^2|D$. (This is the strong form of the Nagell-Lutz theorem.)

2.12. For each of the following cubic curves, determine all of the points of finite order. Also determine the structure of the group formed by these points. (You may need to complete the square on the left before you can use the Nagell-Lutz theorem. Feel free to use the strong form of the Nagell-Lutz theorem described in the previous exercise. The results proven in Chapter IV, Section 3, might also help in limiting the amount of computation you need to do.)

(a) $y^2 = x^3 - 2$

(b) $y^2 = x^3 + 8$

(c) $y^2 = x^3 + 4$

(d) $y^2 = x^3 + 4x$

(e) $y^2 - y = x^3 - x^2$

(f) $y^2 = x^3 + 1$

(g) $y^2 = x^3 - 43x + 166$

(h) $y^2 + 7xy = x^3 + 16x$

(i) $y^2 + xy + y = x^3 - x^2 - 14x + 29$

(j) $y^2 + xy = x^3 - 45x + 81$

(k) $y^2 + 43xy - 210y = x^3 - 210x^2$

(l) $y^2 = x^3 - 4x$

(m) $y^2 + xy - 5y = x^3 - 5x^2$

(n) $y^2 + 5xy - 6y = x^3 - 3x^2$

(o) $y^2 + 17xy - 120y = x^3 - 60x^2$

CHAPTER III

The Group of Rational Points

1. Heights and Descent

In this chapter we will prove Mordell's theorem that the group of rational points on a non-singular cubic is finitely generated. There is a tool used in the proof called the *height*. In brief, the height of a rational point measures how complicated the point is from the viewpoint of number theory.

We begin by defining the height of a rational number. Let $x = \dfrac{m}{n}$ be a rational number written in lowest terms. Then we define the height $H(x)$ to be the maximum of the absolute values of the numerator and the denominator:

$$H(x) = H\left(\frac{m}{n}\right) = \max\{|m|, |n|\}.$$

The height of a rational number is a positive integer.

Why is the height a good way of measuring how complicated a rational number is? For example, why not just take the absolute value $|x|$? Consider the two rational numbers 1 and $\dfrac{99999}{100000}$. They both have about the same absolute value, but the latter is clearly much more "complicated" than the former, at least if one is interested in doing number theory. If this reason is not convincing enough, then possibly the following property of the height will explain why it is a useful notion.

> **Finiteness Property of the Height.** The set of all rational numbers whose height is less than some fixed number is a finite set.

The proof of this fact is easy. If the height of $x = \dfrac{m}{n}$ is less than some fixed constant, then both $|m|$ and $|n|$ are less than that constant, so there are only finitely many possibilities for m and n.

If

$$y^2 = f(x) = x^3 + ax^2 + bx + c$$

is a non-singular cubic curve with integer coefficients a, b, c, and if $P = (x, y)$ is a rational point on the curve, we will define the *height of P* to be simply the height of its x coordinate,

$$H(P) = H(x).$$

We will see that the height behaves somewhat multiplicatively; for example, we will be comparing $H(P + Q)$ with the product $H(P)H(Q)$. For notational reasons it is often more convenient to have a function that behaves additively, so we also define the "small h" height by taking the logarithm:

$$h(P) = \log H(P).$$

So $h(P)$ is always a non-negative real number.

Note that the rational points on C also have the finiteness property. If M is any positive number, then

$$\{P \in C(\mathbb{Q}) : H(P) \leq M\}$$

is a finite set; and the same holds if we use $h(P)$ in place of $H(P)$. This is true because points in the set have only finitely many possibilities for their x coordinate; and for each x coordinate, there are only two possibilities for the y coordinate. That takes care of the points which are not at infinity. There is one point $\mathcal{O}$ at infinity, and we will define its height to be

$$H(\mathcal{O}) = 1, \quad \text{or equivalently, } h(\mathcal{O}) = 0.$$

Our ultimate goal is to prove that the group of rational points $C(\mathbb{Q})$ is finitely generated. This fact will follow from four lemmas. We are going to state the lemmas now and use them to prove the finite generation of $C(\mathbb{Q})$. After that, we will see about proving the lemmas.

Lemma 1. *For every real number M, the set*

$$\{P \in C(\mathbb{Q}) : h(P) \leq M\}$$

is finite.

Lemma 2. *Let P_0 be a fixed rational point on C. There is a constant κ_0, depending on P_0 and on a, b, c, so that*

$$h(P + P_0) \leq 2h(P) + \kappa_0 \quad \text{for all } P \in C(\mathbb{Q}).$$

Lemma 3. *There is a constant κ, depending on a, b, c, so that*

$$h(2P) \geq 4h(P) - \kappa \quad \text{for all } P \in C(\mathbb{Q}).$$

Notice Lemma 3 says that when you double a point, the height goes up quite a bit. So as soon as you get a point with large height, doubling it makes a much larger height. Notice also that Lemmas 2 and 3 relate the group law on C, which is defined geometrically, to the height of points, which is a number theoretic device. So in some sense one can think of the height as a tool to translate geometric information into number theoretic information.

Lemma 4. *The index $\big(C(\mathbb{Q}) : 2C(\mathbb{Q})\big)$ is finite.*

We are using the notation $2C(\mathbb{Q})$ to denote the subgroup of $C(\mathbb{Q})$ consisting of the points which are twice other points. For any commutative group Γ, the multiplication-by-m map

$$\Gamma \longrightarrow \Gamma, \qquad P \longmapsto \underbrace{P + \cdots + P}_{m \text{ terms}} = mP$$

is a homomorphism; and the image of that homomorphism is a subgroup $m\Gamma$ of Γ. The fourth lemma states that for $\Gamma = C(\mathbb{Q})$, the subgroup 2Γ has finite index in Γ.

These lemmas are in increasing order of difficulty. We have already proven Lemma 1. The middle two lemmas are related to the theory of heights of rational numbers; and if you know the formulas for adding and doubling points, then they can be proven without further reference to the curve C. Lemma 4 is much subtler to prove; and since we want to restrict ourselves to working with rational numbers, we will only be able to prove it for a certain, fairly large, class of cubic curves.

We will now show how these four lemmas imply that $C(\mathbb{Q})$ is a finitely generated group. If you like, you can completely forget about rational points on a curve. Just suppose that we are given a commutative group Γ, written additively, and a height function

$$h : \Gamma \longrightarrow [0, \infty)$$

from Γ to the non-negative real numbers. Suppose further that Γ and h satisfy the four lemmas. We will now restate our hypotheses and prove that Γ must be finitely generated.

Descent Theorem. *Let Γ be a commutative group. Suppose that there is a function*

$$h : \Gamma \longrightarrow [0, \infty)$$

with the following three properties.
(a) For every real number M, the set $\{P \in \Gamma : h(P) \le M\}$ is finite.
(b) For every $P_0 \in \Gamma$, there is a constant κ_0 so that

$$h(P + P_0) \le 2h(P) + \kappa_0 \qquad \text{for all } P \in \Gamma.$$

(c) There is a constant κ so that

$$h(2P) \ge 4h(P) - \kappa \qquad \text{for all } P \in \Gamma.$$

Suppose further that
(d) The subgroup 2Γ has finite index in Γ.
Then Γ is finitely generated.

PROOF. The first thing we do is take a representative for each coset of 2Γ in Γ. We know that there are only finitely many cosets, say n of them;

let $Q_1, Q_2, \ldots, Q_n$ be representatives for the cosets. This means that for any element $P \in \Gamma$ there is an index i_1, depending on P, such that

$$P - Q_{i_1} \in 2\Gamma.$$

After all, P has to be in one of the cosets. This means that we can write

$$P - Q_{i_1} = 2P_1$$

for some $P_1 \in \Gamma$. Now we do the same thing with P_1. Continuing this process, we find we can write

$$P_1 - Q_{i_2} = 2P_2$$
$$P_2 - Q_{i_3} = 2P_3$$
$$\vdots$$
$$P_{m-1} - Q_{i_m} = 2P_m,$$

where $Q_{i_1}, \ldots, Q_{i_m}$ are chosen from the coset representatives $Q_1, \ldots, Q_n$, and $P_1, \ldots, P_m$ are elements of Γ.

The basic idea is that because P_i is more-or-less equal to $2P_{i+1}$, the height of P_{i+1} is more-or-less one fourth the height of P_i. So the sequence of points $P, P_1, P_2, \ldots$ should have decreasing height, and eventually we will end up in a set of points having bounded height. From property (a), that set will be finite, which will complete the proof. Now we have to turn these vague remarks into a valid proof.

From the first equation we have

$$P = Q_{i_1} + 2P_1.$$

Now substitute the second equation $P_1 = Q_{i_2} + 2P_2$ into this to get

$$P = Q_{i_1} + 2Q_{i_2} + 4P_2.$$

Continuing in this fashion, we obtain

$$P = Q_{i_1} + 2Q_{i_2} + 4Q_{i_3} + \cdots + 2^{m-1}Q_{i_m} + 2^m P_m.$$

In particular, this says that P is in the subgroup of Γ generated by the Q_i's and P_m. We are going to show that by choosing m large enough, we can force P_m to have height less than a certain fixed bound. Then the finite set of points with height less than that bound, together with the Q_i's, will generate Γ.

Let's take one of the P_j's in the sequence of points $P, P_1, P_2, \ldots$ and examine the relation between the height of P_{j-1} and the height of P_j. We want to show that the height of P_j is considerably smaller. To do that, we need to specify some constants. If we apply (b) with $-Q_i$ in place of P_0, then we get a constant κ_i such that

$$h(P - Q_i) \le 2h(P) + \kappa_i \qquad \text{for all } P \in \Gamma.$$

We do this for each Q_i, $1 \le i \le n$. Let κ' be the largest of the κ_i's. Then

$$h(P - Q_i) \le 2h(P) + \kappa' \qquad \text{for all } P \in \Gamma \text{ and all } 1 \le i \le n.$$

We can do this because there are only finitely many Q_i's. This is one place we are using the property (d) that 2Γ has finite index in Γ.

Let κ be the constant from (c). Then we can calculate

$$4h(P_j) \le h(2P_j) + \kappa = h(P_{j-1} - Q_{i_j}) + \kappa \le 2h(P_{j-1}) + \kappa' + \kappa.$$

We rewrite this as

$$h(P_j) \le \frac{1}{2}h(P_{j-1}) + \frac{\kappa' + \kappa}{4}$$
$$= \frac{3}{4}h(P_{j-1}) - \frac{1}{4}\big(h(P_{j-1}) - (\kappa' + \kappa)\big).$$

From this we see that if $h(P_{j-1}) \ge \kappa' + \kappa$, then

$$h(P_j) \le \frac{3}{4}h(P_{j-1}).$$

So in the sequence of points $P, P_1, P_2, P_3, \ldots$, as long as the point P_j satisfies the condition $h(P_j) \ge \kappa' + \kappa$, then the next point in the sequence has much smaller height, namely, $h(P_{j+1}) \le \frac{3}{4}h(P_j)$. But if you start with a number and keep multiplying it by 3/4, then it approaches zero. So eventually we will find an index m so that $h(P_m) \le \kappa' + \kappa$.

We have now shown that every element $P \in \Gamma$ can be written in the form

$$P = a_1Q_1 + a_2Q_2 + \cdots + a_nQ_n + 2^m R$$

for certain integers $a_1, \ldots, a_n$ and some point $R \in \Gamma$ satisfying the inequality $h(R) \le \kappa' + \kappa$. Hence, the set

$$\{Q_1, Q_2, \ldots, Q_n\} \cup \big\{R \in \Gamma : h(R) \le \kappa' + \kappa\big\}$$

generates Γ. From (a) and (d), this set is finite, which completes the proof that Γ is finitely generated.   □

We have called this a Descent Theorem because the proof is very much in the style of Fermat's method of infinite descent. One starts with an arbitrary point, in our case a point $P \in C(\mathbb{Q})$, and by some clever manipulation one produces (descends to) a smaller point. Of course, one needs to have a way of measuring the size of a point; we have used the height. If one is lucky, repeated application of this idea leads to one of two possible conclusions. In our case we were led to a finite set of generating points; then all of the points arise from this finite generating set by reversing the descent procedure. In other cases, one is led to a contradiction, usually the existence of an integer strictly between zero and one. Then one can conclude that there are no solutions. This is the method that Fermat used to show that $x^4 + y^4 = 1$ has no rational solutions with $xy \neq 0$; and it is undoubtedly the idea he had in mind to prove the same thing for $x^n + y^n = 1$ for any $n \geq 3$. Unfortunately, additional complications arise as n increases, so no one has yet been able to verify Fermat's claim.

In view of the Descent Theorem, and the proof of Lemma 1 already given above, it remains to prove Lemmas 2, 3, and 4. This will occupy us for the next several sections.

2. The Height of $P + P_0$

In this section we will prove Lemma 2, which gives a relationship between the heights of P, P_0, and $P + P_0$. Before beginning we make a couple of remarks.

The first remark is that if $P = (x, y)$ is a rational point on our curve, then x and y have the form

$$x = \frac{m}{e^2} \qquad \text{and} \qquad y = \frac{n}{e^3}$$

for integers m, n, e with $e > 0$ and $\gcd(m, e) = \gcd(n, e) = 1$. In other words, if you write the coordinates of a rational point in lowest terms, then the denominator of x is the square of a number whose cube is the denominator of y. We essentially proved this in Chapter II, because we showed that if p^ν divides the denominator of x, then ν is even and $p^{3\nu/2}$ divides the denominator of y. However, since what we want to know is so easy to prove, we will prove it again without resorting to studying one prime at a time.

Thus, suppose we write

$$x = \frac{m}{M} \qquad \text{and} \qquad y = \frac{n}{N}$$

in lowest terms with $M > 0$ and $N > 0$. Substituting into the equation of the curve gives

$$\frac{n^2}{N^2} = \frac{m^3}{M^3} + a\frac{m^2}{M^2} + b\frac{m}{M} + c;$$

and then clearing denominators yields

$$M^3 n^2 = N^2 m^3 + aN^2 M m^2 + bN^2 M^2 m + cN^2 M^3. \qquad (*)$$

Since N^2 is a factor of all terms on the right, we see that $N^2 | M^3 n^2$. But $\gcd(n, N) = 1$, so $N^2 | M^3$.

Now we want to prove the converse, that is, $M^3 | N^2$. This is done in three steps. First, from $(*)$ we immediately see that $M | N^2 m^3$; and since $\gcd(m, M) = 1$, we find $M | N^2$. Using this fact back in $(*)$, we find that $M^2 | N^2 m^3$, so $M | N$. Finally, using $(*)$ once again, we see that this implies that $M^3 | N^2 m^3$, so $M^3 | N^2$.

We have now shown that $N^2 | M^3$ and $M^3 | N^2$, so $M^3 = N^2$. Further, during the proof we showed that $M | N$. So if we let $e = \dfrac{N}{M}$, then we find that

$$e^2 = \frac{N^2}{M^2} = \frac{M^3}{M^2} = M \quad \text{and} \quad e^3 = \frac{N^3}{M^3} = \frac{N^3}{N^2} = N.$$

Therefore $x = \dfrac{m}{e^2}$ and $y = \dfrac{n}{e^3}$ have the desired form.

Our second remark concerns how we defined the height of the rational points on our curve. We just took the height of the x coordinate. If the point P is given in lowest terms as $P = \left(\dfrac{m}{e^2}, \dfrac{n}{e^3}\right)$, then the height of P is the maximum of $|m|$ and e^2. In particular, $|m| \leq H(P)$ and $e^2 \leq H(P)$. We claim that we can also bound the numerator of the y coordinate in terms of $H(P)$. Precisely, there is a constant $K > 0$, depending on a, b, c, such that

$$|n| \leq K H(P)^{3/2}.$$

To prove this we just use the fact that the point satisfies the equation. Substituting into the equation and multiplying by e^6 to clear denominators gives

$$n^2 = m^3 + ae^2 m^2 + be^4 m + ce^6.$$

Now take absolute values and use the triangle inequality.

$$\begin{aligned} |n^2| &\leq |m^3| + |ae^2 m^2| + |be^4 m| + |ce^6| \\ &\leq H(P)^3 + |a| H(P)^3 + |b| H(P)^3 + |c| H(P)^3. \end{aligned}$$

So we can take $K = \sqrt{1 + |a| + |b| + |c|}$.

We are now ready to prove Lemma 2, which we restate.

Lemma 2. Let P_0 be a fixed rational point on C. There is a constant κ_0, depending on P_0 and on a, b, c, so that

$$h(P + P_0) \leq 2h(P) + \kappa_0 \qquad \text{for all } P \in C(\mathbb{Q}).$$

PROOF. The proof is really nothing more than writing out the formula for the sum of two points and using the triangle inequality. We first remark that the lemma is trivial if $P_0 = \mathcal{O}$; so we can take $P_0 \neq \mathcal{O}$, say $P_0 = (x_0, y_0)$. Next we note that in proving the existence of κ_0, it is enough to prove that the inequality holds for all P except those in a some fixed finite set. This is true because, for any finite number of P, we just look at the differences $h(P + P_0) - 2h(P)$ and take κ_0 larger than the finite number of values that occur. Having said this, it suffices to prove Lemma 2 for all points P with $P \notin \{P_0, -P_0, \mathcal{O}\}$.

We write $P = (x, y)$. The reason for avoiding both P_0 and $-P_0$ is to have $x \neq x_0$, because then we can avoid using the duplication formula. We write

$$P + P_0 = (\xi, \eta).$$

To get the height of $P + P_0$, we need to calculate the height of ξ; so we need a formula for ξ in terms of (x, y) and (x_0, y_0). The formula we derived earlier looks this way:

$$\xi + x + x_0 = \lambda^2 - a, \qquad \lambda = \frac{y - y_0}{x - x_0}.$$

We need to write this out a little bit.

$$\begin{aligned}
\xi &= \frac{(y - y_0)^2}{(x - x_0)^2} - a - x - x_0 \\
&= \frac{(y - y_0)^2 - (x - x_0)^2(x + x_0 + a)}{(x - x_0)^2}.
\end{aligned}$$

If we multiply all of this out, then in the numerator we find $y^2 - x^3$ appearing. Since P is on the curve, the expression $y^2 - x^3$ can be replaced by $ax^2 + bx + c$. What we end up with is an expression

$$\xi = \frac{Ay + Bx^2 + Cx + D}{Ex^2 + Fx + G},$$

where A, B, C, D, E, F, G are certain rational numbers which can be expressed in terms of a, b, c, and (x_0, y_0). Further, multiplying the numerator and the denominator by the least common denominator of $A, \ldots, G$, we may assume that $A, \ldots, G$ are all integers.

In summary, we have integers $A, \ldots, G$, which depend only on a, b, c, and (x_0, y_0), so that for any point $P = (x, y) \notin \{P_0, -P_0, \mathcal{O}\}$, the x coordinate of $P + P_0$ is equal to

$$\xi = \frac{Ay + Bx^2 + Cx + D}{Ex^2 + Fx + G}.$$

The important point is that once the curve and the point P_0 are fixed, then this expression is correct for all points P. So it will be all right for our constant κ_0 to depend on $A, \ldots, G$, as long as it does not depend on (x, y).

Now substitute $x = \dfrac{m}{e^2}$ and $y = \dfrac{n}{e^3}$ and clear the denominators by multiplying the numerator and denominator by e^4. We find

$$\xi = \frac{Ane + Bm^2 + Cme^2 + De^4}{Em^2 + Fme^2 + Ge^4},$$

and now the result we want is almost evident. Notice that we have an expression for ξ as an integer divided by an integer. We do not know that it is in lowest terms, but cancellation will only make the height smaller. Thus,

$$H(\xi) \leq \max\{|Ane + Bm^2 + Cme^2 + De^4|, |Em^2 + Fme^2 + Ge^4|\}.$$

Further, from above we have the estimates

$$e \leq H(P)^{1/2}, \qquad n \leq KH(P)^{3/2}, \qquad m \leq H(P),$$

where K depends only on a, b, c. Using these and the triangle inequality gives

$$\begin{aligned} |Ane + Bm^2 + Cme^2 + De^4| &\leq |Ane| + |Bm^2| + |Cme^2| + |De^4| \\ &\leq (|AK| + |B| + |C| + |D|)H(P)^2; \end{aligned}$$

and

$$\begin{aligned} |Em^2 + Fme^2 + Ge^4| &\leq |Em^2| + |Fme^2| + |Ge^4| \\ &\leq (|E| + |F| + |G|)H(P)^2. \end{aligned}$$

Therefore,

$$H(P + P_0) = H(\xi) \leq \max\{|AK| + |B| + |C| + |D|, |E| + |F| + |G|\}H(P)^2.$$

Taking the logarithm of both sides gives

$$h(P + P_0) \leq 2h(P) + \kappa_0,$$

where the constant $\kappa_0 = \log\max\{|AK| + |B| + |C| + |D|, |E| + |F| + |G|\}$ depends only on a, b, c, and (x_0, y_0) and does not depend on $P = (x, y)$. This completes the proof of Lemma 2. $\qquad\square$

3. The Height of $2P$

In the last section we proved that the height of a sum $P + P_0$ is (essentially) less than twice the height of P. In this section we want to prove Lemma 3, which says that the height of $2P$ is (essentially) greater than four times the height of P. This is harder, because to get the height large, we will need to know that there is not too much cancellation in a certain rational number.

We now restate Lemma 3 and give the proof.

Lemma 3.   *There is a constant* κ, *depending on* a, b, c, *so that*

$$h(2P) \geq 4h(P) - \kappa \qquad \text{for all } P \in C(\mathbb{Q}).$$

PROOF. Just as in our proof of Lemma 2, it is all right to ignore any finite set of points, since we can always take κ larger than $4h(P)$ for all points in that finite set. So we will discard the finitely many points satisfying $2P = \mathcal{O}$.

Let $P = (x, y)$, and write $2P = (\xi, \eta)$. The duplication formula we derived earlier states that

$$\xi + 2x = \lambda^2 - a, \qquad \text{where} \quad \lambda = \frac{f'(x)}{2y}.$$

Putting everything over a common denominator and using $y^2 = f(x)$, we obtain an explicit formula for ξ in terms of x:

$$\xi = \frac{\left(f'(x)\right)^2 - (8x + 4a)f(x)}{4f(x)} = \frac{x^4 + \cdots}{4x^3 + \cdots}.$$

Note that $f(x) \neq 0$ because $2P \neq \mathcal{O}$.

Thus, ξ is the quotient of two polynomials in x with integer coefficients. Since the cubic $y^2 = f(x)$ is non-singular by assumption, we know that $f(x)$ and $f'(x)$ have no common (complex) roots. It follows that the polynomials in the numerator and the denominator of ξ also have no common roots.

Since $h(P) = h(x)$ and $h(2P) = h(\xi)$, we are trying to prove that

$$h(\xi) \geq 4h(x) - \kappa.$$

Thus, we are reduced to proving the following general lemma about heights and quotients of polynomials. Notice this lemma has nothing at all to do with cubic curves.

Lemma 3′.   *Let* $\phi(X)$ *and* $\psi(X)$ *be polynomials with integer coefficients and no common (complex) roots. Let* d *be the maximum of the degrees of* ϕ *and* ψ.
(a) *There is an integer* $R \geq 1$, *depending on* ϕ *and* ψ, *so that for all rational numbers* $\dfrac{m}{n}$,

$$\gcd\left(n^d \phi\left(\frac{m}{n}\right), n^d \psi\left(\frac{m}{n}\right)\right) \quad \text{divides } R.$$

(b) *There are constants* κ_1 *and* κ_2, *depending on* ϕ *and* ψ, *so that for all rational numbers* $\dfrac{m}{n}$ *which are not roots of* ψ,

$$dh\left(\frac{m}{n}\right) - \kappa_1 \leq h\left(\frac{\phi(m/n)}{\psi(m/n)}\right) \leq dh\left(\frac{m}{n}\right) + \kappa_2.$$

Proof. (a) First we observe that since ϕ and ψ have degree at most d, the quantities $n^d\phi\left(\frac{m}{n}\right)$ and $n^d\psi\left(\frac{m}{n}\right)$ are both integers, so it makes sense to talk about their greatest common divisor. The result we are trying to prove says that there is not much cancellation when one takes the quotient of these two integers.

Next we note that ϕ and ψ are interchangable, so for concreteness, we will take $\deg(\phi) = d$ and $\deg(\psi) = e \leq d$. Then we can write

$$n^d\phi.\left(\frac{m}{n}\right) = a_0 m^d + a_1 m^{d-1} n + \cdots + a_d n^d,$$

$$n^d\psi\left(\frac{m}{n}\right) = b_0 m^e n^{d-e} + b_1 m^{e-1} n^{d-e+1} + \cdots + b_e n^d.$$

To ease notation, we will let

$$\Phi(m,n) = n^d\phi\left(\frac{m}{n}\right) \qquad \text{and} \qquad \Psi(m,n) = n^d\psi\left(\frac{m}{n}\right).$$

So we need to find an estimate for $\gcd\big(\Phi(m,n), \Psi(m,n)\big)$ which does not depend on m or n.

Since $\phi(X)$ and $\psi(X)$ have no common roots, they are relatively prime in the Euclidean ring $\mathbb{Q}[X]$. Thus, they generate the unit ideal, so we can find polynomials $F(X)$ and $G(X)$ with rational coefficients satisfying

$$F(X)\phi(X) + G(X)\psi(X) = 1. \tag{$*$}$$

Let A be a large enough integer so that $AF(X)$ and $AG(X)$ have integer coefficients. Further, let D be the maximum of the degrees of F and G. Note that A and D do not depend on m or n.

Now we evaluate the identity $(*)$ at $X = \frac{m}{n}$ and multiply both sides by An^{D+d}. This gives

$$n^D AF\left(\frac{m}{n}\right) \cdot n^d\phi\left(\frac{m}{n}\right) + n^D AG\left(\frac{m}{n}\right) \cdot n^d\psi\left(\frac{m}{n}\right) = An^{D+d}.$$

Let $\gamma = \gamma(m,n)$ be the greatest common divisor of $\Phi(m,n)$ and $\Psi(m,n)$. We have

$$\left\{n^D AF\left(\frac{m}{n}\right)\right\} \Phi(m,n) + \left\{n^D AG\left(\frac{m}{n}\right)\right\} \Psi(m,n) = An^{D+d}.$$

Since the quantities in braces are integers, we see that γ divides An^{D+d}.

This is not good enough because we need to show that γ divides one fixed number which does not depend on n. We will show that γ actually divides Aa_0^{D+d}, where a_0 is the leading coefficient of $\phi(X)$. To prove this, we observe that since γ divides $\Phi(m,n)$, it certainly divides

$$An^{D+d-1}\Phi(m,n) = Aa_0 m^d n^{D+d-1} + Aa_1 m^{d-1} n^{D+d} + \cdots + Aa_d n^{D+2d-1}.$$

But in the sum, every term after the first one contains An^{D+d} as a factor; and we just proved that γ divides An^{D+d}. It follows that γ also divides the first term $Aa_0 m^d n^{D+d-1}$. Thus, γ divides $\gcd(An^{D+d}, Aa_0 m^d n^{D+d-1})$; and because m and n are relatively prime, we conclude that γ divides $Aa_0 n^{D+d-1}$. Notice we have reduced the power of n at the cost of multiplying by a_0.

Now using the fact that γ divides $Aa_0 n^{D+d-2} \Phi(m, n)$ and repeating the above argument shows that γ divides $Aa_0^2 n^{D+d-2}$. The pattern is clear, and eventually we reach the conclusion that γ divides Aa_0^{D+d}, which finishes our proof of (a).

(b) There are two inequalities to be proven. The upper bound is the easier; the proof is similar to the proof of Lemma 2. We will just prove the lower bound and leave the upper bound for you to do as an exercise.

As usual, it is all right to exclude some finite set of rational numbers when we prove an inequality of this sort; we need merely adjust the constant κ_1 to take care of the finitely many exceptions. So we may assume that the rational number $\dfrac{m}{n}$ is not a root of ϕ.

If r is any non-zero rational number, it is clear from the definition that $h(r) = h\left(\dfrac{1}{r}\right)$. So reversing the role of ϕ and ψ if necessary, we may make the same assumption as in (a), namely, that ϕ has degree d and ψ has degree e with $e \le d$.

Continuing with the notation from (a), the rational number whose height we want to estimate is

$$\xi = \frac{\phi\left(\dfrac{m}{n}\right)}{\psi\left(\dfrac{m}{n}\right)} = \frac{n^d \phi\left(\dfrac{m}{n}\right)}{n^d \psi\left(\dfrac{m}{n}\right)} = \frac{\Phi(m, n)}{\Psi(m, n)}.$$

This gives an expression for ξ as a quotient of integers, so the height $H(\xi)$ would be the maximum of the integers $|\Phi(m, n)|$ and $|\Psi(m, n)|$ except for the possibility that they may have common factors.

We proved in (a) that there is some integer $R \ge 1$, independent of m and n, so that the greatest common divisor of $\Phi(m, n)$ and $\Psi(m, n)$ divides R. This bounds the possible cancellation, and we find that

$$
\begin{aligned}
H(\xi) &\ge \frac{1}{R} \max\left\{|\Phi(m, n)|, |\Psi(m, n)|\right\} \\
&= \frac{1}{R} \max\left\{\left|n^d \phi\left(\frac{m}{n}\right)\right|, \left|n^d \psi\left(\frac{m}{n}\right)\right|\right\} \\
&\ge \frac{1}{2R}\left(\left|n^d \phi\left(\frac{m}{n}\right)\right| + \left|n^d \psi\left(\frac{m}{n}\right)\right|\right).
\end{aligned}
$$

For the last line we have used the trivial observation $\max\{a, b\} \ge \frac{1}{2}(a + b)$.

In multiplicative notation, we want to compare $H(\xi)$ to the quantity $H\left(\dfrac{m}{n}\right)^d = \max\{|m|^d, |n|^d\}$, so we consider the quotient

$$\frac{H(\xi)}{H(m/n)^d} \geq \frac{1}{2R} \cdot \frac{\left(\left|n^d\phi\left(\dfrac{m}{n}\right)\right| + \left|n^d\psi\left(\dfrac{m}{n}\right)\right|\right)}{\max\{|m|^d, |n|^d\}}$$

$$= \frac{1}{2R} \cdot \frac{\left(\left|\phi\left(\dfrac{m}{n}\right)\right| + \left|\psi\left(\dfrac{m}{n}\right)\right|\right)}{\max\left\{\left|\dfrac{m}{n}\right|^d, 1\right\}}.$$

This suggests that we look at the function p of a real variable t defined by

$$p(t) = \frac{|\phi(t)| + |\psi(t)|}{\max\{|t|^d, 1\}}.$$

Since ϕ has degree d and ψ has degree at most d, we see that p has a non-zero limit as $|t|$ approaches infinity. This limit is either $|a_0|$, if ψ has degree less than d, or $|a_0| + |b_0|$, if ψ has degree equal to d. So outside some closed interval, the function $p(t)$ is bounded away from zero.

But inside a closed interval, we are looking at a continuous function which never vanishes because by assumption $\phi(X)$ and $\psi(X)$ have no common zeros. And a continuous function on a compact set (such as a closed interval) actually assumes its maximum and minimum values. In particular, since we know that our function is never equal to zero, its minimum value must be positive. This proves that there is a constant $C_1 > 0$ so that $p(t) > C_1$ for all real numbers t.

Using this fact in the inequality we derived above, we find that

$$H(\xi) \geq \frac{C_1}{2R} H\left(\frac{m}{n}\right)^d.$$

The constants C_1 and R do not depend on m and n, so taking logarithms gives the desired inequality

$$h(\xi) \geq dh\left(\frac{m}{n}\right) - \kappa_1$$

with $\kappa_1 = \log(2R/C_1)$.

This concludes the proof of Lemma 3', and also Lemma 3. Notice that there are two ideas in the proof: one is to bound the amount of cancellation, and the other is to look at $\dfrac{H(\phi(x)/\psi(x))}{H(x)^d}$ as a function on something compact. $\qquad\square$

4. A Useful Homomorphism

To complete the proof of Mordell's theorem, we need to prove Lemma 4, which says that the subgroup $2C(\mathbb{Q})$ has finite index inside $C(\mathbb{Q})$. This is the subtlest part of the proof of Mordell's theorem. To ease notation a little bit, we will write Γ instead of $C(\mathbb{Q})$:

$$\Gamma = C(\mathbb{Q}).$$

Unfortunately, we do not know how to prove Lemma 4 for all cubic curves without using some algebraic number theory, and we want to stick to the rational numbers. So we are going to make the additional assumption that the polynomial $f(x)$ has at least one rational root, which amounts to saying that the curve has at least one rational point of order two. The same method of proof works in general if you take a root of the equation $f(x) = 0$ and work in the field generated by that root over the rationals. But ultimately we would need to know some basic facts about the unit group and the ideal class group of this field, topics which we prefer to avoid. So we will prove Lemma 4 in the case $f(x)$ has a rational root x_0. In this section we will develop some tools which we will need for the proof, and then in the next section we will give the proof of Lemma 4, thereby completing the proof of Mordell's theorem.

Since $f(x_0) = 0$, and f is a polynomial with integer coefficients and leading coefficient 1, we know that x_0 is an integer. Making a change of coordinates, we can move the point $(x_0, 0)$ to the origin. This obviously does not affect the group Γ. The new equation again has integer coefficients; in the new coordinates the curve will have the form

$$C : y^2 = f(x) = x^3 + ax^2 + bx,$$

where a and b are integers. Then

$$T = (0, 0)$$

is a rational point on C and satisfies $2T = \mathcal{O}$.

The formula for the discriminant of f given earlier becomes, in this case,

$$D = b^2(a^2 - 4b).$$

We always assume our curve is non-singular, which means that $D \neq 0$, and so neither $a^2 - 4b$ nor b is zero.

Since we are interested in the index $(\Gamma : 2\Gamma)$, or equivalently in the order of the factor group $\Gamma/2\Gamma$, it is extremely helpful to know that the duplication map $P \mapsto 2P$ can be broken down into two simpler operations. The duplication map is in some sense of degree four because the rational function giving the x coordinate of $2P$ is of degree four in the x coordinate of P. We will write the map $P \mapsto 2P$ as a composition of two maps of degree two, each of which will be easier to handle. However, the two maps will not be from C to itself, but rather from C to another curve $\overline{C}$ and then back again to C.

The other curve $\overline{C}$ that we will consider is the curve given by the equation

$$\overline{C} \, : \, y^2 = x^3 + \overline{a}x^2 + \overline{b}x,$$

where

$$\overline{a} = -2a \quad \text{and} \quad \overline{b} = a^2 - 4b.$$

For reasons we will see in a moment, these two curves are intimately related; and it is natural if you are studying C to also study $\overline{C}$. One can play C and $\overline{C}$ off against one another, and that is just what we are planning to do.

Suppose we apply the procedure again and look at

$$\overline{\overline{C}} \, : \, y^2 = x^3 + \overline{\overline{a}}x^2 + \overline{\overline{b}}x.$$

Here

$$\overline{\overline{a}} = -2\overline{a} = 4a \quad \text{and} \quad \overline{\overline{b}} = \overline{a}^2 - 4\overline{b} = 4a^2 - 4(a^2 - 4b) = 16b,$$

so the curve $\overline{\overline{C}}$ is the curve $y^2 = x^3 + 4ax^2 + 16bx$. This is essentially the same as C; we just need to replace y by $8y$ and x by $4x$, and then divide the equation by 64. Thus, the group $\overline{\overline{\Gamma}}$ of rational points on $\overline{\overline{C}}$ is isomorphic to the group Γ of rational points on C.

We are now going to define a map $\phi : C \to \overline{C}$ which will be a group homomorphism and will carry the rational points Γ into the rational points $\overline{\Gamma}$. And then, by the same procedure, we will define a map $\psi : \overline{C} \to \overline{\overline{C}}$. In view of the isomorphism $\overline{\overline{C}} \cong C$, the composition $\psi \circ \phi$ is a homomorphism of C into C which turns out to be multiplication by 2.

The map $\phi : C \to \overline{C}$ is defined in the following way. If $P = (x, y) \in C$ is a point with $x \neq 0$, then the point $\phi(x, y) = (\overline{x}, \overline{y})$ is given by the formulas

$$\overline{x} = x + a + \frac{b}{x} = \frac{y^2}{x^2} \quad \text{and} \quad \overline{y} = y \left(\frac{x^2 - b}{x^2} \right).$$

To see that ϕ is well-defined, we just have to check that $\overline{x}$ and $\overline{y}$ satisfy the equation of $\overline{C}$, which is easy:

$$\begin{aligned}
\overline{x}^3 + \overline{a}\overline{x}^2 + \overline{b}\overline{x} &= \overline{x}(\overline{x}^2 - 2a\overline{x} + (a^2 - 4b)) \\
&= \frac{y^2}{x^2} \left(\frac{y^4}{x^4} - 2a\frac{y^2}{x^2} + (a^2 - 4b) \right) \\
&= \frac{y^2}{x^2} \left(\frac{(y^2 - ax^2)^2 - 4bx^4}{x^4} \right) \\
&= \frac{y^2}{x^6} \left((x^3 + bx)^2 - 4bx^4 \right) \\
&= \left(\frac{y(x^2 - b)}{x^2} \right)^2 = \overline{y}^2.
\end{aligned}$$

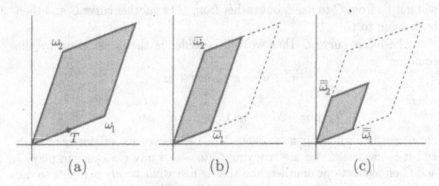

The Map ϕ Described Analytically

Figure 3.1

This defines the map ϕ at all points except $T = (0,0)$ and $\mathcal{O}$. We complete the definition by setting

$$\phi(T) = \overline{\mathcal{O}} \qquad \text{and} \qquad \phi(\mathcal{O}) = \overline{\mathcal{O}}.$$

This ad hoc definition of ϕ looks like magic; we reached into our top hat and out came an amazing map. The reason we presented ϕ in this way is to emphasize that everything about ϕ follows from a little elementary algebra and arithmetic; there is no need to use any analysis. However, if you are willing to think in terms of complex points and the uniformization of the curve C by the complex variable u, then x and y are elliptic functions of u and you can see ϕ quite clearly. Namely, the complex points on our curve can be represented by the points in the period parallelogram for suitable periods ω_1, ω_2. (See Figure 3.1a.)

If we cut that parallelogram in half by a line parallel to one of the sides, then we get a new parallelogram with sides $\overline{\omega}_1$ and $\overline{\omega}_2$ (Figure 3.1b), where $\overline{\omega}_1 = \frac{1}{2}\omega_1$ and $\overline{\omega}_2 = \omega_2$. This parallelogram corresponds to the curve $\overline{C}$. To divide the parallelogram we had to pick a point of order two on C, that is the point T in the figure. There is a natural map of C onto $\overline{C}$ in which the point

$$u = c_1\omega_1 + c_2\omega_2 \qquad \text{is sent to} \qquad \overline{u} = c_1\omega_1 + c_2\omega_2 = 2c_1\overline{\omega}_1 + c_2\overline{\omega}_2.$$

Now if we slice the parallelogram the other way, we get $\overline{\overline{C}}$ which has the period parallelogram $\overline{\overline{\omega}}_1, \overline{\overline{\omega}}_2$ given by $\overline{\overline{\omega}}_1 = \frac{1}{2}\omega_1$ and $\overline{\overline{\omega}}_2 = \frac{1}{2}\omega_2$ (Figure 3.1c). Clearly the curve in Figure 3.1a is isomorphic to the curve in Figure 3.1c via the map $u \mapsto \frac{1}{2}u$, so the elliptic functions with periods $\overline{\overline{\omega}}_1, \overline{\overline{\omega}}_2$ are essentially the same as those with periods ω_1, ω_2. From an analytic point of view, this is the procedure we are using.

What is the kernel of ϕ? From the picture it is clear that the kernel of ϕ consists of the two points $\mathcal{O}$ and T; and if you look at the algebraic formula for ϕ that we gave earlier, you will see that the only two points of C which are sent to $\overline{\mathcal{O}}$ are $\mathcal{O}$ and T. In books on elliptic functions one can find formulas which express elliptic functions with periods $\frac{1}{2}\omega_1$ and ω_2 rationally in terms of elliptic functions with periods ω_1 and ω_2; these are exactly the formulas for $\overline{x}$ and $\overline{y}$ in terms of x and y. Hopefully this explanation helps to make the curve $\overline{C}$ and the map ϕ less mysterious.

We can also consider everything from a highbrow point of view. Since C is an abelian group and $\{\mathcal{O}, T\}$ is a subgroup of C, we might say that $\overline{C}$ is created by forming the quotient group $C/\{\mathcal{O}, T\}$. Unfortunately, it is not obvious that the elements of this quotient group actually correspond to the points on some elliptic curve $\overline{C}$. And even if we know that the quotient is an elliptic curve, it is not obvious that the natural homomorphism from C to $\overline{C}$ is given by rational functions.

However, all of this follows from general theorems on algebraic groups. It is even true that the group of points on an elliptic curve modulo any finite subgroup is again the group of points on an elliptic curve. Granting this, and knowing that any elliptic curve can be written in Weierstrass form, it is not difficult to guess the explicit formulas which we gave earlier.

Both the analytic viewpoint and the "highbrow" approach tell us that the map ϕ is a homomorphism, but we can also prove this directly using explicit formulas. To remind you where we are, and for future reference, we will state this as a formal proposition.

Proposition. *Let C and $\overline{C}$ be the elliptic curves given by the equations*

$$C : y^2 = x^3 + ax^2 + bx \qquad and \qquad \overline{C} : y^2 = x^3 + \overline{a}x^2 + \overline{b}x,$$

where

$$\overline{a} = -2a \qquad and \qquad \overline{b} = a^2 - 4b.$$

Let $T = (0,0) \in C$.

(a) There is a homomorphism $\phi : C \to \overline{C}$ defined by

$$\phi(P) = \begin{cases} \left(\dfrac{y^2}{x^2}, \dfrac{y(x^2 - b)}{x^2} \right), & if\ P = (x,y) \neq \mathcal{O}, T, \\ \overline{\mathcal{O}}, & if\ P = \mathcal{O}\ or\ P = T. \end{cases}$$

The kernel of ϕ is $\{\mathcal{O}, T\}$.

(b) Applying the same process to $\overline{C}$ gives a map $\overline{\phi} : \overline{C} \to \overline{\overline{C}}$. The curve $\overline{\overline{C}}$ is isomorphic to C via the map $(x,y) \to (x/4, y/8)$. There is thus a homomorphism $\psi : \overline{C} \to C$ defined by

$$\psi(\overline{P}) = \begin{cases} \left(\dfrac{\overline{y}^2}{4\overline{x}^2}, \dfrac{\overline{y}(\overline{x}^2 - \overline{b})}{8\overline{x}^2} \right), & if\ \overline{P} = (\overline{x}, \overline{y}) \neq \overline{\mathcal{O}}, \overline{T}, \\ \mathcal{O}, & if\ \overline{P} = \overline{\mathcal{O}}\ or\ \overline{P} = \overline{T}. \end{cases}$$

The composition $\psi \circ \phi : C \to C$ is multiplication by two: $\psi \circ \phi(P) = 2P$.

PROOF. (a) We checked earlier that ϕ maps points of C to points of $\overline{C}$; and once we know that ϕ is a homomorphism, it is obvious that the kernel of ϕ consists of $\mathcal{O}$ and T. So we need to prove that ϕ is a homomorphism. This is somewhat tedious because there are many exceptional cases; so we will do a lot of it, and leave a few cases for you.

We have to prove that

$$\phi(P_1 + P_2) = \phi(P_1) + \phi(P_2) \qquad \text{for all } P_1, P_2 \in C.$$

Note that the first plus sign is addition on C, whereas the second one is addition on $\overline{C}$.

If P_1 or P_2 is $\mathcal{O}$, there is nothing to prove. If one of P_1 or P_2 is T, say, $P_1 = T$, then the thing to be proved is $\phi(T + P) = \phi(P)$. This is not hard to see. Using the explicit formula for the addition law, one easily checks that if $P = (x, y)$, then

$$P + T = \left(\frac{b}{x}, -\frac{by}{x^2} \right).$$

Writing $P + T = \big(x(P+T), y(P+T) \big)$ and $\phi(P+T) = \big(\overline{x}(P+T), \overline{y}(P+T) \big)$, we find

$$\overline{x}(P + T) = \left(\frac{y(P+T)}{x(P+T)} \right)^2 = \frac{(-by/x^2)^2}{(b/x)^2} = \frac{y^2}{x^2} = \overline{x}(P).$$

In the same way we compute

$$\overline{y}(P + T) = \frac{y(P+T)(x(P+T)^2 - b)}{x(P+T)^2} = \frac{-\dfrac{by}{x^2}\left(\left(\dfrac{b}{x}\right)^2 - b \right)}{\left(\dfrac{b}{x}\right)^2} = \overline{y}(P).$$

This shows that $\phi(P + T) = \phi(P)$, except that the argument breaks down if $P = T$. But in that case we obviously have $\phi(T + T) = \phi(T) + \phi(T)$ because everything is $\mathcal{O}$.

Next we observe that ϕ takes negatives to negatives:

$$\phi(-P) = \phi(x, -y) = \left(\left(\frac{-y}{x} \right)^2, \frac{-y(x^2 - b)}{x^2} \right) = -\phi(x, y) = -\phi(P).$$

So in order to prove that ϕ is a homomorphism, it now suffices to show that if $P_1 + P_2 + P_3 = \mathcal{O}$, then $\phi(P_1) + \phi(P_2) + \phi(P_3) = \overline{\mathcal{O}}$; because once we know this, then

$$\phi(P_1 + P_2) = \phi(-P_3) = -\phi(P_3) = \phi(P_1) + \phi(P_2).$$

Further, from what we have already done, we may assume that none of the points P_1, P_2, or P_3 is equal to $\mathcal{O}$ or T.

From the definition of the group law on a cubic curve, the condition $P_1 + P_2 + P_3 = \mathcal{O}$ is equivalent to the statement that P_1, P_2, and P_3 are colinear, so let $y = \lambda x + \nu$ be the line through them. (If two or three of them coincide, then the line should be appropriately tangent to the curve.) We must show that $\phi(P_1)$, $\phi(P_2)$, and $\phi(P_3)$ are the intersection of some line with $\overline{C}$.

Note that $\nu \neq 0$, because $\nu = 0$ would mean that the line $y = \lambda x + \nu$ goes through T, contrary to our assumption that P_1, P_2, P_3 are distinct from T. The line intersecting $\overline{C}$ that we take is

$$y = \overline{\lambda}x + \overline{\nu}, \qquad \text{where} \quad \overline{\lambda} = \frac{\nu\lambda - b}{\nu} \quad \text{and} \quad \overline{\nu} = \frac{\nu^2 - a\nu\lambda + b\lambda^2}{\nu}.$$

To check, say, that $\phi(P_1) = \phi(x_1, y_1) = (\overline{x}_1, \overline{y}_1)$ is on the line $y = \overline{\lambda}x + \overline{\nu}$, we just substitute and compute

$$\begin{aligned}
\overline{\lambda}\overline{x}_1 + \overline{\nu} &= \frac{\nu\lambda - b}{\nu}\left(\frac{y_1}{x_1}\right)^2 + \frac{\nu^2 - a\nu\lambda + b\lambda^2}{\nu} \\
&= \frac{(\nu\lambda - b)y_1^2 + (\nu^2 - a\nu\lambda + b\lambda^2)x_1^2}{\nu x_1^2} \\
&= \frac{\nu\lambda(y_1^2 - ax_1^2) - b(y_1 - \lambda x_1)(y_1 + \lambda x_1) + \nu^2 x_1^2}{\nu x_1^2};
\end{aligned}$$

and now using $y_1^2 - ax_1^2 = x_1^3 + bx_1$ and $y_1 - \lambda x_1 = \nu$, we get

$$\begin{aligned}
&= \frac{\lambda(x_1^3 + bx_1) - b(y_1 + \lambda x_1) + \nu x_1^2}{x_1^2} \\
&= \frac{x_1^2(\lambda x_1 + \nu) - by_1}{x_1^2} \\
&= \frac{(x_1^2 - b)y_1}{x_1^2} = \overline{y}_1.
\end{aligned}$$

The computation for $\phi(P_2)$ and $\phi(P_3)$ is exactly the same.

Notice, however, that strictly speaking it is not enough to show that the three points $\phi(P_1), \phi(P_2), \phi(P_3)$ lie on the line $y = \overline{\lambda}x + \overline{\nu}$. It is enough if $\phi(P_1), \phi(P_2), \phi(P_3)$ are distinct; but in general we really have to show that $\overline{x}(P_1), \overline{x}(P_2), \overline{x}(P_3)$ are the three roots of the cubic $(\overline{\lambda}x + \overline{\nu})^2 = \overline{f}(x)$, whether or not those roots are distinct. We will leave it to you to verify this if there are multiple roots. As an alternative, we might note that ϕ is continuous as a map from the complex points of C to the complex points of $\overline{C}$; so once we know that ϕ is a homomorphism for distinct points, we get by continuity that it is a homomorphism in general.

(b) We noted above that the curve $\overline{\overline{C}}$ is given by the equation

$$\overline{\overline{C}} : y^2 = x^3 + 4ax^2 + 16bx,$$

so it is clear that the map $(x, y) \to (x/4, y/8)$ is an isomorphism from $\overline{\overline{C}}$ to C. From (a) there is a homomorphism $\overline{\phi} : \overline{C} \to \overline{\overline{C}}$ defined by the same equations that define ϕ, but with $\overline{a}$ and $\overline{b}$ in place of a and b. Since the map $\psi : \overline{C} \to C$ is the composition of $\overline{\phi} : \overline{C} \to \overline{\overline{C}}$ with the isomorphism $\overline{\overline{C}} \to C$, we get immediately that ψ is a well-defined homomorphism from $\overline{C}$ to C.

It remains to verify that $\psi \circ \phi$ is multiplication by two, and that is another tedious computation. A little algebra with the explicit formulas we gave earlier yields

$$2P = 2(x, y) = \left(\frac{(x^2 - b)^2}{4y^2}, \frac{(x^2 - b)(x^4 + 2ax^3 + 6bx^2 + 2abx + b^2)}{8y^3} \right).$$

On the other hand, we have

$$\phi(x, y) = \left(\frac{y^2}{x^2}, \frac{y(x^2 - b)}{x^2} \right), \qquad \psi(\overline{x}, \overline{y}) = \left(\frac{\overline{y}^2}{4\overline{x}^2}, \frac{\overline{y}(\overline{x}^2 - \overline{b})}{8\overline{x}^2} \right);$$

so we can compute

$$\psi \circ \phi(x, y) = \psi \left(\frac{y^2}{x^2}, \frac{y(x^2 - b)}{x^2} \right)$$

$$= \left(\frac{\left(\frac{y(x^2 - b)}{x^2} \right)^2}{4 \left(\frac{y^2}{x^2} \right)^2}, \frac{\frac{y(x^2 - b)}{x^2} \left(\left(\frac{y^2}{x^2} \right)^2 - (a^2 - 4b) \right)}{8 \left(\frac{y^2}{x^2} \right)^2} \right)$$

$$= \left(\frac{(x^2 - b)^2}{4y^2}, \frac{(x^2 - b)(y^4 - (a^2 - 4b)x^4)}{8y^3 x^2} \right).$$

Now subsituting $y^4 = x^2(x^2 + ax + b)^2$ and doing a little algebra gives the desired result: $\psi \circ \phi(x, y) = 2(x, y)$.

A similar computation gives $\phi \circ \psi(\overline{x}, \overline{y}) = 2(\overline{x}, \overline{y})$. Or we can argue as follows. Since ϕ is a homomorphism, we know that

$$\phi(2P) = \phi(P + P) = \phi(P) + \phi(P) = 2\phi(P).$$

We just proved that $2P = \psi \circ \phi(P)$, so we get $\phi \circ \psi(\phi(P)) = 2(\phi(P))$. Now $\phi : C \to \overline{C}$ is onto as a map of complex points, so for any $\overline{P} \in \overline{C}$ we can find $P \in C$ with $\phi(P) = \overline{P}$. Therefore $\phi \circ \psi(\overline{P}) = 2\overline{P}$.

Of course, we have really only proven that $\psi \circ \phi = 2$ for points with $x \neq 0$ and $y \neq 0$ because the formulas we used above are not valid if x or y is zero. So we really should check that $\psi \circ \phi(P) = \mathcal{O}$ in the cases that P is a point of order two. We will leave that for you to check explicitly, although again, we could argue that it must be true by continuity.   □

5. Mordell's Theorem

In this section we will complete the proof of Lemma 4, and with it the proof of Mordell's theorem. Continuing with the notation from the last section, we recall that we have two curves

$$C : y^2 = x^3 + ax^2 + bx \quad\text{and}\quad \overline{C} : y^2 = x^3 + \overline{a}x^2 + \overline{b}x,$$

where $\overline{a} = -2a$ and $\overline{b} = a^2 - 4b$; and we have homomorphisms

$$\phi : C \longrightarrow \overline{C} \quad\text{and}\quad \psi : \overline{C} \longrightarrow C$$

such that the compositions $\phi \circ \psi : \overline{C} \to \overline{C}$ and $\psi \circ \phi : C \to C$ are each multiplication by two. Further, the kernel of ϕ consists of the two points $\mathcal{O}$ and $T = (0,0)$; and the kernel of ψ consists of $\overline{\mathcal{O}}$ and $\overline{T} = (0,0)$.

The images of ϕ and ψ are extremely interesting. From the complex point of view, it is obvious that given any point in $\overline{C}$, there is a point in C which maps onto it. In other words, on complex points, the map ϕ is onto. But now we examine what happens to the rational points.

It is clear from the formulas that ϕ maps Γ into $\overline{\Gamma}$; but if you are given a rational point in $\overline{\Gamma}$, it is not at all clear if it comes from a rational point in Γ. If we apply the map ϕ to the rational points Γ, we get a subgroup of the set of rational points $\overline{\Gamma}$; we denote this subgroup by $\phi(\Gamma)$ and call it the *image of Γ by ϕ*. We make the following three claims, which taken together, provide a good description of the image.

(i) $\overline{\mathcal{O}} \in \phi(\Gamma)$.

(ii) $\overline{T} = (0,0) \in \phi(\Gamma)$ if and only if $\overline{b} = a^2 - 4b$ is a perfect square.

(iii) Let $\overline{P} = (\overline{x}, \overline{y}) \in \overline{\Gamma}$ with $\overline{x} \neq 0$. Then $\overline{P} \in \phi(\Gamma)$ if and only if $\overline{x}$ is the square of a rational number.

Statement (i) is obvious, because $\overline{\mathcal{O}} = \phi(\mathcal{O})$. Let's check statement (ii). From the formula for ϕ we see that $\overline{T} \in \phi(\Gamma)$ if and only if there is a rational point $(x,y) \in \Gamma$ such that $\dfrac{y^2}{x^2} = 0$. Note $x \neq 0$, because $x = 0$ means that $(x,y) = T$, and $\phi(T)$ is $\overline{\mathcal{O}}$, not $\overline{T}$. So $\overline{T} \in \phi(\Gamma)$ if and only if there is a rational point $(x,y) \in \Gamma$ with $x \neq 0$ and $y = 0$. Putting $y = 0$ in the equation for Γ gives

$$0 = x^3 + ax^2 + bx = x(x^2 + ax + b).$$

This equation has a non-zero rational root if and only if the quadratic equation $x^2 + ax + b$ has a rational root, which happens if and only if its discriminant $a^2 - 4b$ is a perfect square. This proves statement (ii).

Now we check statement (iii). If $(\overline{x}, \overline{y}) \in \phi(\Gamma)$ is a point with $\overline{x} \neq 0$, then the defining formula for ϕ shows that $\overline{x} = y^2/x^2$ is the square of a rational number. Suppose conversely that $\overline{x} = w^2$ for some rational number w. We want to find a rational point on C that maps to $(\overline{x}, \overline{y})$.

The homomorphism ϕ has two elements in its kernel, $\mathcal{O}$ and T. Thus if $(\overline{x}, \overline{y})$ lies in $\phi(\Gamma)$, there will be two points of Γ that map to it. Let

$$x_1 = \frac{1}{2}\left(w^2 - a + \frac{\overline{y}}{w}\right), \qquad y_1 = x_1 w;$$

$$x_2 = \frac{1}{2}\left(w^2 - a - \frac{\overline{y}}{w}\right), \qquad y_2 = -x_2 w.$$

We claim that the points $P_i = (x_i, y_i)$ are on C and that $\phi(P_i) = (\overline{x}, \overline{y})$ for $i = 1, 2$. Since P_1 and P_2 are clearly rational points, this will prove that $(\overline{x}, \overline{y}) \in \phi(\Gamma)$.

The most efficient way to check that P_1 and P_2 are on C is to do them together, rather than working with them one at a time. First we compute

$$\begin{aligned}
x_1 x_2 &= \frac{1}{4}\left((w^2 - a)^2 - \frac{\overline{y}^2}{w^2}\right) \\
&= \frac{1}{4}\left((\overline{x} - a)^2 - \frac{\overline{y}^2}{\overline{x}}\right) \\
&= \frac{1}{4}\left(\frac{\overline{x}^3 - 2a\overline{x}^2 + a^2\overline{x} - \overline{y}^2}{\overline{x}}\right) \\
&= b.
\end{aligned}$$

The last line follows because $\overline{y}^2 = \overline{x}^3 - 2a\overline{x}^2 + (a^2 - 4b)\overline{x}$.

To show that $P_i = (x_i, y_i)$ lies on C amounts to showing that

$$\frac{y_i^2}{x_i^2} = x_i + a + \frac{b}{x_i}.$$

Since we just proved that $b = x_1 x_2$, and since from the definition of y_1 and y_2 we have $y_i/x_i = \pm w$, this is the same as showing that

$$w^2 = x_1 + a + x_2.$$

This last equality is obvious from the definition of x_1 and x_2.

It remains to check that $\phi(P_i) = (\overline{x}, \overline{y})$, so we must show that

$$\frac{y_i^2}{x_i^2} = \overline{x} \qquad \text{and} \qquad \frac{y_i(x_i^2 - b)}{x_i^2} = \overline{y}.$$

The first equality is clear from the definitions $y_i = \pm x_i w$ and $\overline{x} = w^2$. For the second, we use $b = x_1 x_2$ and the definition of y_i to compute

$$\frac{y_1(x_1^2 - b)}{x_1^2} = \frac{x_1 w(x_1^2 - x_1 x_2)}{x_1^2} = w(x_1 - x_2) \qquad \text{and}$$

$$\frac{y_2(x_2^2 - b)}{x_2^2} = \frac{-x_2 w(x_2^2 - x_1 x_2)}{x_2^2} = w(x_1 - x_2).$$

So we are left to verify that $w(x_1 - x_2) = \bar{y}$, which is obvious from the definition of x_1 and x_2. This completes the verification of statement (iii).

Recall that our aim is to prove Lemma 4, which says that the subgroup 2Γ has finite index inside Γ. As we will see shortly, this will follow if we can prove that the index $(\overline{\Gamma} : \phi(\Gamma))$ is finite and also the index $(\Gamma : \psi(\overline{\Gamma}))$ is finite. In fact, we will now show that $(\overline{\Gamma} : \phi(\Gamma)) \leq 2^{s+1}$, where s is the number of distinct prime factors of $\bar{b} = a^2 - 4b$, and also that $(\Gamma : \psi(\overline{\Gamma})) \leq 2^{r+1}$, where r is the number of distinct prime factors of b.

It is clearly enough to prove one of these statements, so we will just prove the second. From statements (i), (ii), and (iii), we know that $\psi(\overline{\Gamma})$ is the set of points $(x, y) \in \Gamma$ such that x is a non-zero rational square, together with $\mathcal{O}$, and also T if b is a perfect square. The idea of the proof is to find a one-to-one homomorphism from the quotient group $\Gamma/\psi(\overline{\Gamma})$ into a finite group.

Let $\mathbb{Q}^*$ be the multiplicative group of non-zero rational numbers, and let $\mathbb{Q}^{*2}$ denote the subgroup of squares of elements of $\mathbb{Q}^*$:

$$\mathbb{Q}^{*2} = \{u^2 : u \in \mathbb{Q}^*\}.$$

We introduce a map α from Γ to $\mathbb{Q}^*/\mathbb{Q}^{*2}$ defined by

$$\alpha(\mathcal{O}) = 1 \pmod{\mathbb{Q}^{*2}},$$
$$\alpha(T) = b \pmod{\mathbb{Q}^{*2}},$$
$$\alpha(x, y) = x \pmod{\mathbb{Q}^{*2}} \qquad \text{if } x \neq 0.$$

We claim that α is a homomorphism and that the kernel of α is precisely the image of ψ. Further, we are able to say a lot about the image of α. Because this result is so important, we state it formally, and then give the proof. In particular, we want to draw your attention to part (c) of the following proposition. It says that, modulo squares, there are only a finite number of possibilities for the x coordinate of a point on the curve. This miraculous fact is really the crux of the proof that the index $(\Gamma : 2\Gamma)$ is finite.

Proposition. (a) *The map* $\alpha : \Gamma \to \mathbb{Q}^*/\mathbb{Q}^{*2}$ *described above is a homomorphism.*

(b) *The kernel of* α *is the image* $\psi(\overline{\Gamma})$. *Hence* α *induces a one-to-one homomorphism*

$$\frac{\Gamma}{\psi(\overline{\Gamma})} \hookrightarrow \frac{\mathbb{Q}^*}{\mathbb{Q}^{*2}}.$$

(c) *Let* $p_1, p_2, \ldots, p_t$ *be the distinct primes dividing* b. *Then the image of* α *is contained in the subgroup of* $\mathbb{Q}^*/\mathbb{Q}^{*2}$ *consisting of the elements*

$$\{\pm p_1^{\varepsilon_1} p_2^{\varepsilon_2} \cdots p_t^{\varepsilon_t} : \text{each } \varepsilon_i \text{ equals } 0 \text{ or } 1\}.$$

(d) *The index $(\Gamma : \psi(\overline{\Gamma}))$ is at most 2^{t+1}.*

PROOF. (a) First we observe that α sends inverses to inverses, because

$$\alpha(-P) = \alpha(x, -y) = x \equiv \frac{1}{x} = \alpha(x,y)^{-1} = \alpha(P)^{-1} \pmod{\mathbb{Q}^{*2}}.$$

Hence, in order to prove that α is a homomorphism, it is enough to show that whenever $P_1 + P_2 + P_3 = \mathcal{O}$, then $\alpha(P_1)\alpha(P_2)\alpha(P_3) \equiv 1 \pmod{\mathbb{Q}^{*2}}$.

The triples of points which add to zero consist of the intersections of the curve with a line. If the line is $y = \lambda x + \nu$ and the x coordinates of the intersections are x_1, x_2, x_3, we saw in Chapter I, Section 4, that x_1, x_2, x_3 are the roots of the equation

$$x^3 + (a - \lambda^2)x^2 + (b - 2\lambda\nu)x + (c - \nu^2) = 0.$$

This is for the cubic $y^2 = x^3 + ax^2 + bx + c$. Thus,

$$x_1 + x_2 + x_3 = \lambda^2 - a,$$
$$x_1x_2 + x_2x_3 + x_1x_3 = b - 2\lambda\nu,$$
$$x_1x_2x_3 = \nu^2 - c.$$

The last equation is the one we want. We are looking at a curve with $c = 0$, so we find that

$$x_1x_2x_3 = \nu^2 \in \mathbb{Q}^{*2}.$$

Therefore

$$\alpha(P_1)\alpha(P_2)\alpha(P_3) = x_1x_2x_3 = \nu^2 \equiv 1 \pmod{\mathbb{Q}^{*2}}.$$

This completes the proof in the case that P_1, P_2, P_3 are distinct from $\mathcal{O}$ and T. We will leave it as an exercise to check the remaining cases. [N.B. Here we cannot argue by "continuity." Even were we to topologize $C(\mathbb{Q})$ by using the inclusion of $C(\mathbb{Q})$ into the real points of C, there is no way to put a topology onto $\mathbb{Q}^*/\mathbb{Q}^{*2}$ so that the map α would be continuous. Up until now all of the maps we have looked at have been defined geometrically, but the homomorphism α is completely arithmetic in nature.]

(b) Comparing the definition of α with the description of $\psi(\overline{\Gamma})$ given in statements (i), (ii), and (iii), it is clear that the kernel of α is precisely $\psi(\overline{\Gamma})$.

(c) We want to know what rational numbers x can occur as the x coordinate of a point in Γ. We know that such points have coordinates of the form $x = m/e^2$ and $y = n/e^3$. Substituting into the equation and clearing denominators gives

$$n^2 = m^3 + am^2e^2 + bme^4 = m(m^2 + ame^2 + be^4).$$

This equation contains the whole secret. It expresses the square n^2 as a product of two integers. If m and $m^2 + ame^2 + be^4$ were relatively prime, then each of them would be plus or minus a square, and so $x = m/e^2$ would be plus or minus the square of a rational number. In the general case, let

$$d = \gcd(m, m^2 + ame^2 + be^4).$$

Then d divides both m and be^4. But m and e are relatively prime, since we assumed that x was written in lowest terms. Therefore, d divides b.

Thus, the greatest common divisor of m and $m^2 + ame^2 + be^4$ divides b. Since also $n^2 = m(m^2 + ame^2 + be^4)$, we deduce that every prime dividing m appears to an even power except possibly for the primes dividing b. Therefore

$$m = \pm(\text{integer})^2 \cdot p_1^{\varepsilon_1} p_2^{\varepsilon_2} \cdots p_t^{\varepsilon_t},$$

where each ε_i is either 0 or 1, and $p_1, \ldots, p_t$ are the distinct primes dividing b. This proves that

$$\alpha(P) = x = \frac{m}{e^2} \equiv \pm p_1^{\varepsilon_1} p_2^{\varepsilon_2} \cdots p_t^{\varepsilon_t} \pmod{\mathbb{Q}^{*2}};$$

and so the image of α is contained in the indicated set.

If $x = 0$, and hence $m = 0$, our argument breaks down. But then the definition $\alpha(T) = b \pmod{\mathbb{Q}^{*2}}$ shows that the conclusion is still valid because up to squares, b can be written in the indicated form.

(d) The subgroup described in (c) has precisely 2^{t+1} elements. On the other hand, (b) says that the quotient group $\Gamma/\psi(\overline{\Gamma})$ maps one-to-one into this subgroup. Hence, the index of $\psi(\overline{\Gamma})$ inside Γ is at most 2^{t+1}.   □

It has been a long journey, but we now have all the tools needed to prove Lemma 4. Let us remind you what we now know. We have homomorphisms $\phi : \Gamma \to \overline{\Gamma}$ and $\psi : \overline{\Gamma} \to \Gamma$ such that the compositions $\phi \circ \psi$ and $\psi \circ \phi$ are multiplication by two and such that the indices $(\Gamma : \psi(\overline{\Gamma}))$ and $(\overline{\Gamma} : \phi(\Gamma))$ are finite. We want to prove that 2Γ has finite index inside Γ. So the following exercise about abelian groups finishes the proof of Lemma 4.

Lemma. *Let A and B be abelian groups, and consider two homomorphisms $\phi : A \to B$ and $\psi : B \to A$. Suppose that*

$$\psi \circ \phi(a) = 2a \quad \text{for all } a \in A \qquad \text{and} \qquad \phi \circ \psi(b) = 2b \quad \text{for all } b \in B.$$

Suppose further that $\phi(A)$ has finite index in B, and $\psi(B)$ has finite index in A. Then $2A$ has finite index in A. More precisely, the index satisfies

$$(A : 2A) \leq (A : \psi(B))(B : \phi(A)).$$

PROOF. Since $\psi(B)$ has finite index in A, we can find elements $a_1, \ldots, a_n$ representing the finitely many cosets. Similarly, since $\phi(A)$ has finite index in B, we can choose elements $b_1, \ldots, b_m$ representing the finitely many cosets. We claim that the set

$$\{a_i + \psi(b_j) : 1 \leq i \leq n, \ 1 \leq j \leq m\}$$

includes a complete set of representatives for the cosets of $2A$ inside A.

To see this, let $a \in A$. We need to show that a can be written as the sum of an element of this set plus an element of $2A$. Since $a_1, \ldots, a_n$ are representatives for the cosets of $\psi(B)$ inside A, we can find some a_i so that $a - a_i \in \psi(B)$, say $a - a_i = \psi(b)$. Next, since $b_1, \ldots, b_m$ are representatives for the cosets of $\phi(A)$ inside B, we can find some b_j so that $b - b_j \in \phi(A)$, say $b - b_j = \phi(a')$. Then

$$a = a_i + \psi(b) = a_i + \psi\big(b_j + \phi(a')\big)$$
$$= a_i + \psi(b_j) + \psi\big(\phi(a')\big) = a_i + \psi(b_j) + 2a'.$$

$\square$

To celebrate the completion of our proof of Mordell's theorem, we will restate the version that we have proven:

Mordell's Theorem.   (for Curves with a Rational Point of Order Two)
Let C be a non-singular cubic curve given by an equation

$$C : y^2 = x^3 + ax^2 + bx,$$

where a and b are integers. Then the group of rational points $C(\mathbb{Q})$ is a finitely generated abelian group.

PROOF. We saw in Section 1 that Lemmas 1, 2, 3, and 4 imply that $C(\mathbb{Q})$ is finitely generated. We proved Lemma 1 in Section 1, Lemma 2 in Section 2, Lemma 3 in Section 3, and Lemma 4 (for curves with a rational point of order two) in this section. $\square$

Mordell's theorem tells us that we can produce all of the rational points on C by starting from some finite set and using geometry (i.e., using the group law). The following question arises: Given a particular cubic curve, how can one find this generating set? Our proof of Mordell's theorem gives us some tools which often allow us to answer this question. We will do a number of examples in the next section. But at present no one knows a procedure which is guaranteed to work for all cubic curves!

6. Examples and Further Developments

In this section we are going to illustrate Mordell's theorem by working out some numerical examples. First we discuss some consequences of what we have already proven. We have shown that the group Γ of rational points on the curve

$$C : y^2 = x^3 + ax^2 + bx$$

is a finitely generated abelian group. It follows from the fundamental theorem on abelian groups that Γ is isomorphic, as an abstract group, to a direct sum of infinite cyclic groups and finite cyclic groups of prime power order. We will let $\mathbb{Z}$ denote the additive group of integers, and we will let $\mathbb{Z}_m$ denote the cyclic group $\mathbb{Z}/m\mathbb{Z}$ of integers mod m. Then the structure theorem tells us that Γ looks like

$$\Gamma \cong \underbrace{\mathbb{Z} \oplus \mathbb{Z} \oplus \cdots \oplus \mathbb{Z}}_{r \text{ times}} \oplus \mathbb{Z}_{p_1^{\nu_1}} \oplus \mathbb{Z}_{p_2^{\nu_2}} \oplus \cdots \oplus \mathbb{Z}_{p_s^{\nu_s}}.$$

More naively, this says that there are generators

$$P_1, \ldots, P_r, Q_1, \ldots, Q_s \in \Gamma$$

such that every $P \in \Gamma$ can be written in the form

$$P = n_1 P_1 + \cdots + n_r P_r + m_1 Q_1 + \cdots + m_s Q_s.$$

Here the integers n_i are uniquely determined by P, whereas the integers m_j are determined modulo $p_j^{\nu_j}$.

The integer r is called the *rank of* Γ. The group Γ will be finite if and only if it has rank $r = 0$. The subgroup

$$\mathbb{Z}_{p_1^{\nu_1}} \oplus \mathbb{Z}_{p_2^{\nu_2}} \oplus \cdots \oplus \mathbb{Z}_{p_s^{\nu_s}}$$

corresponds to the elements of finite order in Γ; it has order $p_1^{\nu_1} p_2^{\nu_2} \cdots p_s^{\nu_s}$.

Of course, the points $P_1, \ldots, P_r, Q_1, \ldots, Q_s$ are not unique. There are many possible choices of generators for Γ.

We have already studied how to compute the elements of finite order in Γ in a finite number of steps. It is much harder to get hold of the rank. We want to give some illustrations of how to do this in special cases. First we will do a bit more theory, which will help us in doing the computations.

The proof of Mordell's theorem, if we are lucky, will allow us to determine the quotient group $\Gamma/2\Gamma$. From above, the subgroup 2Γ looks like

$$2\Gamma \cong 2\mathbb{Z} \oplus \cdots \oplus 2\mathbb{Z} \oplus 2\mathbb{Z}_{p_1^{\nu_1}} \oplus \cdots \oplus 2\mathbb{Z}_{p_s^{\nu_s}};$$

so the quotient group has the form

$$\frac{\Gamma}{2\Gamma} \cong \frac{\mathbb{Z}}{2\mathbb{Z}} \oplus \cdots \oplus \frac{\mathbb{Z}}{2\mathbb{Z}} \oplus \frac{\mathbb{Z}_{p_1^{\nu_1}}}{2\mathbb{Z}_{p_1^{\nu_1}}} \oplus \cdots \oplus \frac{\mathbb{Z}_{p_s^{\nu_s}}}{2\mathbb{Z}_{p_s^{\nu_s}}}.$$

Now $\mathbb{Z}/2\mathbb{Z} = \mathbb{Z}_2$ is cyclic of order two, whereas

$$\frac{\mathbb{Z}_{p_i^{\nu_i}}}{2\mathbb{Z}_{p_i^{\nu_i}}} \cong \begin{cases} \mathbb{Z}_2, & \text{if } p_i = 2, \\ 0, & \text{if } p_i \neq 2. \end{cases}$$

Thus,

$$(\Gamma : 2\Gamma) = 2^{r+(\text{number of } j \text{ with } p_j=2)}.$$

On the other hand, let $\Gamma[2]$ denote the subgroup of all $Q \in \Gamma$ such that $2Q = \mathcal{O}$. What does $\Gamma[2]$ look like? We need to know when

$$2(n_1 P_1 + \cdots + n_r P_r + m_1 Q_1 + \cdots + m_s Q_s) = 0.$$

This happens if $n_i = 0$ for each i, and $2m_j \equiv 0 \pmod{p_j^{\nu_j}}$. Now if p is odd and $2m \equiv 0 \pmod{p^\nu}$, then $m \equiv 0 \pmod{p^\nu}$. However, if $p = 2$ and $2m \equiv 0 \pmod{p^\nu}$, then we only conclude that $m \equiv 0 \pmod{p^{\nu-1}}$. So the order of the subgroup $\Gamma[2]$ is

$$\#\Gamma[2] = 2^{(\text{number of } j \text{ with } p_j=2)}.$$

Combining these two formulas, we obtain the useful result

$$(\Gamma : 2\Gamma) = 2^r \cdot \#\Gamma[2].$$

This formula holds for any finitely generated abelian group of rank r.

In our case what are the possibilities for $\#\Gamma[2]$? How many points can we have with $2Q = \mathcal{O}$? Aside from $\mathcal{O}$, these are the points with $y = 0$, so it is clear from the equation for the curve that the answer is

$$\#\Gamma[2] = \begin{cases} 2, & \text{if } a^2 - 4b \text{ is not a square,} \\ 4, & \text{if } a^2 - 4b \text{ is a square.} \end{cases}$$

Now we have only to recall the last step of the proof of Mordell's theorem to get a formula for the rank which makes it computable in some cases if we are lucky. Remember we have homomorphisms $\phi : \Gamma \to \overline{\Gamma}$ and $\psi : \overline{\Gamma} \to \Gamma$ such that the composition $\psi \circ \phi$ is multiplication by two. Thus

$$(\Gamma : 2\Gamma) = (\Gamma : \psi \circ \phi(\Gamma)).$$

We have an inclusion of subgroups $\Gamma \supseteq \psi(\overline{\Gamma}) \supseteq 2\Gamma$, and thus

$$(\Gamma : 2\Gamma) = (\Gamma : \psi(\overline{\Gamma}))(\psi(\overline{\Gamma}) : \psi \circ \phi(\Gamma)).$$

We want to analyze this last index $(\psi(\overline{\Gamma}) : \psi \circ \phi(\Gamma))$. We start with an abstract remark. Let A be an abelian group, let B be a subgroup of finite index in A, and let $\psi : A \to A'$ be a homomorphism of A into some group A'. We are interested in the index $(\psi(A) : \psi(B))$.

Using the standard isomorphism theorems from elementary group theory, we find

$$\frac{\psi(A)}{\psi(B)} \cong \frac{A}{(B + \ker(\psi))} \cong \frac{A/B}{(B + \ker(\psi))/B} \cong \frac{A/B}{\ker(\psi)/(\ker(\psi) \cap B)}.$$

Hence,

$$(\psi(A) : \psi(B)) = \frac{(A : B)}{(\ker(\psi) : \ker(\psi) \cap B)}.$$

If you do not like this abstract argument, you can check the equality of indices directly in our case, because $\ker(\psi)$ consists of the two elements $\overline{O}, \overline{T}$, and thus $\ker(\psi) \cap \phi(\Gamma)$ is either $\overline{O}$ or $\ker(\psi)$.

We now apply this abstract formula with $A = \overline{\Gamma}$ and $B = \phi(\Gamma)$. This and the formula for $(\Gamma : 2\Gamma)$ we derived earlier gives

$$(\Gamma : 2\Gamma) = \frac{(\overline{\Gamma} : \psi(\overline{\Gamma})) \cdot (\overline{\Gamma} : \phi(\Gamma))}{(\ker(\psi) : \ker(\psi) \cap \phi(\Gamma))}.$$

But we have seen that $\overline{T} \in \phi(\Gamma)$ if and only if $\overline{b} = a^2 - 4b$ is a square; so

$$(\ker(\psi) : \ker(\psi) \cap \phi(\Gamma)) = \begin{cases} 2, & \text{if } \overline{b} \text{ is not a square,} \\ 1, & \text{if } \overline{b} \text{ is a square.} \end{cases}$$

Now everything falls out nicely, and we find

$$2^r = \frac{(\Gamma : 2\Gamma)}{\#\Gamma[2]} = \frac{(\overline{\Gamma} : \psi(\overline{\Gamma})) \cdot (\overline{\Gamma} : \phi(\Gamma))}{4}.$$

Of course, each of the indices in the numerator is a power of 2.

How should we compute these indices? Recall the method we used to prove that they are finite. We had a homomorphism

$$\alpha : \Gamma \longrightarrow \frac{\mathbb{Q}^*}{\mathbb{Q}^{*2}} \qquad \text{defined by} \qquad \begin{aligned} \alpha(x, y) &= x \pmod{\mathbb{Q}^{*2}} \\ \alpha(T) &= b \pmod{\mathbb{Q}^{*2}}. \end{aligned}$$

We showed that the kernel of α equals the image $\psi(\overline{\Gamma})$; and so the image of α is isomorphic to

$$\alpha(\Gamma) \cong \frac{\Gamma}{\ker(\alpha)} \cong \frac{\Gamma}{\psi(\overline{\Gamma})}.$$

Hence, $(\Gamma : \psi(\overline{\Gamma})) = \#\alpha(\Gamma)$.

Similarly, using the analogous homomorphism $\overline{\alpha} : \overline{\Gamma} \to \mathbb{Q}^*/\mathbb{Q}^{*2}$, we find that $(\overline{\Gamma} : \phi(\Gamma)) = \#\overline{\alpha}(\overline{\Gamma})$. This gives the following alternative formula for the rank of Γ:

$$2^r = \frac{\#\alpha(\Gamma) \cdot \#\overline{\alpha}(\overline{\Gamma})}{4}.$$

It is this formula we will use to try to compute the rank.

In order to determine the image $\alpha(\Gamma)$, we have to find out which rational numbers, modulo squares, can occur as the x coordinates of points in Γ. The way we do that is to write

$$x = \frac{m}{e^2}, \qquad y = \frac{n}{e^3},$$

in lowest terms with $e > 0$.

If $m = 0$, then $(x,y) = T$ and $\alpha(T) = b$. Thus $b \pmod{\mathbb{Q}^{*2}}$ is always in $\alpha(\Gamma)$. If $a^2 - 4b$ is a square, say $a^2 - 4b = d^2$, then Γ has two other points of order two, namely

$$\left(\frac{-a+d}{2}, 0\right) \qquad \text{and} \qquad \left(\frac{-a-d}{2}, 0\right).$$

So if $a^2 - 4b = d^2$, then $\alpha(\Gamma)$ contains $\dfrac{-a \pm d}{2}$.

Now we look at the points with $m, n \neq 0$. These points satisfy

$$n^2 = m^3 + am^2 e^2 + bme^4 = m(m^2 + ame^2 + be^4).$$

In Section 5 we showed that m and $m^2 + ame^2 + be^4$ are practically relatively prime, so m and $m^2 + ame^2 + be^4$ are both more or less squares. Now we do things systematically.

Let $b_1 = \pm \gcd(m, b)$, where we choose the sign so that $mb_1 > 0$. Then we can write

$$m = b_1 m_1, \qquad b = b_1 b_2, \qquad \text{with } \gcd(m_1, b_2) = 1 \text{ and } m_1 > 0.$$

If we substitute in the equation of the curve, we get

$$n^2 = b_1 m_1 (b_1^2 m_1^2 + ab_1 m_1 e^2 + b_1 b_2 e^4) = b_1^2 m_1 (b_1 m_1^2 + am_1 e^2 + b_2 e^4).$$

Thus, $b_1^2 | n^2$, so $b_1 | n$ and we can write $n = b_1 n_1$. Hence,

$$n_1^2 = m_1 (b_1 m_1^2 + am_1 e^2 + b_2 e^4).$$

Since $\gcd(b_2, m_1) = 1$ and $\gcd(e, m_1) = 1$, we see that the quantities m_1 and $b_1 m_1^2 + am_1 e^2 + b_2 e^4$ are relatively prime. Their product is a square, and $m_1 > 0$, so we conclude that each of them is a square. Hence we can factor n_1 as $n_1 = MN$ so that

$$M^2 = m_1 \qquad \text{and} \qquad N^2 = b_1 m_1^2 + am_1 e^2 + b_2 e^4.$$

Eliminating m_1, we obtain

$$N^2 = b_1 M^4 + aM^2 e^2 + b_2 e^4;$$

and this tells the whole story: If you have a point $(x, y) \in \Gamma$ with $y \neq 0$, then you can put that point in the form

$$x = \frac{b_1 M^2}{e^2}, \qquad y = \frac{b_1 M N}{e^3}.$$

Thus modulo squares, the x coordinate of any point on the curve is one of the values of b_1; and since b_1 is a divisor of the non-zero integer b, there are only a finite number of possibilities for b_1.

It is now very "easy" to find the order of $\alpha(\Gamma)$. We take the integer b and factor it as a product $b = b_1 b_2$ in all possible ways. For each way of factoring, we write down the equation

$$N^2 = b_1 M^4 + a M^2 e^2 + b_2 e^4.$$

Here a, b_1, b_2 are fixed, and M, e, N are the variables. Then $\alpha(\Gamma)$ consists of $b \pmod{\mathbb{Q}^{*2}}$, together with those $b_1 \pmod{\mathbb{Q}^{*2}}$ such that the equation has a solution with $M \neq 0$.

In addition, the fact that x and y are in lowest terms implies

$$\gcd(M, e) = \gcd(N, e) = \gcd(b_1, e) = 1;$$

and the assumption that $\gcd(b_2, m_1) = 1$ implies

$$\gcd(b_2, M) = \gcd(M, N) = 1.$$

All admissible solutions must also satisfy these side conditions. Notice that if we find a solution M, e, N, then we get a point in Γ by the formulas for x and y given earlier.

If you are observant, you will have noticed that we appear to have forgotten two elements of $\alpha(\Gamma)$. We noted above that if $a^2 - 4b$ is a square, say $a^2 - 4b = d^2$, then there are points of order two whose images by α are the values $\dfrac{-a \pm d}{2} \in \dfrac{\mathbb{Q}^*}{\mathbb{Q}^{*2}}$. However, notice that there is then a factorization of b given by

$$b = \frac{-a + d}{2} \cdot \frac{-a - d}{2};$$

so in applying the above procedure, we would consider the equation

$$N^2 = \left(\frac{-a \pm d}{2} \right) M^4 + a M^2 e^2 + \left(\frac{-a \mp d}{2} \right) e^4.$$

This equation has the obvious solution $(M, e, N) = (1, 1, 0)$, so our general procedure takes care of these values automatically.

To summarize, in order to determine the order of $\alpha(\Gamma)$, we will write down several equations of the form

$$N^2 = b_1 M^4 + a M^2 e^2 + b_2 e^4,$$

one for each factorization $b = b_1 b_2$. We will need to decide whether or not each of these equations has a solution in integers with $M \neq 0$; and each time we find an equation with a solution (M, e, N), then we get a new point on the curve by the formula

$$x = \frac{b_1 M^2}{e^2}, \qquad y = \frac{b_1 M N}{e^3}.$$

The only trouble with all this is that at present, there is no known method for deciding whether or not an equation like this has a solution. Except for this "little" difficulty, we now have a method for computing the rank.

We can hope to get some results as follows: For each b_1, b_2, either exhibit a solution or show that the equation has no solutions by considering it as a congruence or as an equation in real numbers. In this way we may get enough information to decide the whole story. We will now illustrate this procedure with several examples.

Example 1. $\boxed{C : y^2 = x^3 - x, \qquad \overline{C} : y^2 = x^3 + 4x}$

We will start with a modest example. In this case $a = 0$ and $b = -1$. The first step is to factor b in all possible ways. There are two factorizations:

$$-1 = -1 \times 1 \qquad \text{and} \qquad -1 = 1 \times -1.$$

Thus b_1 can only be ± 1. Since $\alpha(\mathcal{O}) = 1$ and $\alpha(T) = b = -1$, we see that

$$\alpha(\Gamma) = \left\{ \pm 1 \pmod{\mathbb{Q}^{*2}} \right\}$$

is a group of two elements.

Next we must compute $\overline{\alpha}(\overline{\Gamma})$, so we need to apply our procedure to the curve $\overline{C} : y^2 = x^3 + 4x$. Now $\overline{b} = 4$ has lots of factorizations; we can choose

$$b_1 = 1, -1, 2, -2, 4, -4.$$

But $4 \equiv 1 \pmod{\mathbb{Q}^{*2}}$ and $-4 \equiv -1 \pmod{\mathbb{Q}^{*2}}$, so $\overline{\alpha}(\overline{\Gamma})$ consists of at most the four elements $\{1, -1, 2, -2\}$. Of course, we always have $\overline{b} \in \overline{\alpha}(\overline{\Gamma})$, but in this case $\overline{b} = 4$ is a square, so that does not help us.

The four equations we must consider are

(i) $N^2 = M^4 + 4e^4$,

(ii) $N^2 = -M^4 - 4e^4$,

(iii) $N^2 = 2M^4 + 2e^4$,

(iv) $N^2 = -2M^4 - 2e^4$.

Since $N^2 \geq 0$, and we do not allow solutions with $M = 0$, we see that equations (ii) and (iv) have no solutions in integers; in fact, they have no solutions in real numbers with $M \neq 0$, since the right-hand side would be strictly negative.

Equation (i) has the obvious solution $(M, e, N) = (1, 0, 1)$, which corresponds to the fact that $1 \in \overline{\alpha}(\overline{\Gamma})$, so that is nothing new. Finally, our theorem tells us that $\#\alpha(\Gamma) \cdot \#\overline{\alpha}(\overline{\Gamma})$ is at least 4, so for this example we know that $\overline{\alpha}(\overline{\Gamma})$ must have order at least two. Thus, equation (iii) must have a solution. Of course, we needn't rely on this fancy reasoning, because (iii) has the obvious solution

$$2^2 = 2 \cdot 1^4 + 2 \cdot 1^4.$$

So we conclude that $\overline{\alpha}(\overline{\Gamma})$ has order two. Thus, the rank of Γ is zero, and the same for the rank of $\overline{\Gamma}$. This proves that the group of rational points on C and $\overline{C}$ are both finite, and so all rational points have finite order.

To find the points of finite order, we can use the Nagell-Lutz theorem. Thus, if $P = (x, y)$ is a point of finite order in Γ, then either $y = 0$ or y divides $b^2(a^2 - 4b) = 4$. The points with $y = 0$ are $(0, 0)$ and $(\pm 1, 0)$; and it is a simple matter to check that there are no points with $y = \pm 1$, $y = \pm 2$, or $y = \pm 4$. We have thus proved that the group of rational points on the curve $C : y^2 = x^3 - x$ is precisely

$$C(\mathbb{Q}) = \{\mathcal{O}, (0, 0), (1, 0), (-1, 0)\} \cong \frac{\mathbb{Z}}{2\mathbb{Z}} \oplus \frac{\mathbb{Z}}{2\mathbb{Z}}.$$

So here is the first explicit cubic equation for which we have actually found all of the rational solutions.

Similarly, the points of finite order in $\overline{\Gamma}$ satisfy either $y = 0$ or y divides $\overline{b}^2(\overline{a}^2 - 4\overline{b}) = -256$. After some work, one finds four points of finite order:

$$\overline{C}(\mathbb{Q}) = \{\mathcal{O}, (0, 0), (2, 4), (2, -4)\} \cong \frac{\mathbb{Z}}{4\mathbb{Z}}.$$

In this case the group of rational points is a cyclic group of order four, because one easily checks that $(2, 4) + (2, 4) = (0, 0)$.

Example 2. $\boxed{C : y^2 = x^3 + x, \qquad \overline{C} : y^2 = x^3 - 4x}$
The situation here is a slight variant of the previous example, so we will leave the details to you. Again one finds that the rank is zero; the finite groups of rational points being given by

$$C(\mathbb{Q}) = \{\mathcal{O}, T\} \cong \frac{\mathbb{Z}}{2\mathbb{Z}};$$

$$\overline{C}(\mathbb{Q}) = \{\mathcal{O}, (0, 0), (2, 0), (-2, 0)\} \cong \frac{\mathbb{Z}}{2\mathbb{Z}} \oplus \frac{\mathbb{Z}}{2\mathbb{Z}}.$$

As a by-product of the calculation, we get the answer to an interesting question. Any integer solution of the equation $N^2 = M^4 + e^4$ with $e \neq 0$ gives a rational point on the curve C, namely, the point $\left(\dfrac{M^2}{e^2}, \dfrac{MN}{e^3}\right)$. So once we know that Γ has only the two elements $\mathcal{O}$ and $(0,0)$, it follows that the equation $N^2 = M^4 + e^4$ has no solutions in which M, N, e are all non-zero. This means, in particular, that the Fermat equation $Z^4 = X^4 + Y^4$ has no solutions in non-zero integers. Of course, there are more elementary proofs of this fact.

Example 3. $\boxed{C : y^2 = x^3 - 5x, \qquad \overline{C} : y^2 = x^3 + 20x}$

For the curve C, we have $a = 0$ and $b = -5$, so the possibilities for b_1 are $1, -1, 5, -5$. The corresponding equations are

(i) $\qquad N^2 = M^4 - 5e^4$,

(ii) $\qquad N^2 = -M^4 + 5e^4$,

(iii) $\qquad N^2 = 5M^4 - e^4$,

(iv) $\qquad N^2 = -5M^4 + e^4$.

Notice that equations (i) and (ii) are the same as equations (iii) and (iv) with the variables M and e reversed. Since the solutions that we find will satisfy $Me \neq 0$, it is enough to consider the first two equations.

After a little trial-and-error, we find solutions to (i) and (ii):

$$1^2 = 3^4 - 5 \cdot 2^4,$$
$$2^2 = -(1^4) + 5 \cdot 1^4.$$

Hence, all b_1's occur; and as a by-product of the method, we can use the formulas

$$x = \frac{b_1 M^2}{e^2}, \qquad y = \frac{b_1 M N}{e^3}$$

to get the rational points $\left(\frac{9}{4}, \frac{3}{8}\right)$ and $(-1, -2)$ on C. This proves that

$$\alpha(\Gamma) = \{\pm 1, \pm 5\} \left(\text{mod } \mathbb{Q}^{*2}\right),$$

which is the Four Group.

What about $\overline{\alpha}(\overline{\Gamma})$? Since $\overline{b} = a^2 - 4b = 20$, the possibilities for $\overline{b}_1$ are

$$\overline{b}_1 = \pm 1, \pm 2, \pm 4, \pm 5, \pm 10, \pm 20.$$

But $\pm 4 = \pm(2^2)$ and $\pm 20 = \pm(5 \cdot 2^2)$, so the possibilities for $\overline{b}_1$ modulo squares are $\pm 1, \pm 2, \pm 5, \pm 10$.

Next we observe that since $\overline{b}_1 \overline{b}_2 = \overline{b} = 20$, both $\overline{b}_1$ and $\overline{b}_2$ will have the same sign. If they are negative, then the equation

$$N^2 = \overline{b}_1 M^4 + \overline{b}_2 e^4$$

can have no non-zero rational solutions, because it will not even have non-zero real solutions. So we are down to

$$\overline{\alpha}(\overline{\Gamma}) \subseteq \{1, 2, 5, 10 \,(\text{mod } \mathbb{Q}^{*2})\}.$$

Now $\overline{\alpha}(\overline{\mathcal{O}}) = 1$ and $\overline{\alpha}(\overline{T}) = \overline{b} = 20 \equiv 5 \pmod{\mathbb{Q}^{*2}}$ are both in $\overline{\alpha}(\overline{\Gamma})$. How do we eliminate $\overline{b}_1 = 2$ and $\overline{b}_1 = 10$?

We have to decide whether the equation

$$N^2 = 2M^4 + 10e^4$$

has a solution in integers. Looking back at the relative primality conditions satisfied by M, N, e, it is enough to show that there are no solutions with $\gcd(M, 10) = 1$. Suppose that there is a solution. Since M is relatively prime to 5, we know from Fermat's Little Theorem that $M^4 \equiv 1 \pmod 5$. So reducing the equation modulo 5, we see that N satisfies

$$N^2 \equiv 2 \pmod 5.$$

But this congruence has no solutions, from which we conclude that the equation $N^2 = 2M^4 + 10e^4$ has no solutions in integers with $\gcd(M, 10) = 1$. Therefore $2 \notin \overline{\alpha}(\overline{\Gamma})$.

A similar calculation would show that $10 \notin \overline{\alpha}(\overline{\Gamma})$, but there is an easier way. Since $\overline{\alpha}(\overline{\Gamma})$ is a subgroup of $\mathbb{Q}^*/\mathbb{Q}^{*2}$, and we already know that 5 is in this subgroup and 2 is not, it is immediate that 10 is not. So now we know that

$$\overline{\alpha}(\overline{\Gamma}) = \{1, 5\} \pmod{\mathbb{Q}^{*2}}.$$

Putting all this together, we find that

$$2^r = \frac{\#\alpha(\Gamma) \cdot \#\overline{\alpha}(\overline{\Gamma})}{4} = \frac{4 \cdot 2}{4} = 2;$$

and so the rank of $C(\mathbb{Q})$ is 1.

There is a general principle involved here. In eliminating the equations $N^2 = \overline{b}_1 M^4 + \overline{b}_2 e^4$ with $\overline{b}_1$ and $\overline{b}_2$ negative, we viewed it as an equation in real numbers. This point of view was no help in eliminating $\overline{b}_1 = 2$ and $\overline{b}_1 = 10$, but from the point of view of congruences modulo the prime $p = 5$, we saw that there are no solutions to the congruence $N^2 \equiv 2M^4 + 10e^4 \pmod 5$. Thus, for the equation $y^2 = x^3 - 5x$, we could settle the whole issue by taking certain equations and looking at them as equations in the real field and as congruences.

Life gets much rougher when we find a curve for which we do our best to eliminate the b_1's by real and congruence considerations, and still there remain some b_1's which we cannot eliminate and for which we cannot find a solution to $N^2 = b_1 M^4 + b_2 e^4$. Such curves do occur in nature, and the problems in such a situation are of a much higher order of difficulty. We will exhibit an equation of this sort in the next example, although we will not give a proof.

Example 4. $\boxed{C_p : y^2 = x^3 + px}$
It is curious that $y^2 = x^3 + 20x$ has infinitely many rational solutions, whereas $y^2 = x^3 + x$ and $y^2 = x^3 + 4x$ have only a finite number. In

general, it is difficult to predict the rank from the equation of the curve. For example, let's look at the curves $C_p : y^2 = x^3 + px$, where p is a prime. In this case, $b = p$ and $\bar{b} = -4p$, and it is not too hard to show that the rank of C_p is either 0, 1, or 2.

If $p \equiv 7$ or $11 \,(\mathrm{mod}\ 16)$, then an argument similar to the ones we gave above can be used to show that C_p has rank 0. Next, if

$$p \equiv 3 \text{ or } 5 \text{ or } 13 \text{ or } 15 \quad (\mathrm{mod}\ 16),$$

then it is conjectured but not yet proven that the rank is always equal to 1. Finally, in the remaining case $p \equiv 1 \,(\mathrm{mod}\ 8)$, it is believed that the rank is always either 0 or 2, never 1. Both of these can occur, because the curves C_{73} and C_{89} both have rank 2, whereas the curves C_{17} and C_{41} both have rank 0.

The last two curves give examples of the hard problem mentioned above. In trying to compute the rank of C_{17}, for example, one needs to check if the equation $N^2 = 17M^4 - 4e^4$ has a non-trivial solution in integers. It turns out that there are no such solutions, even though one can check that there are real solutions and also solutions modulo m for every integer m! So the proof that there are no integer solutions is of necessity somewhat indirect.

We cannot resist mentioning one more C_p, studied by Bremner and Cassels [1]. They show that the innocuous looking curve $y^2 = x^3 + 877x$ has rank 1, as it should by the conjecture mentioned above. They further show that its group of rational points is generated by the points $T = (0,0)$ and $P = (x_0, y_0)$, where x_0 has the value

$$x_0 = \left(\frac{612776083187947368101}{7884153586063900210} \right)^2 .$$

So even cubic curves with comparatively small coefficients may require points of extremely large height to generate their group of rational points.

We have now seen cubic curves whose rational points have rank 0 and 1, and it is not too hard to find examples with rank 2 or 3 or even 4. But it is extremely difficult to find curves with very large rank. In fact, it is still an open question as to whether or not there exist curves with arbitrarily large rank. Mestre [2] has shown that the curve

$$y^2 + xy = x^3 + bx + c,$$
$$b = -2098119445112830964947539994\underline{8}5,$$
$$c = 36653992551590286206010035905960909459942897,$$

has rank at least 15. As of April 14, 1992, there are no cubic curves known with rank 16 or larger.

7. Singular Cubic Curves

As promised earlier, we will now briefly look at singular cubic curves. We will show that the rational points on singular cubic curves and on non-singular cubic curves behave completely differently.

Let C be a cubic curve with a singular point $S \in C$. Then any line through S will intersect C at S with multiplicity at least two. If there were a second singular point $S' \in C$, then the line L connecting S and S' would intersect C at least twice at S and at least twice at S', so L would intersect C at least four times. But a line and a cubic intersect in only three points counting multiplicities. Thus, a cubic curve C can have at most one singular point.

Even if C is singular, we would like to make the points of C into a group, just as we did for non-singular cubics. It turns out that this can be done quite easily provided we discard the singular point S. So for any cubic curve C we will define

$$C_{\text{ns}} = \{P \in C : P \text{ is not a singular point}\}.$$

(The subscript ns stands for "non-singular.") Similarly, we let $C_{\text{ns}}(\mathbb{Q})$ denote the subset of C_{ns} consisting of points with rational coordinates. As usual, we also fix a point $\mathcal{O} \in C_{\text{ns}}$ to be the origin. Then to add two points $P, Q \in C_{\text{ns}}$, we use the same geometric procedure that worked for non-singular curves. First we draw the line L connecting P and Q, and let R be the other intersection point of $L \cap C$. Then we draw the line L' through R and $\mathcal{O}$. The third intersection point of $L' \cap C$ is defined to be the sum $P + Q$. Then one can check that C_{ns} is an abelian group; and if $\mathcal{O}$ is in $C_{\text{ns}}(\mathbb{Q})$, then $C_{\text{ns}}(\mathbb{Q})$ is a subgroup of C_{ns}.

This describes the group law geometrically, but we can also give explicit equations. In fact, if we make a change of variables so that the singular cubic curve is given by a Weierstrass equation

$$y^2 = x^3 + ax^2 + bx + c,$$

then all of the formulas for the addition law derived in Chapter I, Section 4 are still true. For example, on the singular cubic curve

$$y^2 = x^3$$

with singular point $S = (0,0)$, the addition law becomes

$$(x_1, y_1) + (x_2, y_2) = \left(\frac{\nu^2}{x_1 x_2}, \frac{-\nu^3}{y_1 y_2} \right), \qquad \text{where } \nu = \frac{y_1 x_2 - x_1 y_2}{x_2 - x_1}.$$

If C is non-singular, the Mordell-Weil theorem tells us that $C(\mathbb{Q})$ is a finitely generated group. We are now going to describe exactly what the group $C_{ns}(\mathbf{Q})$ looks like in the case that C is singular. The answer and the proof will be much easier than the Mordell-Weil theorem. The only slight complication is that there are two different answers, depending on what the singularity looks like.

We observed in Chapter I that there are two possible pictures for the singularity S. Either there are distinct tangent directions at S (cf. Figure 1.13) or there is a cusp (cf. Figure 1.14). A typical example of the first case is the equation

$$C : y^2 = x^3 + x^2$$

and a typical example of the second case is

$$C' : y^2 = x^3.$$

We also saw in Chapter I that it is very easy to parametrize all of the rational points on C and C'. For example, the maps

$$
\begin{array}{ccccc}
\mathbf{Q} & \longrightarrow & C(\mathbf{Q}) & \longrightarrow & \mathbf{Q} \\
r & \longmapsto & (r^2 - 1, r^3 - r) & & \\
& & (x, y) & \longmapsto & \dfrac{y}{x}
\end{array}
$$

are inverses to one another, so the set $C(\mathbf{Q})$ essentially looks like the set $\mathbf{Q}$. Similarly the map $r \to (r^2, r^3)$ shows that the set $C'(\mathbf{Q})$ also looks like $\mathbf{Q}$. However, we are now interested in $C_{ns}(\mathbf{Q})$ and $C'_{ns}(\mathbf{Q})$ not as sets, but as groups. It turns out that by using slightly different maps, we actually get group homomorphisms.

Theorem. (a) *Let C be the singular cubic curve $y^2 = x^3 + x^2$. Then the map*

$$\phi : C_{ns}(\mathbf{Q}) \longrightarrow \mathbf{Q}^*, \qquad \phi(P) = \begin{cases} \dfrac{y-x}{y+x} & \text{if } P = (x, y), \\ 1 & \text{if } P = \mathcal{O}, \end{cases}$$

is a group isomorphism from $C_{ns}(\mathbf{Q})$ to the multiplicative group of non-zero rational numbers.

(b) *Let C be the singular cubic curve $y^2 = x^3$. Then the map*

$$\phi : C_{ns}(\mathbf{Q}) \longrightarrow \mathbf{Q}, \qquad \phi(P) = \begin{cases} \dfrac{x}{y} & \text{if } P = (x, y), \\ 0 & \text{if } P = \mathcal{O}, \end{cases}$$

is a group isomorphism from $C_{ns}(\mathbf{Q})$ to the additive group of all rational numbers.

PROOF. (a) First we observe that ϕ is well defined. The only possible problem would be if we had a point $(x, y) \in C_{ns}(\mathbf{Q})$ with $y + x = 0$. But then the equation of C would imply that

$$x^3 = y^2 - x^2 = (y+x)(y-x) = 0,$$

so $x = 0$; and then also $y = 0$. Since $(0,0)$ is the singular point on C, we see that $y + x \neq 0$ for all points $(x, y) \in C_{\text{ns}}$.

Next, if we set $t = \dfrac{y - x}{y + x}$ and solve for $y = \left(\dfrac{1 + t}{1 - t}\right) x$, we can substitute into $y^2 = x^3 + x^2$ and solve for x in terms of t:

$$x = \frac{4t}{(1 - t)^2}.$$

Thus, we get a map

$$\psi : \mathbb{Q}^* \longrightarrow C_{\text{ns}}(\mathbb{Q}), \qquad \psi(t) = \begin{cases} \left(\dfrac{4t}{(1 - t)^2}, \dfrac{4t(1 + t)}{(1 - t)^3}\right) & \text{if } t \neq 1, \\ \mathcal{O} & \text{if } t = 1. \end{cases}$$

It is easy to check that $\phi(\psi(t)) = t$ and $\psi(\phi(P)) = P$, which proves that ϕ and ψ are inverse maps. Hence ϕ and ψ are one-to-one and onto as maps of sets. It remains to show that they are homomorphisms.

First we check that ψ sends inverses to inverses.

$$\psi\left(\frac{1}{t}\right) = \left(\frac{4t^{-1}}{(1 - t^{-1})^2}, \frac{4t^{-1}(1 + t^{-1})}{(1 - t^{-1})^3}\right)$$

$$= \left(\frac{4t}{(1 - t)^2}, -\frac{4t(1 + t)}{(1 - t)^3}\right) = -\psi(t).$$

Next let $P_1, P_2, P_3 \in C_{\text{ns}}$ be any three points on C_{ns}. We know that their sum is zero if and only if they are collinear. If we use coordinates $P_i = (x_i, y_i)$, then the line through P_1 and P_2 has the equation

$$(x_2 - x_1)(y - y_1) = (y_2 - y_1)(x - x_1).$$

Substituting $(x, y) = (x_3, y_3)$ and multiplying out both sides, we find that P_1, P_2, P_3 are collinear if and only if their coordinates satisfy the condition

$$x_1 y_2 - x_2 y_1 + x_2 y_3 - x_3 y_2 + x_3 y_1 - x_1 y_3 = 0. \qquad (*)$$

Now we need to verify that if three elements $t_1, t_2, t_3 \in \mathbb{Q}^*$ multiply to 1, then their images $\psi(t_1), \psi(t_2), \psi(t_3) \in C_{\text{ns}}(\mathbb{Q})$ sum to $\mathcal{O}$. The formula for ψ given above says that

$$\psi(t) = \left(\frac{4t}{(1 - t)^2}, \frac{4t(1 + t)}{(1 - t)^3}\right).$$

Letting $P_1 = \psi(t_1)$, $P_2 = \psi(t_2)$, and $P_3 = \psi(t_3)$, and substituting into the left-hand-side of $(*)$, we find after some algebra that

$$x_1 y_2 - x_2 y_1 + x_2 y_3 - x_3 y_2 + x_3 y_1 - x_1 y_3$$
$$= \frac{32(t_1 - t_2)(t_1 - t_3)(t_2 - t_3)(t_1 t_2 t_3 - 1)}{(1 - t_1)^3 (1 - t_2)^3 (1 - t_3)^3}.$$

This proves that

$$t_1 t_2 t_3 = 1 \implies \psi(t_1), \psi(t_2), \text{ and } \psi(t_3) \text{ are collinear}$$
$$\implies \psi(t_1) + \psi(t_2) + \psi(t_3) = \mathcal{O},$$

at least provided t_1, t_2, t_3 are distinct and not equal to 1. The remaining cases can be dealt with similarly, or we could define the group law on all of the real points in C_{ns} and argue that because $\psi : \mathbb{R}^* \to C_{\text{ns}}(\mathbb{R})$ is a homomorphism for distinct points, it is a homomorphism for all points by continuity.

(b) The proof for this curve is similar to the proof in (a), but easier, so we will leave it for you as an exercise.   □

The Mordell-Weil theorem tells us that if C is a non-singular cubic curve, then the group $C(\mathbb{Q})$ is finitely generated. On the other hand, it is easy to see that the groups $(\mathbb{Q}^*, *)$ and $(\mathbb{Q}, +)$ are not finitely generated. So our theorem implies that the group of rational points $C_{\text{ns}}(\mathbb{Q})$ on a singular cubic curve is not finitely generated, at least for the two curves covered in the theorem. In the exercises we will explain how to show that $C_{\text{ns}}(\mathbb{Q})$ is not finitely generated for all singular cubic curves. So the rational points on singular and non-singular cubic curves behave quite differently; and further, the rational points on the singular curves form groups ($\mathbb{Q}^*$ and $\mathbb{Q}$) with which we are very familiar. We hope that this explains why we have devoted most of our attention to studying rational points on the more interesting and mysterious non-singular cubic curves.

EXERCISES

3.1. (a) Prove that the set of rational numbers x with height $H(x)$ less than κ contains at most $2\kappa^2 + \kappa$ elements.

(b) * Let $R(\kappa)$ be the set of rational numbers x with height $H(x)$ less than κ. Prove that

$$\lim_{\kappa \to \infty} \frac{\#R(\kappa)}{\kappa^2} = \frac{12}{\pi^2}.$$

3.2. Let $P_1 = (x_1, y_1)$ and $P_2 = (x_2, y_2)$ be points on the non-singular cubic curve

$$y^2 = x^3 + ax^2 + bx + c,$$

where a, b, c are integers. Let

$$P_3 = (x_3, y_3) = P_1 + P_2 \qquad \text{and} \qquad P_4 = (x_4, y_4) = P_1 - P_2.$$

(a) Derive formulas for the quantities $x_3 + x_4$ and $x_3 x_4$ in terms of x_1 and x_2. (Note that you should be able to eliminate y_1 and y_2 from these formulas.)

(b) Prove that there is a constant κ, which depends only on a, b, c, so that for all rational points P_1 and P_2,

$$h(P_1 + P_2) + h(P_1 - P_2) \le 2h(P_1) + 2h(P_2) + \kappa.$$

Notice that this greatly strengthens the inequality given in Lemma 2.

(c) Prove that if κ is replaced by a suitably large negative number, then the opposite inequality in (b) will hold. In other words, prove that there is a constant κ, depending only on a, b, c, so that for all rational points P_1 and P_2,

$$-\kappa \le h(P_1 + P_2) + h(P_1 - P_2) - 2h(P_1) - 2h(P_2) \le \kappa.$$

(*Hint.* In (b), replace P_1 and P_2 by $P_1 + P_2$ and $P_1 - P_2$ and use the lower bound $h(2P) \ge 4h(P) - \kappa_0$ provided by Lemma 3.)

(d) Prove that for any integer m there is a constant κ_m, depending on a, b, c, m, so that for all rational points P,

$$-\kappa_m \le h(mP) - m^2 h(P) \le \kappa_m.$$

3.3. * Let C be a rational cubic curve given by the usual Weierstrass equation.

(a) Prove that for any rational point $P \in C(\mathbb{Q})$, the limit

$$\hat{h}(P) = \lim_{n \to \infty} \frac{1}{4^n} h(2^n P)$$

exists. The quantity $\hat{h}(P)$ is called the *canonical height of P*. (*Hint.* Try to prove that the sequence is Cauchy.)

(b) Prove that there is a constant κ, depending only on a, b, c, so that for all rational points P we have

$$-\kappa \le \hat{h}(P) - h(P) \le \kappa.$$

(c) Prove that for any integer m and any rational point P,

$$\hat{h}(mP) = m^2 \hat{h}(P).$$

(d) Prove that $\hat{h}(P) = 0$ if and only if P is a point of finite order.

3.4. Prove the upper bound in Lemma 3′, Section 3, whose proof was omitted in the text.

3.5. Let $\alpha : \Gamma \to \mathbb{Q}^*/\mathbb{Q}^{*2}$ be the map defined in Section 5 by the rule

$$\alpha(\mathcal{O}) = 1 \quad (\text{mod } \mathbb{Q}^{*2}),$$
$$\alpha(T) = b \quad (\text{mod } \mathbb{Q}^{*2}),$$
$$\alpha(x, y) = x \quad (\text{mod } \mathbb{Q}^{*2}) \qquad \text{if } x \neq 0.$$

Prove that if $P_1 + P_2 + T = \mathcal{O}$, then $\alpha(P_1)\alpha(P_2)\alpha(T) \equiv 1 \ (\text{mod } \mathbb{Q}^{*2})$.
(Except for a few trivial cases, this completes the proof that α is a homomorphism.)

3.6. Let A and B be abelian groups and let $\phi : A \to B$ and $\psi : B \to A$ be homomorphisms. Suppose that there is an integer $m \geq 2$ so that

$$\psi \circ \phi(a) = ma \quad \text{for all } a \in A,$$
$$\phi \circ \psi(b) = mb \quad \text{for all } b \in B.$$

Suppose further that $\phi(A)$ has finite index in B, and $\psi(B)$ has finite index in A.
(a) Prove that mA has finite index in A, and that the index satisfies the inequality

$$(A : mA) \leq \big(A : \psi(B)\big)\big(B : \phi(A)\big).$$

(b) Give an example to show that it is possible for the inequality in (a) to be a strict inequality. More generally, show that the ratio

$$\frac{\big(A : \psi(B)\big)\big(B : \phi(A)\big)}{(A : mA)}$$

is an integer and give a good description of what this ratio represents.

3.7. This exercise describes a variant of the Nagell-Lutz theorem which often simplifies calculations on curves with a rational point of order two.
(a) Let C be a non-singular cubic curve given in Weierstrass form by an equation
$$C : y^2 = x^3 + ax^2 + bx,$$

where a and b are integers. Let $P = (x, y) \in C(\mathbb{Q})$ be a point of finite order with $y \neq 0$. Prove that x divides b and that the quantity

$$x + a + \frac{b}{x}$$

is a perfect square. [Note that if this quantity is a square, say equal to N^2, then (x, xN) is a rational point on C; but such a point need not have finite order. So this exercise gives a necessary condition for P to have finite order, but not a sufficient condition.]

(b) Let p be a prime. Prove that the only points of finite order on the curve $C : y^2 = x^3 + px$ are $\mathcal{O}$ and T.

(c) ** Let $D \neq 0$ be an integer. Prove that the points of finite order on the curve $y^2 = x^3 + Dx$ are as described in the following table:

$$\left\{ P \in C(\mathbb{Q}) : \begin{array}{l} P \text{ has} \\ \text{finite order} \end{array} \right\} \cong \begin{cases} \dfrac{\mathbb{Z}}{4\mathbb{Z}}, & \text{if } D = 4d^4 \text{ for some } d, \\[2ex] \dfrac{\mathbb{Z}}{2\mathbb{Z}} \oplus \dfrac{\mathbb{Z}}{2\mathbb{Z}}, & \text{if } D = -d^2 \text{ for some } d, \\[2ex] \dfrac{\mathbb{Z}}{2\mathbb{Z}}, & \text{otherwise.} \end{cases}$$

3.8. For primes p, let C_p be the cubic curve $y^2 = x^3 + px$ discussed in Section 6.

   (a) Prove that the rank of C_p is either 0, 1, or 2.

   (b) If $p \equiv 7 \pmod{16}$, prove that C_p has rank 0.

   (c) If $p \equiv 3 \pmod{16}$, prove that C_p has rank either 0 or 1.

3.9. Using the method developed in Section 6, find the rank of each of the following curves.

   (a) $y^2 = x^3 + 3x$

   (b) $y^2 = x^3 + 5x$

   (c) $y^2 = x^3 + 7x$

   (d) ** $y^2 = x^3 + 17x$

   (e) $y^2 = x^3 + 73x$

   (f) * $y^2 = x^3 - 82x$

3.10. (a) Let C be the singular cubic curve $y^2 = x^3$. Prove that the group law on C_{ns} is given by the formula

$$(x_1, y_1) + (x_2, y_2) = \left(\frac{\nu^2}{x_1 x_2}, \frac{-\nu^3}{y_1 y_2} \right), \qquad \text{where } \nu = \frac{y_1 x_2 - x_1 y_2}{x_2 - x_1}.$$

(b) Let C be the singular cubic curve $y^2 = x^3 + x^2$. Find a formula for the group law on C_{ns} similar to the formula in (a).

3.11. Let C be the singular cubic curve $y^2 = x^3$. Prove that the map

$$\phi : C_{\text{ns}}(\mathbb{Q}) \longrightarrow \mathbb{Q}, \qquad \phi(P) = \begin{cases} \dfrac{x}{y} & \text{if } P = (x, y), \\[2ex] 0 & \text{if } P = \mathcal{O}, \end{cases}$$

is a group isomorphism from $C_{\text{ns}}(\mathbb{Q})$ to the additive group of all rational numbers.

3.12. Let $P_1 = (x_1, y_1)$, $P_2 = (x_2, y_2)$, and $P_3 = (x_3, y_3)$ be three points in the plane. Prove that P_1, P_2, and P_3 are collinear if and only if

$$\det \begin{vmatrix} x_1 & y_1 & 1 \\ x_2 & y_2 & 1 \\ x_3 & y_3 & 1 \end{vmatrix} = 0.$$

3.13. (a) Prove that the additive group of rational numbers $(\mathbb{Q}, +)$ is not a finitely generated group.
(b) Prove that the multiplicative group of non-zero rational numbers $(\mathbb{Q}^*, *)$ is not a finitely generated group.

3.14. Let C be a cubic curve given by an equation

$$C : y^2 = x^3 + ax^2 + bx + c$$

with $a, b, c \in \mathbb{Q}$. Suppose that C is singular, and let $S = (x_0, y_0)$ be the singular point.
(a) Prove that x_0 and y_0 are in $\mathbb{Q}$.
(b) Prove that the change of coordinates $x = X + x_0$, $y = Y$ gives a new equation for C which has the form

$$Y^2 = X^3 + AX^2 \qquad \text{for some } A \in \mathbb{Q}.$$

(c) Suppose that $A = B^2 \neq 0$ for some $B \in \mathbb{Q}$. Prove that $C_{\mathrm{ns}}(\mathbb{Q})$ is isomorphic (as a group) to the multiplicative group $\mathbb{Q}^*$ of non-zero rational numbers.

3.15. This is a continuation of the previous exercise. Let A be a non-zero rational number which is not a perfect square (i.e., $\sqrt{A} \notin \mathbb{Q}$).
(a) Let H be the conic $u^2 - Av^2 = 1$. If (u_1, v_1) and (u_2, v_2) are two points in $H(\mathbb{Q})$, we define their "product" by the formula

$$(u_1, v_1) * (u_2, v_2) = (u_1 u_2 + Av_1 v_2, u_1 v_2 + u_2 v_1).$$

Prove that with this operation, $H(\mathbb{Q})$ is an abelian group.
(b) Prove that $H(\mathbb{Q})$ is not a finitely generated group.
(c) Let C be the singular cubic curve $y^2 = x^3 + Ax^2$. Prove that the map

$$C_{\mathrm{ns}}(\mathbb{Q}) \longrightarrow H(\mathbb{Q}), \qquad P \longmapsto \begin{cases} \left(\dfrac{y^2 + Ax^2}{x^3}, \dfrac{-2y}{x^2} \right) & \text{if } P = (x, y), \\ (1, 0) & \text{if } P = \mathcal{O}, \end{cases}$$

is an isomorphism of groups. Deduce that $C_{\mathrm{ns}}(\mathbb{Q})$ is not a finitely generated group.

(*Hint.* If you have studied field theory, it might help to reformulate this problem in terms of the field $K = \mathbb{Q}(\sqrt{A})$. First show that $H(\mathbb{Q})$ is isomorphic to a certain subgroup of K^*. Then the map in (c) becomes $(x, y) \to (y - x\sqrt{A})/(y + x\sqrt{A})$.)

CHAPTER IV

Cubic Curves over Finite Fields

1. Rational Points over Finite Fields

In this chapter we will look at cubic equations over a finite field, the field of integers modulo p. We will denote this field by $\mathbb{F}_p$. Of course, now we cannot visualize things; but we can look at polynomial equations

$$C : F(x, y) = 0$$

with coefficients in $\mathbb{F}_p$ and ask for solutions (x, y) with $x, y \in \mathbb{F}_p$. More generally, we can look for solutions $x, y \in \mathbb{F}_q$, where $\mathbb{F}_q$ is an extension field of $\mathbb{F}_p$ containing $q = p^e$ elements. We call such a solution a point on the curve C. If the coordinates x and y of the solution lie in $\mathbb{F}_p$, we call it a *rational point*.

If we have a cubic curve which is non-singular, then we can define an addition law on it, and the points form a commutative group. There is no need to use any pictures; the procedures and formulas that we described earlier make perfect sense for any field.

For example, consider the curve

$$y^2 = x^3 + ax^2 + bx + c$$

for some $a, b, c \in \mathbb{F}_p$. This curve is non-singular if and only if $p \neq 2$ and the discriminant $D = -4a^3c + a^2b^2 + 18abc - 4b^3 - 27c^2$ of the cubic is not zero (as an element of $\mathbb{F}_p$). Given points $P_1 = (x_1, y_1)$ and $P_2 = (x_2, y_2)$, we define the sum $P_1 + P_2$ by the usual rules. Ignoring a few exceptional cases (namely $P_1 = \mathcal{O}$, $P_2 = \mathcal{O}$, and $P_1 + P_2 = \mathcal{O}$), we take $y = \lambda x + \nu$ to be the line through P_1 and P_2, so

$$\lambda = \begin{cases} \dfrac{y_2 - y_1}{x_2 - x_1}, & \text{if } x_1 \neq x_2, \\[2mm] \dfrac{3x_1^2 + 2ax_1 + b}{2y_1}, & \text{if } P_1 = P_2, \end{cases}$$

and let $\nu = y_1 - \lambda x_1 = y_2 - \lambda x_2$. Then $P_3 = (x_3, y_3) = P_1 + P_2$ is given by the formulas

$$x_3 = \lambda^2 - a - x_1 - x_2 \quad \text{and} \quad y_3 = -\lambda x_3 - \nu.$$

All of this makes perfect sense if $a, b, c, x_1, y_1, x_2, y_2$ are in the finite field $\mathbb{F}_p$. Of course, it would be a lot of work to verify that this addition law defines a group, since there are a lot of special cases to check. In particular, the associative law would require some lengthy calculations. But we have given you explicit formulas to work with; so if you have any doubts, feel free to do the necessary checking.

If the curve C is given by an equation

$$C : F(x, y) = 0,$$

we will denote the set of rational points by

$$C(\mathbb{F}_p) = \big\{ (x, y) : x, y \in \mathbb{F}_p \quad \text{and} \quad F(x, y) = 0 \big\}.$$

Actually, just as with our cubic curves, we may also include one or more points "at infinity." These extra points come from making F into a homogenous polynomial in three variables. We will see an example in the next section.

Before doing more general theory, let's look at an example. Consider the curve

$$y^2 = x^3 + x + 1$$

over the field $\mathbb{F}_5$. How can we find the rational points? Since x and y are supposed to be in $\mathbb{F}_5$, we can just take each of the five possibilities for x, plug them into the polynomial $x^3 + x + 1$, and see when the result is a square in $\mathbb{F}_5$. Doing this, we find nine points (including the point $\mathcal{O}$ at infinity):

$$C(\mathbb{F}_5) = \big\{ \mathcal{O}, (0, \pm 1), (2, \pm 1), (3, \pm 1), (4, \pm 2) \big\}.$$

Thus, $C(\mathbb{F}_5)$ is an abelian group of order nine, so it is either cyclic or a product of two groups of order three. We can determine which one by making a group table. Let $P = (0, 1)$. Then using the formulas given above, we find that

$$2P = (4, 2), \quad 3P = (2, 1), \quad 4P = (3, -1), \ldots .$$

Hence, $C(\mathbb{F}_5)$ is a cyclic group of order nine. The two points of order three in $C(\mathbb{F}_5)$ are $(2, \pm 1)$; all of the other non-zero points have order nine.

As this example makes clear, there is never a problem about the group $C(\mathbb{F}_p)$ being finitely generated. Since there are only a finite number of possibilities for x and y, $C(\mathbb{F}_p)$ is actually a finite group. A natural question to ask is how big is it? Can we make any estimate as to the number of points in $C(\mathbb{F}_p)$?

To get an idea of what might be true, let's consider some simpler cases. First, how many points are there on a straight line? If the line is $y = ax + b$, we can take any value for x, and then the value for y is determined. So that gives p points. But we really want to count projective points, and a line always has one additional point "at infinity." (In homogeneous coordinates, the line has the equation $Y = aX + bZ$, so it contains the extra point $[1, a, 0]$. See Appendix A, Sections 1 and 2.) Thus a line has $p + 1$ points.

Next we might look at a conic C which is the set of solutions $x, y \in \mathbb{F}_p$ to a quadratic equation

$$ax^2 + bxy + cy^2 + dx + ey + f = 0.$$

In Chapter I, Section 1 we discussed the solutions to such equations with x and y in the field of rational numbers $\mathbb{Q}$; and everything we said works equally well if we replace $\mathbb{Q}$ by the finite field $\mathbb{F}_p$. Further, it turns out that $C(\mathbb{F}_p)$ is never empty, so as long as C is non-singular, $C(\mathbb{F}_p)$ will always contain exactly $p + 1$ points.

We now turn our attention to the curve C given by the equation

$$C : y^2 = f(x),$$

where $f(x)$ is a polynomial with coefficients in $\mathbb{F}_p$. How many points would we expect C to have? We suppose $p \neq 2$. As we observed above, among the non-zero elements $1, 2, \ldots, p-1$ of the field $\mathbb{F}_p$, half of them are squares (the quadratic residues) and half of them are non-squares (the quadratic nonresidues).

Now think of substituting the different values $x = 0, 1, \ldots, p-1$ into the equation $y^2 = f(x)$. If $f(x) = 0$, there is one solution, $y = 0$. If $f(x) \neq 0$, then for half the possible non-zero values of $f(x)$, there are two solutions for y; and for the other possible values of $f(x)$, there are no solutions y. So if the $f(x)$'s were randomly distributed among the squares and the non-squares, we would expect again to get approximately $p + 1$ points. Of course, this does not constitute a proof. But intuitively, each value for x either yields one solution (if $f(x) = 0$), or else it has a 50% chance of producing two solutions and a 50% chance of producing no solutions. So the p possible values for x should give approximately p solutions, and then including the point $\mathcal{O}$ at infinity gives $p + 1$. Thus, the number of solutions should look like

$$\#C(\mathbb{F}_p) = p + 1 + (\text{error term}),$$

where we expect the "error term" to be fairly small compared to p.

It turns out that this is true; as long as the polynomial $f(x)$ has distinct roots, there is no tendency for the values of $f(x)$ to be squares or non-squares. So it is true that the number of points on a curve does not differ too much from the number of points on a line. These rough remarks are made precise by the following theorem.

Theorem. (Hasse, Weil) *If C is a non-singular irreducible curve of genus g defined over the finite field $\mathbb{F}_p$, then the number of points on C with coordinates in $\mathbb{F}_p$ is $p + 1 + \varepsilon$, where the "error term" ε satisfies $|\varepsilon| \leq 2g\sqrt{p}$.*

It would take us too far afield to actually define the genus, but that will not matter. Let us just say that whenever you have a curve $F(x, y) = 0$, there is a non-negative integer g associated to it called its genus; and as long as the curve is not too singular, the genus increases as the degree of F increases. For example, the Fermat curve $x^n + y^n = 1$ has genus equal to $\frac{1}{2}(n-1)(n-2)$ (assuming p does not divide n). In particular, the cubic curve $x^3 + y^3 = 1$ which we will be studying in Section 2 is a curve of genus 1. More generally, any non-singular curve given by a cubic equation is a curve of genus 1; so an alternative title for this book would have been "Rational Points on Curves of Genus 1"! (But that might have sounded too forbidding to the uninitiated.)

For an elliptic curve C over the finite field $\mathbb{F}_p$, the Hasse-Weil theorem thus gives the estimate

$$-2\sqrt{p} \leq \#C(\mathbb{F}_p) - p - 1 \leq 2\sqrt{p}.$$

The Hasse-Weil theorem is also called the Riemann hypothesis for curves over finite fields because there is an alternative way to state it which is analogous to the famous, as yet unsolved, Riemann hypothesis. The theorem was conjectured by E. Artin in his thesis, was proved by Hasse [1] in the case $g = 1$ (i.e., for elliptic curves), and was proved by Weil [1] for arbitrary g. An amazingly deep generalization to higher dimensions, suggested by Weil [2], was proved by Deligne [1].

For some special cubic curves, the result is due to Gauss. In the next section we will give Gauss' proof of one of these special cases.

2. A Theorem of Gauss

In the last section we described an estimate for the number of solutions to a cubic equation over a finite field. Certain special cases of that theorem were proved by Gauss. We are going to discuss one of those cases, the Fermat curve

$$x^3 + y^3 = 1.$$

This comes from Gauss' *Disquisitiones Arithmeticae*, Article 358; it is the first non-trivial case of the theorem ever treated. If you want, you can read about it in Latin in the *Disquisitiones*. (It's easy Latin. Or you can read it in the language of your choice — there are several translations available.)

We take the curve in homogeneous form

$$x^3 + y^3 + z^3 = 0$$

and consider solutions in the projective sense. That is, we do not count the trivial solution $(0, 0, 0)$; and we identify a solution (x, y, z) with all of its non-zero multiples (ax, ay, az). With these conventions, we can now state the theorem of Gauss.

Theorem. (Gauss) *Let M_p be the number of projective solutions to the equation*

$$x^3 + y^3 + z^3 = 0$$

with x, y, z in the finite field $\mathbb{F}_p$.
(a) *If $p \not\equiv 1 \,(\mathrm{mod}\ 3)$, then $M_p = p + 1$.*
(b) *If $p \equiv 1 \,(\mathrm{mod}\ 3)$, then there are integers A and B such that*

$$4p = A^2 + 27B^2.$$

A and B are unique up to changing their signs, and if we fix the sign of A so that $A \equiv 1 \,(\mathrm{mod}\ 3)$, then

$$M_p = p + 1 + A.$$

Note that if $p \equiv 1 \,(\mathrm{mod}\ 3)$, then the equation $4p = A^2 + 27B^2$ implies that $A^2 \equiv 1 \,(\mathrm{mod}\ 3)$. So $A \equiv \pm 1 \,(\mathrm{mod}\ 3)$, and replacing A by $-A$ if necessary, we can always make $A \equiv 1 \,(\mathrm{mod}\ 3)$.

Since $B^2 > 0$, it follows that $A^2 = 4p - 27B^2 < 4p$; and so $|A| < 2\sqrt{p}$. Since the genus in this case is $g = 1$, the Hasse-Weil theorem says that we should have $|M_p - p - 1| \le 2\sqrt{p}$. But $M_p - p - 1 = A$, so we do indeed have a special case of the theorem.

Before beginning the proof of Gauss' theorem, we make a few remarks about the field $\mathbb{F}_p$. This field consists of p elements, $0, 1, \ldots, p - 1$. The multiplicative group $\mathbb{F}_p^*$ of $\mathbb{F}_p$ consists of the non-zero elements $1, 2, \ldots, p-1$ with the group operation being multiplication.

The multiplicative group $\mathbb{F}_p^*$ is a *cyclic* group of order $p - 1$. Why is it cyclic? Well, if G is a non-cyclic finite abelian group, and if ℓ is the least common multiple of the orders of its elements, then $\ell < \#G$ (strict inequality) and every element of G satisfies the equation $x^\ell = 1$. In other words, the equation $x^\ell = 1$ has more than ℓ solutions in G. But over a field, a polynomial equation never has more roots than its degree. Hence, the multiplicative group of a finite field is cyclic. More generally, if K is any field, and if G is a finite subgroup of its multiplicative group K^*, then G is cyclic. You may have run across this fact when K is the field of complex numbers and G is a finite group of roots of unity.

Using this elementary fact about $\mathbb{F}_p^*$, the first part of Gauss' theorem is easy.

PROOF. (of Gauss' theorem) (a) For this part we suppose that

$$p \not\equiv 1 \,(\mathrm{mod}\ 3).$$

Then 3 does not divide the order $p - 1$ of the cyclic group $\mathbb{F}_p^*$. It follows that the map $x \to x^3$ is an isomorphism from $\mathbb{F}_p^*$ to itself.

For example, if $p = 5$, then in $\mathbb{F}_5$ we have

$$1^3 = 1, \quad 2^3 = 3, \quad 3^3 = 2, \quad 4^3 = 4.$$

And, of course, $0^3 = 0$, so in the case that $p \not\equiv 1 \,(\mathrm{mod}\ 3)$, every element of $\mathbb{F}_p$ has a unique cube root. Thus, the number of solutions of $x^3 + y^3 + z^3 = 0$ is equal to the number of solutions of the linear equation $x + y + z = 0$. This is the equation of a line in the projective plane, so it has exactly $p + 1$ points rational for $\mathbb{F}_p$. Therefore, $M_p = p + 1$. So the case $p \not\equiv 1 \,(\mathrm{mod}\ 3)$ is extremely easy.

(b) Now we consider the case that $p \equiv 1 \,(\mathrm{mod}\ 3)$. Let us write

$$p = 3m + 1.$$

Since 3 divides the order of the group $\mathbb{F}_p^*$, the map $x \to x^3$ is a homomorphism of $\mathbb{F}_p^*$ to itself which is neither one-to-one nor onto. The image of this homomorphism is a subgroup which we will denote R, so

$$R = \{x^3 \,:\, x \in \mathbb{F}_p^*\}.$$

The subgroup R has index 3 inside $\mathbb{F}_p^*$. The kernel of the map $x \to x^3$ consists of three elements $1, u, u^2$ satisfying $u^3 = 1$. For example, if $p = 13$, then $R = \{\pm 1, \pm 5\}$, and the kernel of the homomorphism $x \to x^3$ consists of the numbers $1, 3, 9 \in \mathbb{F}_{13}^*$.

The elements of R are called *cubic residues*. We will let S and T be the other two cosets of R in $\mathbb{F}_p^*$. For example, if we take any $s \in \mathbb{F}_p^*$ which is not in R, then we could let $S = sR$ and $T = s^2 R$. As illustration, if $p = 13$, we can choose $s = 2$, and then $S = 2R = \{\pm 2, \pm 10\}$ and $T = 4R = \{\pm 4, \pm 7\}$.

In general $\mathbb{F}_p$ is a disjoint union,

$$\mathbb{F}_p = \{0\} \cup R \cup S \cup T.$$

The number of elements in each of the sets R, S, T is m. Notice also that $-1 = (-1)^3$ is a cube, so $R = -R$, $S = -S$, and $T = -T$. (This means that if $r \in R$, then $-r \in R$; and similarly for S and T.) Thinking in terms of R, S, and T is the key to finding the number of solutions of $x^3 + y^3 + z^3 = 0$.

We want to express the number of solutions M_p in terms of R, S, and T. It's a question of counting. We need to introduce a symbol. Suppose that X, Y, Z are subsets of the field $\mathbb{F}_p$. We let $[X, Y, Z]$ or $[XYZ]$ denote the number of triples (x, y, z) such that

$$x \in X, \quad y \in Y, \quad z \in Z, \quad \text{and} \quad x + y + z = 0.$$

What is the number of solutions M_p in terms of this symbol? Let's first consider the solutions of $x^3 + y^3 + z^3 = 0$ where none of x, y, z are zero.

The number of ways of writing zero as a sum of three non-zero cubes is obviously $[RRR]$. But for each non-zero cube there are three possible field elements which give that cube. Thus, there are $27[RRR]$ solutions (x, y, z) of $x^3 + y^3 + z^3 = 0$ with x, y, z non-zero. But we have agreed not to distinguish proportional solutions (x, y, z) and (ax, ay, az). There are $p - 1$ choices for the multiplier a. Thus, there are $\frac{27[RRR]}{p-1} = \frac{9[RRR]}{m}$ projective solutions of $x^3 + y^3 + z^3 = 0$ in which none of x, y, z is zero.

How many solutions are there if one of them is zero, say $z = 0$. Then neither x nor y can be zero, because we do not allow $(0, 0, 0)$. So we can pick anything non-zero for x, and once we do that there are three possible values for y, namely the solutions of $y^3 = -x^3$. This has three solutions because as we noted earlier, the group $\mathbb{F}_p^*$ has an element u of order three; and so for a given x, the equation $y^3 = -x^3$ has the three solutions $y = -x, -ux, -u^2x$. Thus, there are $3(p - 1)$ triples $(x, y, 0)$ such that $x^3 + y^3 = 0$. Similarly for $y = 0$ and for $x = 0$, so there are $9(p - 1)$ triples (x, y, z) such that $x^3 + y^3 + z^3 = 0$ and one of x, y, z is zero. Since we do not distinguish proportional triples, we must divide by the $p - 1$ possible multipliers; and so there are $\frac{9(p-1)}{p-1} = 9$ projective solutions with one coordinate zero.

Combining these two calculations, we have shown that

$$M_p = \frac{9[RRR]}{m} + 9 = 9\left(\frac{[RRR]}{m} + 1\right).$$

The symbol $[XYZ]$ has many marvelous properties which are easy to verify, such as the following:

$[XY(Z \cup W)] = [XYZ] + [XYW]$ if $Z \cap W = \emptyset$.

$[XYZ] = [aX, aY, aZ]$ for any $a \neq 0$, where $aX = \{ax : x \in X\}$.

$[XYZ] = [XZY] = [YXZ] = [YZX] = [ZXY] = [ZYX]$.

Thus, since $\mathbb{F}_p = \{0\} \cup R \cup S \cup T$ is a disjoint union, and $[RR\mathbb{F}_p] = m^2$, we have

$$[RR\{0\}] + [RRR] + [RRS] + [RRT] = m^2.$$

Now fix elements $s \in S$ and $t \in T$. Since $[RRS] = [sR, sR, sS] = [SST]$ and $[RRT] = [tR, tR, tT] = [TTS]$, we obtain

$$[RR\{0\}] + [RRR] + [SST] + [TTS] = m^2. \qquad (*)$$

Again using $\mathbb{F}_p = \{0\} \cup R \cup S \cup T$ and the obvious fact $[\mathbb{F}_p TS] = m^2$, we get

$$[\{0\}TS] + [RTS] + [STS] + [TTS] = m^2. \qquad (**)$$

Now $[\{0\}TS] = 0$ because $-S = S$ and $S \cap T = \emptyset$. Also $[RR\{0\}] = m$ because $-R = R$. So if we subtract $(**)$ from $(*)$, then we get

$$m + [RRR] = [RTS];$$

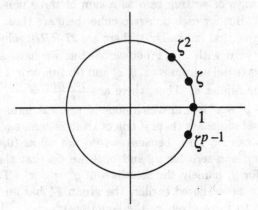

The pth Roots of Unity

Figure 4.1

and so we have the beautiful formula

$$M_p = 9\frac{[RTS]}{m}.$$

Now we just have to find a clever method of getting $[RTS]$. What we are going to do is look at some complex numbers called *cubic Gauss sums*. These complex numbers that we use in the proof are gadgets for keeping track of information about the sets R, S, and T.

We recall a little bit about p^{th} roots of unity. (See Figure 4.1.) Let $\zeta = e^{2\pi i/p}$. The pth roots of unity are then $1 = \zeta^0, \zeta, \zeta^2, \ldots, \zeta^{p-1}$. Now $\zeta^a = \zeta^b$ if and only if $a \equiv b \pmod{p}$, which really means that ζ^a makes sense if a is an element of our finite field $a \in \mathbb{F}_p$. Further, if $a, b \in \mathbb{F}_p$, then $\zeta^{a+b} = \zeta^a\zeta^b$.

We define three complex numbers $\alpha_1, \alpha_2, \alpha_3$ which are sums of powers of ζ:

$$\alpha_1 = \sum_{r \in R} \zeta^r,$$

$$\alpha_2 = \sum_{s \in S} \zeta^s,$$

$$\alpha_3 = \sum_{t \in T} \zeta^t.$$

The complex numbers $\alpha_1, \alpha_2, \alpha_3$ are thus each a sum of m different p^{th} roots of unity; they are called *cubic Gauss sums*. It turns out that they are the three roots of a polynomial equation with integer coefficients, and that equation is what we next need to derive.

To do this, we multiply together two of the α_i's, say $\alpha_2\alpha_3$.

$$\alpha_2\alpha_3 = \sum_{s \in S}\sum_{t \in T} \zeta^s\zeta^t = \sum_{s \in S, t \in T} \zeta^{s+t} = \sum_{x \in \mathbb{F}_p} N_x\zeta^x,$$

where N_x is the number of pairs (s, t) with $s \in S, t \in T$, such that $s+t = x$. We observe that for $r \in R$,

$$N_x = [ST\{-x\}] = [rS, rT, \{-rx\}] = [S, T, \{-rx\}] = N_{rx},$$

so that N_x only depends on the coset R, S, or T that x lies in. Thus,

$$mN_x = [S, T, Rx] = \begin{cases} [STR] & \text{if } x \in R, \\ [STS] & \text{if } x \in S, \\ [STT] & \text{if } x \in T. \end{cases}$$

Define integers a, b, c by

$$[STR] = ma, \qquad [STS] = mb, \qquad [STT] = mc.$$

Then

$$M_p = 9a$$

and

$$\alpha_2\alpha_3 = a\alpha_1 + b\alpha_2 + c\alpha_3.$$

A similar calculation gives

$$\alpha_3\alpha_1 = a\alpha_2 + b\alpha_3 + c\alpha_1,$$
$$\alpha_1\alpha_2 = a\alpha_3 + b\alpha_1 + c\alpha_2.$$

From now on you can relax because everything else is merely substituting one formula into another until we find an expression for the integer a. Since

$$0 = \zeta^p - 1 = (\zeta - 1)(\zeta^{p-1} + \zeta^{p-2} + \cdots + \zeta + 1)$$

and $\zeta \neq 1$, we have $\zeta^{p-1} + \zeta^{p-2} + \cdots + \zeta + 1 = 0$. Hence,

$$\alpha_1 + \alpha_2 + \alpha_3 = -1$$

because the three α_i's include all powers of ζ except ζ^0. Now adding up the three formulas for the $\alpha_i\alpha_j$'s, we find

$$\alpha_1\alpha_2 + \alpha_1\alpha_3 + \alpha_2\alpha_3 = (a + b + c)(\alpha_1 + \alpha_2 + \alpha_3) = -(a + b + c).$$

But

$$m(a + b + c) = [STR] + [STS] + [STT]$$
$$= [ST(R \cup S \cup T)] = [ST\mathbb{F}_p] - [ST\{0\}] = m^2,$$

so we find that

$$\alpha_1\alpha_2 + \alpha_1\alpha_3 + \alpha_2\alpha_3 = -m.$$

This allows us to compute the sum of the squares of the α_i's as

$$\alpha_1^2 + \alpha_2^2 + \alpha_3^2 = (\alpha_1 + \alpha_2 + \alpha_3)^2 - 2(\alpha_1\alpha_2 + \alpha_1\alpha_3 + \alpha_2\alpha_3) = 1 + 2m.$$

What we need to find next is $\alpha_1\alpha_2\alpha_3$. To get that, write

$$\alpha_1(\alpha_2\alpha_3) = \alpha_1(a\alpha_1 + b\alpha_2 + c\alpha_3),$$
$$\alpha_2(\alpha_3\alpha_1) = \alpha_2(a\alpha_2 + b\alpha_3 + c\alpha_1),$$
$$\alpha_3(\alpha_1\alpha_2) = \alpha_3(a\alpha_3 + b\alpha_1 + c\alpha_2).$$

Now adding these and using that we know

$$\alpha_1^2 + \alpha_2^2 + \alpha_3^2 = 1 + 2m \qquad \text{and} \qquad \alpha_1\alpha_2 + \alpha_1\alpha_3 + \alpha_2\alpha_3 = -m,$$

we get

$$3\alpha_1\alpha_2\alpha_3 = a(1 + 2m) + (b + c)(-m) = a + km,$$

where we have introduced a new letter

$$k = 2a - b - c = 3a - m.$$

So if we can find a value for k, then we will also have computed

$$M_p = 9a = 3(k + m) = 3k + p - 1.$$

Let's stop for a moment and review what we are doing. The sets R, S and T are defined multiplicatively in terms of cubing, whereas the symbol $[RTS]$ tells us how many times the sum of three things is zero. We are mixing up multiplication and addition and counting, and out of that mixture we have concocted three complex numbers $\alpha_1, \alpha_2, \alpha_3$ and three integers a, b, c and various algebraic relations among them. Now all we are doing is manipulating those relations until we get what we want because we know that $9a$ is our answer for the number of points on the curve.

The complex numbers $\alpha_1, \alpha_2, \alpha_3$ are the roots of the polynomial

$$F(t) = (t - \alpha_1)(t - \alpha_2)(t - \alpha_3) = t^3 + t^2 - mt - \frac{a + km}{3}.$$

Let D_F be the discriminant of F. Using our formulas for the $\alpha_i\alpha_j$'s, we can calculate a square root of D_F as

$$
\begin{aligned}
\sqrt{D_F} &= (\alpha_1 - \alpha_2)(\alpha_1 - \alpha_3)(\alpha_2 - \alpha_3) \\
&= \alpha_2\alpha_3(\alpha_2 - \alpha_3) + \alpha_3\alpha_1(\alpha_3 - \alpha_1) + \alpha_1\alpha_2(\alpha_1 - \alpha_2) \\
&= (a\alpha_1 + b\alpha_2 + c\alpha_3)(\alpha_2 - \alpha_3) + (a\alpha_2 + b\alpha_3 + c\alpha_1)(\alpha_3 - \alpha_1) \\
&\qquad\qquad\qquad\qquad\qquad + (a\alpha_3 + b\alpha_1 + c\alpha_2)(\alpha_1 - \alpha_2) \\
&= (b - c)(\alpha_1^2 + \alpha_2^2 + \alpha_3^2 - \alpha_1\alpha_2 - \alpha_1\alpha_3 - \alpha_2\alpha_3) \\
&= (b - c)(1 + 3m) \\
&= (b - c)p.
\end{aligned}
$$

Put
$$\beta_i = 1 + 3\alpha_i;$$

then we find that

$$\beta_1 + \beta_2 + \beta_3 = 0,$$
$$\beta_1\beta_2 + \beta_1\beta_3 + \beta_2\beta_3 = -3p,$$
$$\beta_1\beta_2\beta_3 = (3k - 2)p.$$

The polynomial whose roots are $\beta_1, \beta_2, \beta_3$ is

$$G(t) = (t - \beta_1)(t - \beta_2)(t - \beta_3) = t^3 - 3pt - (3k - 2)p.$$

Let $A = 3k - 2$; so as noted above, the number of solutions M_p is given by the formula

$$M_p = 3k + p - 1 = p + 1 + A.$$

Thus, this is the A referred to in the statement of Gauss' theorem. We just need to show that it has all of the necessary properties.

Let D_G be the discriminant of the polynomial $G(t)$. From the formula for the discriminant of a cubic, we have

$$D_G = -4(-3p)^3 - 27(Ap)^2 = 4 \cdot 27p^3 - 27A^2p^2.$$

On the other hand, since $\beta_i - \beta_j = 3(\alpha_i - \alpha_j)$, we have

$$D_G = (27)^2 D_F.$$

Thus,

$$4 \cdot 27p^3 - 27A^2p^2 = D_G = (27)^2 D_F = (27)^2(b - c)^2 p^2.$$

Cancelling $27p^2$, we find

$$4p = A^2 + 27B^2$$

with $B = b - c$ and $A = 3k - 2 \equiv 1 \,(\text{mod } 3)$. So, magically, we obtain the result that $4p$ can be written in the form $4p = A^2 + 27B^2$ with $A \equiv 1 \,(\text{mod } 3)$ and $M_p = p + 1 + A$.

It remains to show that A is uniquely determined by the two conditions $4p = A^2 + 27B^2$ and $A \equiv 1 \,(\text{mod } 3)$. One can argue conceptually or do it with formulas. In keeping with the first part of the proof, we will do it with formulas. So suppose we had another representation $4p = A_1^2 + 27B_1^2$. Then

$$4p(B_1^2 - B^2) = (A^2 + 27B^2)B_1^2 - (A_1^2 + 27B_1^2)B^2 = (AB_1 + A_1B)(AB_1 - A_1B).$$

Since p divides the product on the right-hand side, it divides one of the factors on the left, say $p \,|\, (AB_1 - A_1B)$.

Now multiplying the two formulas for $4p$, we get

$$16p^2 = A^2A_1^2 + 27B^2A_1^2 + 27B_1^2A^2 + (27)^2B^2B_1^2,$$

so that

$$16p^2 - (AA_1 + 27BB_1)^2 = 27(AB_1 - A_1B)^2.$$

Since $p \mid (AB_1 - A_1B)$, it follows that $p \mid (AA_1 + 27BB_1)^2$; and so

$$16 - \left(\frac{AA_1 + 27BB_1}{p}\right)^2 = 27\left(\frac{AB_1 - A_1B}{p}\right)^2.$$

Well now, something is fishy because the left-hand side is less than 16, whereas the right-hand side is 27 times the square of an integer. So both sides must be zero. In particular, $AB_1 - A_1B = 0$; so if we let

$$\lambda = \frac{A_1}{A} = \frac{B_1}{B},$$

then $A_1 = \lambda A$ and $B_1 = \lambda B$. Substituting into $A^2 + 27B^2 = 4p = A_1^2 + 27B_1^2$ gives $\lambda^2 = 1$, so $\lambda = \pm 1$. Finally, the assumption that $A \equiv A_1 \equiv 1 \pmod 3$ forces $\lambda = 1$, which proves the uniqueness and completes the proof of Gauss' theorem. $\square$

We illustrate the theorem with some examples. To find the number of points M_p, we just have to solve the equation $4p = A^2 + 27B^2$; and for smallish p, that is not too hard. Here is a short table:

p	A	B	$M_p = A + p + 1$
7	1	1	9
13	-5	1	9
19	7	1	27
31	4	2	36
4027	-104	14	3924

M_p is always divisible by 9. This is because the group of points on the curve $x^3 + y^3 + z^3 = 0$ has nine points of order three corresponding to the solutions of $x^3 + y^3 + z^3 = 0$. These are the points where one of x, y, or z is zero, and the other two are cube roots of 1 and -1. Note that in the field $\mathbb{F}_p$, there are three distinct cube roots of 1, so we get nine distinct projective points on the curve. We will leave it to you to check that these nine points form a subgroup isomorphic to $\mathbb{Z}/3\mathbb{Z} \oplus \mathbb{Z}/3\mathbb{Z}$, which implies that M_p is always divisible by 9. Of course, all this is only for the case that $p \equiv 1 \pmod 3$.

So we now have this crazy method for computing the number of points on the curve. Take $4p$ and write it as $A^2 + 27B^2$. We know we can do it. If we actually want to compute A and B, it helps to note that M_p is divisible by 9, so $A \equiv -p - 1 \pmod 9$. And in looking for B, we can think of the formula $4p = A^2 + 27B^2$ as a congruence modulo some low primes. This gives us some information, a kind of sieve, with congruences that B must satisfy. This somewhat eases the quest for A and B.

There is a famous problem concerning the roots $\alpha_1, \alpha_2, \alpha_3$. Letting $\zeta = e^{2\pi i/p}$ be the usual p^{th} root of unity, we have the well-defined complex number

$$\alpha_1 = \sum_{r \in R} \zeta^r = \frac{1}{3} \sum_{x \in F_p^*} \zeta^{x^3}.$$

In fact, since ζ^{-r} is the complex conjugate of ζ^r, and $-R = R$, we see that α_1 is actually a real number,

$$\alpha_1 = \frac{1}{3} \sum_{n=1}^{(p-1)/2} \left(\zeta^{n^3} + \zeta^{-n^3} \right) = \frac{2}{3} \sum_{n=1}^{(p-1)/2} \cos\left(\frac{2\pi n^3}{p} \right).$$

Similarly, α_2 and α_3 are real. For a given prime p, we can compute the α_i's very easily by writing $4p = A^2 + 27B^2$, and then observing that the α_i's are the roots of the polynomial

$$F(t) = t^3 + t^2 - \frac{p-1}{3}t - \frac{p(A+3)-1}{27}.$$

Since $D_F \neq 0$, the α_i's are distinct.

Question. For which primes p is α_1 the smallest of the three roots?

The primes $p \equiv 1 \pmod 3$ are mysteriously divided into three types, those types for which α_1 is the smallest root, the middle root, and the largest root of the equation $F(t) = 0$. Let's call these Class 1, Class 2, and Class 3. Kummer [1] made a table for all primes less than 500, and found that there are 7 primes of Class 1, 14 primes of Class 2, and 24 primes of Class 3. Based on this evidence, he suggested that maybe primes fall into the three classes in the ratio 1-to-2-to-3. When early computers became available, Emil Artin suggested this problem to von Neumann and Goldstine to try out on the MANIAC computer. This is a good problem to test on a machine because there is a built in check. On the one hand, it can compute α_1 directly as a sum of cosines, while on the other hand, it can search for A and B and use them to get the polynomial $F(t)$. Then it can substitute α_1 into $F(t)$ and see if it gets zero. They computed for all primes less than 2000 and found that Kummer's table was correct — he had not made any mistakes. This is quite a feat, since when p is around 500 you have to add up 133 cosines to get α_1.

However, tables of this sort for small primes can be quite misleading, and Kummer's guess turned out to be wrong. What is true is that the primes $p \equiv 1 \pmod 3$ are equally distributed among the three types. This beautiful result was proven by Heath-Brown and Patterson [1]. The proof, which is extremely difficult, uses tools from number theory, geometry, and analysis.

Suppose that we take a non-singular cubic with integer coefficients, say

$$ax^3 + bx^2y + cxy^2 + dy^3 + ex^2 + fxy + gy^2 + hx + iy + j = 0,$$

and suppose that we read it as a congruence modulo p for various primes p. If we ask for a formula for the number of solutions M_p, then it is only for very special cubics that we get an answer like the one we obtained for $x^3 + y^3 = 1$. In general, not much is known about the behavior of M_p as a function of p, although there is a conjecture due to Taniyama and Weil which would associate to the collection of M_p's a certain holomorphic function (called a *modular form*) which has some wonderful transformation properties. And of course, we always have the Hasse-Weil estimate

$$|M_p - p - 1| \le 2\sqrt{p};$$

but in general there is no nice formula for M_p like Gauss' result. It is only when the elliptic curve has what is known as *complex multiplication* that such a formula exists. We will discuss complex multiplication (from a somewhat different viewpoint) in Chapter VI.

We conclude this section by describing a conjecture about the distribution of the M_p's in the case that the cubic curve does not have complex multiplication. Since $|M_p - p - 1| \le 2\sqrt{p}$, we can define an angle θ_p between 0 and π by the condition

$$\cos\theta_p = \frac{M_p - p - 1}{2\sqrt{p}}.$$

We also recall the standard notation $\pi(X)$ for the number of primes less than X. The prime number theorem says that $\pi(X)$ is asymptotically equal to $X/\log X$. This means that the ratio $\pi(X)/(X/\log X) \to 1$ as $X \to \infty$.

Conjecture. (Sato-Tate) *Assume that the cubic curve does not have complex multiplication. For any fixed angles $0 \le \alpha \le \beta \le \pi$,*

$$\lim_{X \to \infty} \frac{\#\{p \le X : \alpha \le \theta_p \le \beta\}}{\pi(X)} = \frac{2}{\pi}\int_\alpha^\beta \sin^2 t\, dt.$$

So the angles θ_p, which determine the number of solutions M_p by the formula

$$M_p = p + 1 + 2\sqrt{p}\cos\theta_p,$$

are conjecturally distributed in the interval $[0, \pi]$ according to a $\sin^2$ distribution.

3. Points of Finite Order Revisited

Let C be a cubic curve, given as usual by a Weierstrass equation

$$C : y^2 = x^3 + ax^2 + bx + c$$

with integer coefficients a, b, c. In Chapters II and III we studied the group of rational points $C(\mathbb{Q})$ on this curve, and in particular we showed that this group is finitely generated (Mordell's theorem) and that the points of finite order have integer coordinates (Nagell-Lutz theorem).

In this chapter we have been looking at curves with coefficients in a finite field $\mathbb{F}_p$. Suppose that we write $z \to \tilde{z}$ for the map "reduction modulo p,"

$$\mathbb{Z} \longrightarrow \frac{\mathbb{Z}}{p\mathbb{Z}} = \mathbb{F}_p, \qquad z \longmapsto \tilde{z}.$$

Then we can take the equation for C, which has integer coefficients, and we can reduce those coefficients modulo p to get a new curve with coefficients in $\mathbb{F}_p$:

$$\tilde{C} : y^2 = x^3 + \tilde{a}x^2 + \tilde{b}x + \tilde{c}.$$

When will the curve $\tilde{C}$ be non-singular? It will be non-singular provided $p \geq 3$ and provided the discriminant

$$\tilde{D} = -4\tilde{a}^3\tilde{c} + \tilde{a}^2\tilde{b}^2 + 18\tilde{a}\tilde{b}\tilde{c} - 4\tilde{b}^3 - 27\tilde{c}^2$$

is non-zero. But reduction modulo p from $\mathbb{Z}$ to $\mathbb{F}_p$ is a homomorphism, so $\tilde{D}$ is just the reduction modulo p of the discriminant D of the cubic $x^3 + ax^2 + bx + c$. In other words, the reduced curve $\tilde{C} \pmod{p}$ will be non-singular provided $p \geq 3$ and p does not divide the discriminant D.

Having reduced the curve C, it is natural to try taking points on C and reducing them modulo p to get points on $\tilde{C}$. We can do this provided that the coordinates of the point have no p in their denominator. In particular, if a point has integer coordinates, then we can reduce that point modulo p for any prime p. That is, if $P = (x, y)$ is a point in $C(\mathbb{Q})$ with integer coordinates, then x and y satisfy the equation

$$y^2 = x^3 + ax^2 + bx + c.$$

This equation gives a relation among integers, so we can reduce it modulo p to get the equation

$$\tilde{y}^2 = \tilde{x}^3 + \tilde{a}\tilde{x}^2 + \tilde{b}\tilde{x} + \tilde{c}.$$

This last equation says that $\tilde{P} = (\tilde{x}, \tilde{y})$ is a point in $\tilde{C}(\mathbb{F}_p)$. So we get a map from the points in $C(\mathbb{Q})$ with integer coordinates to $\tilde{C}(\mathbb{F}_p)$.

We proved in Chapter II, Section 4 that points of finite order in $C(\mathbb{Q})$ always have integer coordinates. This was the hard part of the Nagell-Lutz theorem. We are going to study the collection of points of finite order, so let us give it a name:

$$\Phi = \big\{ P = (x,y) \in C(\mathbb{Q}) \; : \; P \text{ has finite order} \big\}.$$

Clearly, Φ is a subgroup of $C(\mathbb{Q})$ because if P_1, P_2 are points of finite order, say $m_1 P_1 = \mathcal{O}$ and $m_2 P_2 = \mathcal{O}$, then $(m_1 m_2)(P_1 \pm P_2) = \mathcal{O}$. So both $P_1 + P_2$ and $P_1 - P_2$ are in Φ.

Since Φ consists of points with integer coordinates, together with $\mathcal{O}$, we can define a *reduction modulo p map*

$$\Phi \longrightarrow \tilde{C}(\mathbb{F}_p), \qquad P \longmapsto \tilde{P} = \begin{cases} (\tilde{x}, \tilde{y}) & \text{if } P = (x,y), \\ \tilde{\mathcal{O}} & \text{if } P = \mathcal{O}. \end{cases}$$

Now Φ is a subgroup of $C(\mathbb{Q})$, so it is a group; and provided p does not divide $2D$, we know that $\tilde{C}(\mathbb{F}_p)$ is a group. So we have a map from the group Φ to the group $\tilde{C}(\mathbb{F}_p)$, and we now want to check that this map is a homomorphism. (For a more general description of the reduction modulo p map $C(\mathbb{Q}) \to \tilde{C}(\mathbb{F}_p)$ and a proof that it is a homomorphism, see Appendix A, Section 5.)

First we note that negatives go to negatives:

$$\widetilde{-P} = \widetilde{(x,-y)} = (\tilde{x}, -\tilde{y}) = -\tilde{P}.$$

So it suffices to show that if $P_1 + P_2 + P_3 = \mathcal{O}$, then $\tilde{P}_1 + \tilde{P}_2 + \tilde{P}_3 = \tilde{\mathcal{O}}$. As usual, there are some special cases to check.

If any of P_1, P_2, or P_3 equals $\mathcal{O}$, then the result we want follows from the fact that negatives go to negatives. So we may assume that P_1, P_2, and P_3 are not equal to $\mathcal{O}$. We write their coordinates as

$$P_1 = (x_1, y_1), \qquad P_2 = (x_2, y_2), \qquad P_3 = (x_3, y_3).$$

From the definition of the group law on C, the condition $P_1 + P_2 + P_3 = \mathcal{O}$ is equivalent to saying that P_1, P_2, and P_3 lie on a line. Let

$$y = \lambda x + \nu$$

be the line through P_1, P_2, P_3. (If two or three of the points coincide, then the line has to satisfy certain tangency conditions.)

Our explicit formula for adding points says that

$$x_3 = \lambda^2 - a - x_1 - x_2, \qquad y_3 = \lambda x_3 + \nu.$$

Since x_1, x_2, x_3, y_3, and a are all integers, we see that λ and ν are also integers. This fact is what we needed because now we can reduce λ and ν modulo p.

Substituting the equation of the line into the equation of the cubic, we know that the equation

$$x^3 + ax^2 + bx + c - (\lambda x + \nu)^2 = 0$$

has x_1, x_2, x_3 as its three roots. In other words, we have the factorization

$$x^3 + ax^2 + bx + c - (\lambda x + \nu)^2 = (x - x_1)(x - x_2)(x - x_3).$$

This is the relation that ensures that $P_1 + P_2 + P_3 = \mathcal{O}$, regardless of whether or not the points are distinct.

Reducing this last equation modulo p, we obtain

$$x^3 + \tilde{a}x^2 + \tilde{b}x + \tilde{c} - (\tilde{\lambda}x + \tilde{\nu})^2 = (x - \tilde{x}_1)(x - \tilde{x}_2)(x - \tilde{x}_3).$$

Of course, we can also reduce the equations $y_i = \lambda x_i + \nu$ to get

$$\tilde{y}_i = \tilde{\lambda}\tilde{x}_i + \tilde{\nu} \qquad \text{for } i = 1, 2, 3.$$

This means that the line $y = \tilde{\lambda}x + \tilde{\nu}$ intersects the curve $\tilde{C}$ at the three points $\tilde{P}_1, \tilde{P}_2$, and $\tilde{P}_3$. Further, if two of the points $\tilde{P}_1, \tilde{P}_2, \tilde{P}_3$ are the same, say $\tilde{P}_1 = \tilde{P}_2$, then the line is tangent to $\tilde{C}$ at $\tilde{P}_1$; and similarly, if all three coincide, then the line has a triple order contact with $\tilde{C}$. Therefore

$$\tilde{P}_1 + \tilde{P}_2 + \tilde{P}_3 = \tilde{\mathcal{O}},$$

which completes the proof that the reduction modulo p map is a homomorphism from Φ to $\tilde{C}(\mathbb{F}_p)$.

Now, lo and behold, we observe that this homomorphism is one-to-one. Why is this true? Because a non-zero point $(x, y) \in \Phi$ is sent to the reduced point $(\tilde{x}, \tilde{y}) \in \tilde{C}(\mathbb{F}_p)$, and that reduced point is clearly not $\tilde{\mathcal{O}}$. So the kernel of the reduction map consists only of $\mathcal{O}$, and hence the map is one-to-one. This means that Φ looks like a subgroup of $\tilde{C}(\mathbb{F}_p)$ for every prime p such that p is relatively prime to $2D$. As we will see, this often lets us determine Φ with very little work. But before giving some examples, we will restate formally the theorem we have just finished proving.

Reduction Modulo p Theorem. *Let C be a non-singular cubic curve*

$$C : y^2 = x^3 + ax^2 + bx + c$$

with integer coefficients a, b, c and let D be the discriminant

$$D = -4a^3c + a^2b^2 + 18abc - 4b^3 - 27c^2.$$

Let $\Phi \subseteq C(\mathbb{Q})$ be the subgroup consisting of all points of finite order. For any prime p, let $P \to \tilde{P}$ be the reduction modulo p map

$$\Phi \longrightarrow \tilde{C}(\mathbb{F}_p), \qquad P \longmapsto \tilde{P} = \begin{cases} (\tilde{x}, \tilde{y}) & \text{if } P = (x, y); \\ \tilde{\mathcal{O}} & \text{if } P = \mathcal{O}. \end{cases}$$

If p does not divide $2D$, then the reduction modulo p map is an isomorphism of Φ onto a subgroup of $\tilde{C}(\mathbb{F}_p)$.

How can we use this theorem to determine the points of finite order? We give three examples to illustrate how it is used.

Example 1. $\boxed{C : y^2 = x^3 + 3}$

The discriminant for this curve is $D = -243 = -3^5$, so there is a one-to-one homomorphism $\Phi \rightarrow \tilde{C}(\mathbb{F}_p)$ for all primes $p \geq 5$. But it is easy to check that

$$\#\tilde{C}(\mathbb{F}_5) = 6 \quad \text{and} \quad \#\tilde{C}(\mathbb{F}_7) = 13.$$

Thus $\#\Phi$ divides both 6 and 13, so $\#\Phi = 1$. In other words, C has no points of finite order other than $\mathcal{O}$. Notice this means that the point $(1, 2) \in C(\mathbb{Q})$ has infinite order, so C has infinitely many rational points.

It is worth comparing this method for determining Φ with the procedure given by the Nagell-Lutz theorem. Using the Nagell-Lutz theorem, we would have to check that there are no points on C with y coordinate in the set

$$\{\pm 1, \pm 3, \pm 9, \pm 27, \pm 81, \pm 243\}.$$

[Using the stronger form of the Nagell-Lutz theorem, we would only need to check $y \in \{\pm 1, \pm 3, \pm 9\}$.] Clearly $y = \pm 1$ gives no rational points. But if y is divisible by 3, then the equation $y^2 = x^3 + 3$ shows that x must also be divisible by 3. Then $3 = y^2 - x^3$ means that 3 would be divisible by 9, which is absurd. So using the Nagell-Lutz theorem, we have again proven that $\#\Phi = 1$. We will let you decide which method you think was more efficient for computing Φ for this curve.

Example 2. $\boxed{C : y^2 = x^3 + x}$

Here the discriminant $D = -4$ is quite small, so it might be easiest to use the Nagell-Lutz theorem, but we will use the reduction theorem. We have a one-to-one map $\Phi \rightarrow \tilde{C}(\mathbb{F}_p)$ for all primes $p \geq 3$. A little computation gives the values

$$\#\tilde{C}(\mathbb{F}_3) = 4, \qquad \#\tilde{C}(\mathbb{F}_5) = 4, \qquad \#\tilde{C}(\mathbb{F}_7) = 8.$$

In fact, it is not hard to check that $\#\tilde{C}(\mathbb{F}_p)$ is divisible by 4 for every prime $p \geq 3$.

But suppose we look at the actual groups.

$$\tilde{C}(\mathbb{F}_3) = \{\tilde{\mathcal{O}}, (0, 0), (2, 1), (2, 2)\},$$
$$\tilde{C}(\mathbb{F}_5) = \{\tilde{\mathcal{O}}, (0, 0), (2, 0), (3, 0)\}.$$

We know that a point in $\tilde{C}$ has order two if and only if its y coordinate is zero. So

$$\tilde{C}(\mathbb{F}_3) \cong \frac{\mathbb{Z}}{4\mathbb{Z}}, \qquad \tilde{C}(\mathbb{F}_5) \cong \frac{\mathbb{Z}}{2\mathbb{Z}} \oplus \frac{\mathbb{Z}}{2\mathbb{Z}}.$$

Now Φ looks like a subgroup of both of these groups, so the only possibilities are that Φ is trivial or cyclic of order two. Since $(0, 0) \in C(\mathbb{Q})$ is a point of order two, we conclude that $\Phi = \{\mathcal{O}, (0, 0)\}$.

Example 3. $\boxed{C : y^2 = x^3 - 43x + 166}$

The discriminant is $D = -425984 = -2^{15} \cdot 13$. Starting to apply the Nagell-Lutz theorem, we soon find the point $P = (3, 8)$ which might be a point of finite order. Using the doubling formula, we can easily compute the x coordinates of $2P$, $4P$, and $8P$, which turn out to be

$$x(P) = 3, \qquad x(2P) = -5, \qquad x(4P) = 11, \qquad x(8P) = 3.$$

Since $x(P) = x(8P)$, it follows that $8P = \pm P$, so P is a point of finite order.

Next we use the reduction theorem. Since $2D$ is relatively prime to 3, we know that Φ is a subgroup of $\tilde{C}(\mathbb{F}_3)$. It is easy to check that $\#\tilde{C}(\mathbb{F}_3) = 7$, so Φ must have order 1 or 7. Since Φ contains the point $P = (3, 8)$, we conclude that Φ has order 7. Therefore, the points of finite order in $C(\mathbb{Q})$ form a cyclic subgroup of order 7, and $(3, 8)$ generates this subgroup. Computing the multiples of $(3, 8)$, we can find all of the points of finite order:

$$\Phi = \big\{ \mathcal{O}, (3, \pm 8), (-5, \pm 16), (11, \pm 32) \big\}.$$

4. A Factorization Algorithm Using Elliptic Curves

In this section we are going to discuss the classical problem of factoring integers. The fundamental theorem of arithmetic says that every integer can be written as a product of primes in an essentially unique way. So suppose we are given a large positive integer n and asked to factor it into primes. First, n itself might be prime, in which case we are done. How can we check? We will see below that it is not difficult to compute $2^k \pmod{n}$, even if n and k are very large. Now if n is prime, then Fermat's little theorem says that $2^{n-1} \equiv 1 \pmod{n}$. So if we compute $2^{n-1} \pmod{n}$ and find it is not equal to 1, then we will know that n is composite. Suppose that this is what happens. Then we will have conclusively proven that n is composite without having any idea how to factor it!

Warning. The converse to Fermat's little theorem is not true. In fact, there are composite numbers n such that $a^{n-1} \equiv 1 \pmod{n}$ for all a relatively prime to n. Such numbers are called *pseudo-primes*. So you cannot use Fermat's theorem to prove that a number is prime, but only to (sometimes) prove that a number is not prime.

Suppose we are given a number n which we know is composite. If n factors as $n = n_1 n_2$, the smaller factor must be less than $\sqrt{n}$. So here is a method which will factor n. First we check if $2|n$. If it does, we have found a factor. If not, then we check if $3|n$, then if $4|n$, then if $5|n$, etc. And by the time we get up to $\sqrt{n}$, we are guaranteed to find a factor. Of course, this procedure is somewhat inefficient. For example, suppose that n has around 100 digits, and suppose that every second we can check 1,000,000 possible divisors. Then we will find a factor of n in no more than 3.2×10^{37} years. And even if we made our calculations a million times faster, it could still take around 3.2×10^{31} years. So we clearly need to find a better procedure.

Why do we want to be able to factor large numbers? From a purely mathematical viewpoint, the fundamental theorem of arithmetic is a beautiful theorem, so it is natural to want to be able to compute the factorization it describes. But there is also a practical reason to factor large numbers. Recently, mathematicians have devised new sorts of codes (really ciphers) based on trap-door functions built around the problem of factoring large integers. We have already seen the underlying principle, namely it is often easy to tell if a number is composite, but hard to actually find a factor. It would take us too far afield to describe these new ciphers, but let us just say that if a message is enciphered using the composite integer n, then you will be able to read the message if you can factor n. So the question of factoring large integers is of great interest to governments and businesses that wish to keep their messages secret.

Before we discuss the problem of factorization, we will consider two other arithmetic problems and describe efficient algorithms to solve them.

Example 1. *Raising to Powers Modulo n*
Suppose we are given three positive integers a, k, and n, and we want to compute
$$a^k \pmod{n}.$$

This means we want to find an integer b satisfying
$$b \equiv a^k \pmod{n} \qquad \text{and} \qquad 0 \le b < n.$$

How long will it take us to do this? The obvious method is to compute $a_2 = a \cdot a$, reduce a_2 modulo n, compute $a_3 = a_2 \cdot a$, reduce a_3 modulo n, etc. When we get to a_k we will have our answer, at the cost of k operations, where each operation consists of one multiplication and one reduction modulo n. Is there a better way?

The answer is that there is a much better way, which we will illustrate for the case $k = 1000$. The first step is to write k as a sum of powers of 2; that is, write it in base 2. Thus

$$1000 = 2^3 + 2^5 + 2^6 + 2^7 + 2^8 + 2^9.$$

Then we observe that a^{1000} can be written as

$$a^{1000} = a^{2^3} \cdot a^{2^5} \cdot a^{2^6} \cdot a^{2^7} \cdot a^{2^8} \cdot a^{2^9}.$$

The number a^{2^i} can be calculated with only i multiplications by successive squaring. Thus we let $A_0 = a$ and calculate

$$A_1 = A_0 \cdot A_0 = a^2 \pmod{n}$$
$$A_2 = A_1 \cdot A_1 = a^4 \pmod{n}$$
$$A_3 = A_2 \cdot A_2 = a^{2^3} \pmod{n}$$
$$\vdots$$
$$A_9 = A_8 \cdot A_8 = a^{2^9} \pmod{n}.$$

Then

$$a^{1000} = A_3 \cdot A_5 \cdot A_6 \cdot A_7 \cdot A_8 \cdot A_9 \pmod{n}.$$

So it takes nine operations to get the A_i's, and then six operations to get a^{1000}. This is much better than the 1000 operations required by our original method. And if k is much larger, say $k \approx 10^{100}$, the savings are enormous.

In general, to compute $a^k \pmod{n}$ we write

$$k = k_0 + k_1 \cdot 2 + k_2 \cdot 2^2 + k_3 \cdot 2^3 + \cdots + k_r \cdot 2^r$$

with each k_i either 0 or 1. Next we calculate

$$A_0 = a, \quad A_1 = A_0^2, \quad A_2 = A_1^2, \quad \ldots, \quad A_r = A_{r-1}^2,$$

all calculations being done modulo n. Finally we get a^k as

$$a^k = (\text{product of the } A_i\text{'s for which } k_i = 1).$$

It takes r operations to get the A_i's, and then at most r operations to get a^k. The speed of this method depends on the size of r. We may assume that $k_r = 1$ because otherwise k has a smaller binary expansion. Then

$$k = k_0 + k_1 \cdot 2 + k_2 \cdot 2^2 + \cdots + k_r \cdot 2^r \geq 2^r, \qquad \text{so} \quad r \leq \log_2 k.$$

We have proved the following result.

Proposition. *It is possible to compute $a^k \pmod{n}$ in at most $2 \log_2 k$ operations, where each operation consists of one multiplication and one reduction modulo n.*

The logarithm function grows very slowly, so this provides an extremely practical method for computing $a^k \pmod{n}$ even for very large k. For example, if $k = 10^{100}$, then the computation takes fewer than 700 steps.

Example 2. *Greatest Common Divisor*

Let a and b be given positive integers. How can we compute the greatest common divisor of a and b, that is, the largest integer which divides both a and b? If we can factor a and b into products of primes, then it is easy; but if a and b are large, this is not a feasible procedure.

An efficient way to compute $\gcd(a, b)$ is the *Euclidean algorithm*, which many of you have probably already seen. The idea is to use division with remainder. Thus, first we divide a by b and get a quotient q and remainder r. In other words,

$$a = bq + r \qquad \text{with } 0 \leq r < b.$$

Next we divide b by r, and so on. This leads to the following sequence of equations.

$$
\begin{aligned}
a &= bq_1 + r_2 &&\text{with } 0 \leq r_2 < b \\
b &= r_2 q_2 + r_3 &&\text{with } 0 \leq r_3 < r_2 \\
r_2 &= r_3 q_3 + r_4 &&\text{with } 0 \leq r_4 < r_3 \\
&\;\;\vdots && \;\;\vdots \\
r_{n-1} &= r_n q_n + r_{n+1} &&\text{with } 0 \leq r_{n+1} < r_n \\
r_n &= r_{n+1} q_{n+1}.
\end{aligned}
$$

(If you let $r_0 = a$ and $r_1 = b$, the numbering system on the r_i's and q_i's will make more sense.) Since $b = r_1 > r_2 > r_3 > \cdots$ and the r_i's are non-negative integers, we eventually get to zero; say $r_{n+2} = 0$. Then it is not hard to check that

$$\gcd(a, b) = r_{n+1}.$$

How many steps does the Euclidean algorithm take in order to compute $\gcd(a, b)$? We claim that the successive remainders satisfy the estimate

$$r_{i+1} \leq \frac{1}{2} r_{i-1}.$$

So every two steps the remainder is at least halved, and the algorithm finishes when we reach a remainder of zero. Switching a and b if necessary, we may assume that $a \geq b$; and we know that at the first step we have $r_2 < b$. Hence,

$$r_4 < \frac{1}{2} b, \quad r_6 < \frac{1}{2} r_4 < \frac{1}{4} b, \quad r_8 < \frac{1}{2} r_6 < \frac{1}{8} b, \quad \ldots, \quad r_{2i} < \frac{1}{2^{i-1}} b.$$

But r_{2i} is a non-negative integer, so as soon as $2^{i-1} \geq b$, we get $r_{2i} < 1$, which means that $r_{2i} = 0$. In other words,

$$i \geq 1 + \log_2 b = \log_2(2b) \qquad \text{implies that} \qquad r_{2i} = 0.$$

So the Euclidean algorithm takes at most $2 \log_2(2b)$ steps to compute the greatest common divisor of a and b. Again, because the logarithm grows so slowly, this makes the Euclidean algorithm practical even for very large values of a and b.

Now we will verify the claim we made above that

$$r_{i+1} \leq \frac{1}{2} r_{i-1}.$$

If $r_i \leq \frac{1}{2} r_{i-1}$, then we are done because $r_{i+1} < r_i$. On the other hand, suppose that $r_i > \frac{1}{2} r_{i-1}$. We know that

$$r_{i-1} = r_i q_i + r_{i+1} \quad \text{with} \quad 0 \leq r_{i+1} < r_i.$$

Now under our assumption that $r_i > \frac{1}{2} r_{i-1}$, we find that

$$r_{i+1} = r_{i-1} - r_i q_i < r_{i-1} \left(1 - \frac{1}{2} q_i\right).$$

Clearly, $q_i \neq 0$ because otherwise $r_{i-1} = r_{i+1}$, which would contradict the fact that the r_i's are strictly decreasing. Hence, $q_i \geq 1$ (in fact, $q_i = 1$,) and so $r_{i+1} < \frac{1}{2} r_{i-1}$.

We have proven:

Proposition. *The Euclidean algorithm will compute the greatest common divisor of a and b in at most $2 \log_2 \max\{2a, 2b\}$ operations, where each operation is one division with remainder.*

Now we turn to the difficult problem of factoring an integer into a product of primes. We saw earlier that it is always possible to find a factor of n in at most $\sqrt{n}$ steps, but that takes too long. We will start by describing a factorization method due to Pollard [1]. Pollard's method does not work for all n's; but when it does work, it is fairly efficient. Further, it is the prototype for the elliptic curve method we will discuss later.

The idea underlying Pollard's algorithm is not difficult. Suppose that n happens to have a prime factor p such that $p-1$ is a product of small primes. From Fermat's little theorem, we know that if p does not divide a, then

$$a^{p-1} \equiv 1 \pmod{p};$$

so p will divide $\gcd(a^{p-1} - 1, n)$.

Of course, at the start we do not know what p is, so we cannot actually compute $a^{p-1} - 1$. Instead, we choose an integer

$$k = 2^{e_2} \cdot 3^{e_3} \cdot 5^{e_5} \cdots r^{e_r},$$

where $2, 3, \ldots, r$ are the first few primes and $e_1, e_2, \ldots, e_r$ are small positive integers. Then we compute

$$\gcd(a^k - 1, n).$$

Note that it is only necessary to compute $a^k - 1$ modulo n. So from Examples 1 and 2 discussed earlier, we can compute $\gcd(a^k - 1, n)$ in fewer than $2\log_2(2kn)$ operations, which is a very reasonable amount of time even if k and n are as large as 10^{1000}.

Now if we are lucky enough that n has a prime factor p with $p - 1 | k$, then p will divide $a^k - 1$.[‡] So in this case we will have

$$\gcd(a^k - 1, n) \geq p > 1.$$

If $\gcd(a^k - 1, n) \neq n$, then this gcd provides a non-trivial factor of n; so we can factor n into two pieces and repeat the procedure for each piece. On the other hand, if the gcd equals n, then we choose a new a and try again. So the idea is to compute $\gcd(a^k - 1, n)$. If it is strictly between 1 and n, we have factored n; if it equals n, we choose a new a; and if it equals 1, we choose a larger k.

Let's see how this works in practice. We will try to factor

$$n = 246082373.$$

The first thing to do is to verify that n is not itself prime. This can be done by computing $2^{n-1} \pmod{n}$ and showing it is not equal to 1. We will leave this for you to check. So now we know that n is composite, and we want to find a factor.

We will take

$$a = 2 \quad \text{and} \quad k = 2^2 \cdot 3^2 \cdot 5 = 180.$$

Since

$$180 = 2^2 + 2^4 + 2^5 + 2^7,$$

we need to compute $2^{2^i} \pmod{n}$ for $0 \leq i \leq 7$. The results are compiled in the following table.

i	$2^{2^i} \pmod{246082373}$
0	2
1	4
2	16
3	256
4	65536
5	111566955
6	166204404
7	214344997

[‡] Actually, if a and n are not relatively prime, then Fermat's little theorem cannot be used. However, if $\gcd(a,n) > 1$, then we have already found a non-trivial factor of n.

Using this table, we can compute

$$2^{180} = 2^{2^2+2^4+2^5+2^7}$$

$$\equiv 16 \cdot 65536 \cdot 111566955 \cdot 28795219 \pmod{246082373}$$

$$\equiv 121299227 \pmod{246082373}.$$

Then a short calculation using the Euclidean algorithm yields

$$\gcd(2^{180} - 1, n) = \gcd(121299226, 246082373) = 1.$$

So the test fails, and n has no prime factors p such that $p - 1$ divides 180.

But all is not lost, we can just go back and choose a larger k. For our new k we will take

$$k = \text{LCM}[2, 3, \ldots, 9] = 2^3 \cdot 3^2 \cdot 5 \cdot 7 = 2520.$$

Since

$$2520 = 2^3 + 2^4 + 2^6 + 2^7 + 2^8 + 2^{11},$$

we need to extend our table a little. This is easily done.

i	2^{2^i} (mod 246082373)
8	111354998
9	82087367
10	7262569
11	104815687

Now we can compute

$$2^{2520} = 2^{2^3+2^4+2^6+2^7+2^8+2^{11}} \equiv 101220672 \pmod{246082373}.$$

Then the Euclidean algorithm yields

$$\gcd(2^{2520} - 1, n) = \gcd(101220671, 246082373) = 2521,$$

so we have found a non-trivial factor of n.

More precisely, we have factored n as

$$n = 246082373 = 2521 \cdot 97613.$$

Further, it is easy to check that each of these factors is prime, so we have completely factored n. (Notice that $\sqrt{97613} = 312.43\ldots$, so one could even check that 97613 is prime by hand using the trivial method of checking all possible factors up to $\sqrt{97613}$.)

Of course, we do not mean to suggest that Pollard's algorithm is needed to factor a number such as $n = 246082373$ because even a micro-computer could easily use the trivial algorithm and check all possible divisors up to $\sqrt{n} \approx 15687$. But this example reveals all the salient features of the algorithm, which we now summarize.

Pollard's $p-1$ Algorithm.   *Let $n \geq 2$ be a composite integer for which we are to find a factor.*

$\boxed{\text{Step 1}}$ *Choose a number k which is a product of small primes to small powers. For example, take*

$$k = \text{LCM}[1, 2, 3, \ldots, K]$$

for some integer K.

$\boxed{\text{Step 2}}$ *Choose an arbitrary integer a satisfying $1 < a < n$.*

$\boxed{\text{Step 3}}$ *Calculate $\gcd(a, n)$. If it is strictly greater than 1, then it is a non-trival factor of n, so we are done. Otherwise, go on to Step 4.*

$\boxed{\text{Step 4}}$ *Calculate $D = \gcd(a^k - 1, n)$. If $1 < D < n$, then D is a non-trivial factor of n and we are done. If $D = 1$, go back to Step 1 and take a larger k. If $D = n$, go back to Step 2 and choose another a.*

Notice that Pollard's algorithm should eventually stop because eventually K in Step 1 will equal $\frac{1}{2}(p-1)$ for some prime p dividing n, so eventually there will be some $p-1$ dividing k. However, if it takes this long to work, then the algorithm will not be practical for large values of n. The algorithm only works in a "reasonable" amount of time if it happens that n has a prime divisor p satisfying

$$p - 1 = \text{product of small primes to small powers.}$$

Now we are ready to describe Lenstra's idea [1] for using elliptic curves to create an algorithm which (conjecturally) does not have this defect. Pollard's algorithm is based on the fact that the non-zero elements in $\mathbb{Z}/p\mathbb{Z}$ form a group $(\mathbb{Z}/p\mathbb{Z})^*$ of order $p-1$; so if $p-1|k$, then $a^k = 1$ in the group. Lenstra's idea is to replace the group $(\mathbb{Z}/p\mathbb{Z})^*$ by the group of points on an elliptic curve $C(\mathbb{F}_p)$, and to replace the integer a by a point $P \in C(\mathbb{F}_p)$. As in Pollard's algorithm, we choose an integer k composed of a product of small primes. Then, if it happens that the number of elements in $C(\mathbb{F}_p)$ divides k, we will have $kP = \mathcal{O}$ in $C(\mathbb{F}_p)$. And just as before, the fact that $kP = \mathcal{O}$ will generally allow us to find a non-trivial factor of n.

What is the advantage of Lenstra's algorithm? If we choose only one curve C with integer coefficients and consider its reductions modulo various primes, then there is no advantage. For a single curve C, we will win if there is some prime p dividing n so that $\#\tilde{C}(\mathbb{F}_p)$ is a product of small primes. Similarly, we win using Pollard's algorithm if there is a prime p dividing n so that $p-1$ is a product of small primes. But suppose now that we do not win. Using Pollard's algorithm, not winning means losing, and the game is over. But with Lenstra's algorithm, there is a new flexibility which allows us to continue playing. Namely, we are free to choose a new elliptic curve and start all over again. Since $\#\tilde{C}(\mathbb{F}_p)$ varies considerably for a fixed prime p and varying curve C, our odds of eventually winning are fairly good.

Now we take these vague comments and turn them into an explicit algorithm. We noted in Section 1 that if C is a non-singular cubic curve with coefficients in $\mathbb{F}_p$, then

$$\#C(\mathbb{F}_p) = p + 1 - \varepsilon_p \qquad \text{with } |\varepsilon_p| \le 2\sqrt{p}.$$

Further, one can show that as C varies over all such curves, the numbers ε_p are quite well spread out over the interval from $-2\sqrt{p}$ to $+2\sqrt{p}$. So it is quite likely (but not yet rigorously proven) that we will fairly rapidly run across a curve C with $\#C(\mathbb{F}_p)$ equal to a product of small primes.

So here, formally laid out, is Lenstra's algorithm.

Lenstra's Elliptic Curve Algorithm. *Let $n \ge 2$ be a composite integer for which we are to find a factor.*

$\boxed{\text{Step 1}}$ *Check that $\gcd(n, 6) = 1$ and that n does not have the form m^r for some $r \ge 2$.*

$\boxed{\text{Step 2}}$ *Choose random integers b, x_1, y_1 between 1 and n.*

$\boxed{\text{Step 3}}$ *Let $c = y_1^2 - x_1^3 - bx_1 \pmod{n}$, let C be the cubic curve*

$$C : y^2 = x^3 + bx + c, \qquad \text{and let } P = (x_1, y_1) \in C.$$

$\boxed{\text{Step 4}}$ *Check that $\gcd(4b^3 + 27c^2, n) = 1$. (If it equals n, go back and choose a new b. If it is strictly between 1 and n, then it is a non-trivial factor of n, so we are done.)*

$\boxed{\text{Step 5}}$ *Choose a number k which is a product of small primes to small powers. For example, take*

$$k = \text{LCM}[1, 2, 3, \dots, K]$$

for some integer K.

$\boxed{\text{Step 6}}$ *Compute*

$$kP = \left(\frac{a_k}{d_k^2}, \frac{b_k}{d_k^3} \right).$$

$\boxed{\text{Step 7}}$ *Calculate $D = \gcd(d_k, n)$. If $1 < D < n$, then D is a non-trivial factor of n and we are done. If $D = 1$, either go back to Step 5 and increase k or go back to Step 2 and choose a new curve. If $D = n$, then go back to Step 5 and decrease k.*

There are two things about this algorithm that we should discuss. First, why does it work? Second, how do we perform Step 6, which seems to involve calculating

$$\underbrace{P + P + \dots + P}_{k \text{ times}} ?$$

To see why the algorithm works, suppose that we are lucky enough to choose a curve C and number k so that, for some prime p dividing n, we have $\#\tilde{C}(\mathbb{F}_p)$ dividing k. Then every element in $\tilde{C}(\mathbb{F}_p)$ has order dividing k, so in particular if we take the point $P \in C(\mathbb{Q})$ and reduce it modulo p, we know that[‡]

$$\widetilde{kP} = k\tilde{P} = \tilde{\mathcal{O}}.$$

In other words, the reduction of kP modulo p is the point $\mathcal{O}$ at infinity, so we must have p dividing d_k. Hence, p will divide $\gcd(d_k, n)$. Further, unless we are extremely unlucky, n will not divide d_k, so we will get a non-trivial factor of n. This explains why Lenstra's algorithm works.

Now how do we compute kP efficiently? Clearly, not as the k-fold sum $P + P + \cdots + P$. Instead, we use the same binary expansion scheme we used to compute a^k. First we write k as

$$k = k_0 + k_1 \cdot 2 + k_2 \cdot 2^2 + k_3 \cdot 2^3 + \cdots + k_r \cdot 2^r$$

with each k_i either 0 or 1. As before, we can do this with $r \leq \log_2 k$. Next we compute

$$
\begin{aligned}
P_0 &= P \\
P_1 &= 2P_0 &&= 2P \\
P_2 &= 2P_1 &&= 2^2 P \\
P_3 &= 2P_2 &&= 2^3 P \\
&\;\;\vdots &&\;\;\vdots \\
P_r &= 2P_{r-1} &&= 2^r P.
\end{aligned}
$$

Finally we calculate

$$kP = (\text{sum of } P_i\text{'s for which } k_i = 1).$$

So we can compute kP in fewer than $2\log_2 k$ steps of doubling and adding points.

Note, however, that we really do not want to compute the coordinates of kP as rational numbers because the numerators and denominators would have approximately k^2 digits. Even for relatively small values of k, such as $k \approx 10^{50}$, this leads to numbers with more digits than there are elementary particles in the known universe. So it is much better to perform all of our computations modulo n.

But n is not prime, so how can we use the formulas for doubling and adding points? Let's consider the problem of adding together two points, say $Q_1 = (x_1, y_1)$ and $Q_2 = (x_2, y_2)$, where x_1, y_1, x_2, y_2 are integers modulo n. Our formula for $Q_3 = Q_1 + Q_2$ says that

$$x_3 = \lambda^2 - x_1 - x_2 \quad \text{and} \quad y_3 = -\lambda x_3 - (y_1 - \lambda x_1), \quad \text{where}$$

$$\lambda = \frac{y_2 - y_1}{x_2 - x_1}.$$

[‡] Actually, we need to know that reduction modulo p gives a homomorphism from all of $C(\mathbb{Q})$ to $C(\mathbb{F}_p)$. In Section 3 we only proved that it is a homomorphism on the subgroup of $C(\mathbb{Q})$ consisting of the points of finite order. We will leave the general case as an exercise, or you can find a proof in Appendix A, Section 5.

The difficulty lies in computing λ because the ring $\mathbb{Z}/n\mathbb{Z}$ is not a field, so $x_2 - x_1$ may not have an inverse. In trying to do this computation, we are faced with three possible outcomes:

(1) $\gcd(x_2 - x_1, n) = 1$

In this case $x_2 - x_1$ does have an inverse in $\dfrac{\mathbb{Z}}{n\mathbb{Z}}$, so we can calculate Q_3 modulo n. [Possibly we should mention that if $\gcd(a, n) = 1$, then the Euclidean algorithm can be modified to give a solution to the equation $ax \equiv 1 \pmod{n}$. So there is a fast way to find the inverse of a modulo n; cf. Exercise 4.18.]

(2) $1 < \gcd(x_2 - x_1, n) < n$

In this case we cannot find Q_3, but we don't care because the integer $\gcd(x_2 - x_1, n)$ gives us the desired factor of n. So the algorithm can be terminated here.

(3) $\gcd(x_2 - x_1, n) = n$

If this case occurs, then we have been very unlucky. The best bet is to go back to Step 5 and reduce the value of k; or we could go back to Step 2 and start with another curve.

Similarly, to double a point $Q = (x, y)$ modulo n, we need to compute the ratio

$$\lambda = \frac{f'(x)}{2y} = \frac{3x^2 + 2ax + b}{2y} \pmod{n}.$$

So we get the same three alternatives: either we can compute $2Q \pmod{n}$; or we get a non-trivial factor of n; or $\gcd(y, n) = n$ and we have to start with a new k or a new curve.

This, in essence, is how Lenstra's elliptic curve algorithm works; although in practice there are several ways of making it more efficient. To illustrate the general procedure, we will now use Lenstra's algorithm to factor the integer

$$n = 1715761513.$$

The first thing to do is to check that n is not prime. Using the successive squaring scheme described earlier, we easily calculate that

$$2^{n-1} \equiv 93082891 \pmod{n}.$$

By Fermat's little theorem, this *proves* that n is not prime; so now we search for a factor.

The first step in Lenstra's algorithm says to check that n is not a perfect power. Using a calculator, we compute

$$\sqrt{n}, \quad \sqrt[3]{n}, \quad \sqrt[4]{n}, \quad \ldots \quad \sqrt[31]{n} \approx 1.9855.$$

None of them are integers, so n is not a perfect power.

Since $\sqrt{1715761513} \approx 42422$, we know that n has some prime factor p less than 42422. We want to choose a value for k so that some integer close to p divides k. We will try

$$k = \text{LCM}[1, 2, 3, \ldots, 17] = 12252240,$$

which has lots of factors less than 42422.

Next we have to choose an elliptic curve and a point on that curve. As indicated in the description of Lenstra's algorithm, it is easiest to fix the point P and one of the coefficients of the curve, and then choose the other coefficient so that the point is on the curve. Thus, we will fix the point

$$P = (2, 1),$$

we will take various values of b, and then we will set $c = -7 - 2b$. To start with we will let $b = 1$, so $c = -9$. Thus, we are looking at the curve C and point P given by

$$C : y^2 = x^3 + x - 9, \qquad P = (2, 1) \in C.$$

Our goal is to compute $kP \pmod{n}$ using successive doubling, so the first thing to do is express k as a sum of powers of 2. This is easily accomplished, yielding

$$k = 12252240$$
$$= 2^4 + 2^6 + 2^{10} + 2^{12} + 2^{13} + 2^{14} + 2^{15} + 2^{17}$$
$$+ 2^{19} + 2^{20} + 2^{21} + 2^{23}.$$

So in order to compute $kP \pmod{n}$, we need to determine $2^i P \pmod{n}$ for $0 \le i \le 23$. This would take a lot of paper if we were to do it by hand, but it is an easy task for even the smallest computer. The results are compiled in the table given on the next page.

Finally, adding up the appropriate points in this table, we find the value of $kP \pmod{n}$:

$$2^4 P = 16P = (385062894, 618628731)$$
$$(2^4 + 2^6)P = 80P = (831572269, 1524749605)$$
$$(2^4 + 2^6 + 2^{10})P = 1104P = (1372980126, 736595454)$$
$$(2^4 + 2^6 + 2^{10} + 2^{12})P = 5200P = (1247661424, 958124008)$$

(previous partial sum) $+ 2^{13}P = 13392P = (1548582473, 1559853215)$

(previous partial sum) $+ 2^{14}P = 29776P = (201510394, 7154559)$

(previous partial sum) $+ 2^{15}P = 62544P = (629067322, 264081696)$

(previous partial sum) $+ 2^{17}P = 193616P = (844665131, 537510825)$

(previous partial sum) $+ 2^{19}P = 717904P = (886345533, 342856598)$

(previous partial sum) $+ 2^{20}P = 1766480P = (370579416, 1254954111)$

(previous partial sum) $+ 2^{21}P = 3863632P = (77302130, 514483068)$

(previous partial sum) $+ 2^{23}P = 12252240P = (1225303014, 142796033)$

i	$2^i P \pmod{1715761513}$
0	$(2,1)$
1	$(1286821173, 1072350709)$
2	$(1334478523, 112522703)$
3	$(912789305, 77695868)$
4	$(385062894, 618628731)$
5	$(866358838, 450284374)$
6	$(904716938, 169383608)$
7	$(808696477, 1201030016)$
8	$(572301268, 107111567)$
9	$(1512647092, 1695275444)$
10	$(1858186, 1224662922)$
11	$(1550404618, 825515387)$
12	$(1519325194, 1657497846)$
13	$(522917322, 524407354)$
14	$(25207285, 1375034461)$
15	$(781360494, 1457273929)$
16	$(1108412304, 25813532)$
17	$(435914774, 323718902)$
18	$(1399483199, 1203611423)$
19	$(778823593, 192206539)$
20	$(853199887, 1012680972)$
21	$(501929966, 910060788)$
22	$(1315182921, 305331854)$
23	$(257200250, 318342966)$

So now we know that on the curve $y^2 = x^3 + x - 9$ considered modulo n, we have

$$kP = 12252240(2,1) \equiv (421401044, 664333727) \pmod{1715761513}.$$

What does this tell us about factors of n? Nothing! The whole point of Lenstra's algorithm is that it gives us a factor of n precisely when the addition law breaks down. So if we are actually able to compute $kP \pmod{n}$, then we have to start over with either a new k, a new P, or a new curve.

We will take the last alternative and vary the curve. So we will stick with $k = 12252240$ and with $P = (2,1)$, but now we will take $b = 2$ and $c = -7 - 2b = -11$. Using this curve and repeating the above calculation, we again find we are able to compute $kP \pmod{n}$. So we go on to $b = 3$ and $c = -13$, etc. This is all perfectly feasible with a small computer, and we find that we are able to compute $kP \pmod{n}$ for all $b = 3, 4, 5, \ldots, 41$.

However, when we try $b = 42$, and $c = -91$, the addition law breaks down and we find a factor of n. What happens is the following. We have no trouble making a table of $2^i P \pmod{n}$ for $0 \le i \le 23$, just as above. Then we start adding up the points in the table to compute $kP \pmod{n}$.

At the penultimate step we find

$$(2^4 + 2^6 + 2^{10} + \cdots + 2^{20} + 2^{21})P = 3863632P$$
$$\equiv (1115004543, 1676196055) \pmod{n}.$$

Next, we read off from the (omitted) table

$$2^{23}P \equiv (1267572925, 848156341) \pmod{n}.$$

So to get kP we need to add these two points,

$$(1115004543, 1676196055) + (1267572925, 848156341) \pmod{n}.$$

To do this we have to take the difference of their x coordinates and find the inverse modulo n. But when we try to do this, we discover that the inverse does not exist because

$$\gcd(1115004543-1267572925, n) = \gcd(-152568382, 1715761513) = 26927.$$

So the attempt to compute $12252240(2, 1)$ on the curve

$$y^2 = x^3 + 42x - 91 \pmod{1715761513}$$

fails, but it leads to the factorization

$$n = 1715761513 = 26927 \times 63719.$$

One easily checks that each of these factors is prime, so this gives the full factorization of n.

Of course, we were lucky that the k we picked was large enough. In practice, if we had gotten up to a large b, say $b = 1000$, without finding a factor of n, then we might have increased k to $\mathrm{LCM}[1, 2, \ldots, 25]$ and started over again. In that case it would probably make sense to also use a new initial point, say $P = (3, 1)$.

EXERCISES

4.1. Let $p \neq 2$ be a prime, let $a, b, c, d \in \mathbb{F}_p$ satisfy $acd \neq 0$, and let C be the conic given by the homogeneous equation

$$C : ax^2 + bxy + cy^2 = dz^2.$$

(a) If $b^2 \neq 4ac$, prove that $\#C(\mathbb{F}_p) = p + 1$.
(b) If $b^2 = 4ac$, prove that either

$$\#C(\mathbb{F}_p) = 1 \quad \text{or} \quad \#C(\mathbb{F}_p) = 2p + 1.$$

Give examples (e.g., with $p = 3$) to show that both possibilities can occur.

4.2. Compute the group $C(\mathbb{F}_p)$ for the curve

$$C : y^2 = x^3 + x + 1$$

and the primes $p = 3$, 7, 11, and 13.

4.3. Let $p \geq 3$ be a prime, and let $m \geq 1$ be an integer which is relatively prime to $p - 1$.
(a) Prove that the map $x \rightarrow x^m$ is an isomorphism from $\mathbb{F}_p^*$ to itself.
(b) Prove that the equation

$$x^m + y^m + z^m = 0$$

has exactly $p + 1$ projective solutions with $x, y, z \in \mathbb{F}_p$.
(c) ** Suppose instead that m divides $p - 1$. Let M_p be the number of projective solutions to the equation given in (b). Prove that M_p satisfies the inequality

$$|M_p - p - 1| \leq (m - 1)(m - 2)\sqrt{p}.$$

This problem is a little easier if you take m to be a prime, so you might try to do that case first. [The Fermat equation $x^m + y^m + z^m = 0$ has genus $\frac{1}{2}(m - 1)(m - 2)$, so (c) is a special case of the Hasse-Weil theorem.]

4.4. Let p be an odd prime and let $\zeta \in \mathbb{C}$ be a root of the equation

$$x^{p-1} + x^{p-2} + \cdots + x + 1 = 0.$$

(Notice that $\zeta^p = 1$ and $\zeta \neq 1$.) We define the set of *quadratic residues* R in $\mathbb{F}_p^*$ by

$$R = \{x^2 : x \in \mathbb{F}_p^*\}.$$

(a) Prove that R is a subgroup of $\mathbb{F}_p^*$ of index 2. We denote the other coset of R in $\mathbb{F}_p^*$ by N and call it the set of *quadratic non-residues*.
(b) Prove that $-1 \in R$ if and only if $p \equiv 1 \pmod 4$.
(c) Define quadratic Gauss sums by the formulas

$$\alpha = \sum_{r \in R} \zeta^r, \qquad \beta = \sum_{n \in N} \zeta^n.$$

Prove that $\alpha + \beta = -1$.
(d) * Prove that

$$\alpha\beta = \begin{cases} -\dfrac{p-1}{4} & \text{if } p \equiv 1 \,(\text{mod } 4), \\ \dfrac{p+1}{4} & \text{if } p \equiv 3 \,(\text{mod } 4). \end{cases}$$

Deduce that

$$2\alpha + 1 = \begin{cases} \pm\sqrt{p} & \text{if } p \equiv 1 \,(\text{mod } 4), \\ \pm\sqrt{-p} & \text{if } p \equiv 3 \,(\text{mod } 4). \end{cases}$$

(e) If we fix $\zeta = e^{2\pi i/p}$, compute the value of α for small values of p, and use your computations to make a conjecture about the correct sign for $2\alpha + 1$.
(f) ** Prove that your conjecture in (e) is correct.

4.5. Let C be the cubic curve given by the equation

$$C : y^2 = x^3 + x + 1.$$

(a) For each prime $p < 1000$, compute the number of points

$$M_p = \#C(\mathbb{F}_p)$$

on C over the field $\mathbb{F}_p$. Don't forget to include the point $\mathcal{O}$. Also compute the angles θ_p determined by the conditions

$$\cos \theta_p = \frac{M_p - p - 1}{2\sqrt{p}}, \qquad 0 \le \theta_p \le \pi.$$

(b) Compare the quantities

$$\#\{p \le 1000 : \alpha \le \theta_p \le \beta\} \qquad \text{and} \qquad 168 \left(\frac{2}{\pi} \int_\alpha^\beta \sin^2 t \, dt \right)$$

for various values of α and β, such as

$$(\alpha, \beta) = \left(0, \frac{\pi}{2}\right) \quad \text{or} \quad \left(\frac{\pi}{3}, \frac{\pi}{2}\right) \quad \text{or} \quad \left(\frac{2\pi}{5}, \frac{3\pi}{5}\right).$$

Note that $168 = \pi(1000)$ is the number of primes less than 1000. Do your computations convince you that the Sato-Tate conjecture is probably true? (For this exercise you will undoubtedly need to use a computer.)

4.6. This exercise describes a special case of a theorem that was originally proven by Eichler and Shimura. The Taniyama-Weil conjecture says that a similar statement is true for every rational elliptic curve.

(a) Let C be the cubic curve given by the equation

$$C : y^2 = x^3 - 4x^2 + 16.$$

As usual, let $M_p = \#C(\mathbb{F}_p)$ be the number of points on C over the field $\mathbb{F}_p$. Calculate M_p for all primes $3 \le p \le 13$. (If you have a computer, calculate up to $p \le 100$.)

(b) Let $F(q)$ be the (formal) power series given by the infinite product

$$F(q) = q \prod_{n=1}^{\infty} (1 - q^n)^2 (1 - q^{11n})^2 = q - 2q^2 - q^3 + 2q^4 + \cdots.$$

Let N_n be the coefficient of q^n in $F(q)$,

$$F(q) = \sum_{n=1}^{\infty} N_n q^n.$$

Calculate N_n for $n \leq 13$. (If you have a computer, calculate up to $n \leq 100$.)

(c) For each prime p, compute the sum $M_p + N_p$ of the quantities calculated in (a) and (b). Formulate a conjecture as to what this value should be in general.

(d) ** Prove that your conjecture in (c) is correct.

(e) If we replace the indeterminate q by the quantity $e^{2\pi i z}$, we obtain a function

$$\Phi(z) = F(e^{2\pi i z}) = e^{2\pi i z} \prod_{n=1}^{\infty} (1 - e^{2\pi i n z})^2 (1 - e^{2\pi i 11 n z})^2.$$

Prove that $\Phi(z)$ is a holomorphic function on the upper half plane $\mathcal{H}$, where $\mathcal{H} = \{z = x + iy \in \mathbb{C} : y > 0\}$. Prove that

$$\lim_{y \to \infty} \Phi(x + iy) = 0.$$

(f) ** Prove that for every prime p except $p = 11$, the function $\Phi(z)$ satisfies the relation

$$N_p \Phi(z) = \Phi(pz) + \sum_{j=0}^{p} \Phi\left(\frac{z+j}{p}\right) \qquad \text{for all } z \in \mathcal{H}.$$

Prove that if a, b, c, d are integers satisfying

$$ad - bc = 1 \qquad \text{and} \qquad c \equiv 0 \,(\mathrm{mod}\ 11),$$

then

$$\Phi\left(\frac{az+b}{cz+d}\right) = (cz+d)^2 \Phi(z) \qquad \text{for all } z \in \mathcal{H}.$$

[These are two of the amazing properties enjoyed by the function $\Phi(z)$, which is called a modular form of weight 2 for the congruence subgroup $\Gamma_0(11)$. Assuming that you have solved part (d), you will see that the coefficients of this modular form completely determine the number of points in $C(\mathbb{F}_p)$ for all primes p.]

4.7. Let b and c be integers satisfying

$$b \equiv 11 \pmod{15} \qquad \text{and} \qquad c \equiv 4 \pmod{15}.$$

Further, assume that $4b^3 + 27c^2 \neq 0$, and let C be the elliptic curve

$$C : y^2 = x^3 + bx + c.$$

Find all points of finite order in $C(\mathbb{Q})$.

4.8. Let $p \equiv 3 \,(\mathrm{mod}\ 4)$ be a prime, and let $b \in \mathbb{F}_p^*$.

(a) Show that the equation

$$v^2 = u^4 - 4b$$

has $p - 1$ solutions (u, v) with $u, v \in \mathbb{F}_p$.

(b) Show that if (u, v) is a solution of the equation in (a), then

$$\phi(u, v) = \left(\tfrac{1}{2}(u^2 + v), \tfrac{1}{2}u(u^2 + v)\right)$$

is a point on the elliptic curve

$$C : y^2 = x^3 + bx.$$

(c) Prove that the curve C defined in (b) satisfies $\#C(\mathbb{F}_p) = p + 1$.

4.9. Let b be a non-zero integer which is *fourth power free*. (This means that there are no primes p with $p^4 | b$.) Let C be the elliptic curve

$$C : y^2 = x^3 + bx,$$

and let $\Phi \subseteq C(\mathbb{Q})$ be the subgroup consisting of all points of finite order.

(a) Prove that $\#\Phi$ divides 4.

(b) More precisely, show that Φ is given by the following table:

$$\Phi \cong \begin{cases} \dfrac{\mathbb{Z}}{4\mathbb{Z}} & \text{if } b = 4, \\[2mm] \dfrac{\mathbb{Z}}{2\mathbb{Z}} \oplus \dfrac{\mathbb{Z}}{2\mathbb{Z}} & \text{if } -b \text{ is a square}, \\[2mm] \dfrac{\mathbb{Z}}{2\mathbb{Z}} & \text{otherwise}. \end{cases}$$

4.10. Let $p \equiv 2 \,(\mathrm{mod}\ 3)$ be a prime, and let $c \in \mathbb{F}_p^*$. Prove that the curve

$$C : y^2 = x^3 + c$$

satisfies $\#C(\mathbb{F}_p) = p + 1$.

4.11. Let c be a non-zero integer which is *sixth power free*. (This means that there are no primes p with $p^6 | c$.) Let C be the elliptic curve .

$$C : y^2 = x^3 + c,$$

and let $\Phi \subseteq C(\mathbb{Q})$ be the subgroup consisting of all points of finite order.

(a) Prove that $\#\Phi$ divides 6.

(b) More precisely, show that Φ is given by the following table:

$$\Phi \cong \begin{cases} \dfrac{\mathbb{Z}}{6\mathbb{Z}} & \text{if } c = 1, \\[2mm] \dfrac{\mathbb{Z}}{3\mathbb{Z}} & \text{if } c \neq 1 \text{ is a square, or if } c = -432, \\[2mm] \dfrac{\mathbb{Z}}{2\mathbb{Z}} & \text{if } c \neq 1 \text{ is a cube}, \\[2mm] \{\mathcal{O}\} & \text{otherwise}. \end{cases}$$

4.12. Let C be a cubic curve given by a Weierstrass equation

$$y^2 = x^3 + ax^2 + bx + c$$

with integer coefficients. Let $p \geq 3$ be a prime not dividing the discriminant, so when we reduce C modulo p we get a non-singular cubic curve $\tilde{C}$ with coefficients in $\mathbb{F}_p$. We define a general *reduction modulo p map* from $C(\mathbb{Q})$ to $\tilde{C}(\mathbb{F}_p)$ as follows: Let

$$P = (x, y) = \left(\frac{a}{d^2}, \frac{b}{d^3} \right) \in C(\mathbb{Q}).$$

If p does not divide d, we choose an integer e such that $de \equiv 1 \pmod{p}$ and set

$$\tilde{P} = (\tilde{a}\tilde{e}^2, \tilde{b}\tilde{e}^3) \in \tilde{C}(\mathbb{F}_p).$$

If p does divide d, we set $\tilde{P} = \tilde{O}$. Show that this map is a homomorphism from $C(\mathbb{Q})$ to $\tilde{C}(\mathbb{F}_p)$, and that its kernel is the subgroup $C(p)$ we discussed in Chapter II, Sections 4 and 5. Thus, there is a one-to-one homomorphism

$$\frac{C(\mathbb{Q})}{C(p)} \longrightarrow \tilde{C}(\mathbb{F}_p).$$

Since we proved in Chapter II that $C(p) \cap \Phi = \{O\}$, this immediately implies the reduction theorem and provides a useful generalization.

4.13. (a) Prove that $561 = 3 \cdot 11 \cdot 17$ is a pseudo-prime. That is, prove that if a is any integer relatively prime to 561, then

$$a^{560} \equiv 1 \pmod{561}.$$

(This can, of course, be checked on a computer. But with a little cleverness, you should be able to verify it by hand in just a few lines.)
(b) Fix an integer $a \geq 2$. Prove that there are infinitely many composite numbers m which satisfy $a^{m-1} \equiv 1 \pmod{m}$.

4.14. Use the successive doubling method described in Section 4, Example 1, to compute the following powers.
(a) $17^{5386} \pmod{26}$.
(b) $2^{35687} \pmod{38521}$.

4.15. Prove that the Euclidean algorithm described in Section 4, Example 2, does compute the greatest common divisor of a and b.

4.16. Use the Euclidean algorithm to compute the gcd of the following pairs of integers a, b. Write out each of the intermediate equations and compare the number of steps needed with the upper bound $2 \log_2(2b)$.
(a) $a = 1187319, \quad b = 438987$.
(b) $a = 4152983, \quad b = 298936$.

4.17. If $a > b > 0$, we proved that the Euclidean algorithm presented in Section 4, Example 2, will compute $\gcd(a, b)$ in no more than $2\log_2(2b)$ steps.

(a) Suppose that we revise the Euclidean algorithm as follows. Each time we do a division with remainder $r_{i-1} = r_i q_i + r_{i+1}$, we choose the remainder to satisfy $-\frac{1}{2}|r_i| < r_{i+1} \le \frac{1}{2}|r_i|$. Prove that the algorithm still computes $\gcd(a, b)$, but now in no more than $\log_2(2b)$ steps.

(b) Using the revised version of the Euclidean algorithm described in (a), prove that the r_i's satisfy

$$|r_{i+2}| \le \tfrac{1}{5}|r_i|.$$

Deduce that the revised algorithm computes $\gcd(a, b)$ in no more than $2\log_5(5b)$ steps. How large does b have to be before this estimate is better than the estimate in (a)?

(c) Compute $\gcd(4152983, 298936)$ using the revised Euclidean algorithm. Compare the actual number of steps needed with the estimate $2\log_5(5b)$ from (b).

4.18. If $\gcd(a, b) = 1$, then we know that there exists integers a', b' such that $aa' + bb' = 1$. The Euclidean algorithm described in Section 4, Example 2, gives a sequence of quotients $q_1, \ldots, q_{n+1}$ and remainders $r_0, \ldots, r_{n+1}$ used in computing $\gcd(a, b)$. Show how one can use the q_i's and r_i's to find a' and b'. (Notice that this gives an efficient way to find the inverse of a modulo b, which is needed in the implementation of Lenstra's algorithm.)

4.19. Let $n = 246082373$.

(a) Write n in the form

$$n = k_0 + k_1 \cdot 2 + k_2 \cdot 2^2 + \cdots + k_r 2^r$$

with each k_i either 0 or 1, and $k_r = 1$.

(b) Extend the table in Section 4 up to $2^{2^r} \pmod{n}$ for the r in (a).

(c) Calculate $2^{n-1} \pmod{n}$, and use the answer to deduce that n is not prime.

4.20. Let $n = 7591548931$.

(a) Calculate $2^{n-1} \pmod{n}$, and use the result to prove that n is not prime.

(b) Let $k = \text{LCM}[2, 3, \cdots, K]$ for various values of K, and compute $2^k \pmod{n}$ and $\gcd(2^k - 1, n)$ until you find a non-trivial factor of n. (You'll probably need to use a computer for this problem.)

4.21. Let $n = 199843247$. Using the elliptic curve

$$C : y^2 = x^3 + 59x - 59,$$

the point $P = (1, 1)$, and the integer $k = 16296$, compute $kP \pmod{n}$ as described in Lenstra's algorithm to find a non-trivial factor of n.

CHAPTER V

Integer Points on Cubic Curves

1. How Many Integer Points?

Let C be a non-singular cubic curve given by an equation

$$ax^3 + bx^2y + cxy^2 + dy^3 + ex^2 + fxy + gy^2 + hx + iy + j = 0$$

with integer coefficients. We have seen that if C has a rational point (possibly at infinity), then the set of all rational points on C forms a finitely generated abelian group. So we can get every rational point on C by starting from some finite set and adding points using the geometrically defined group law.

Another natural number theoretic question would be to describe the solutions (x, y) to the cubic equation with x and y both *integers*. Since the cubic equation may have infinitely many rational points, we are really asking which of these rational points actually have integer coordinates. For a curve given by a Weierstrass equation,

$$C : y^2 = x^3 + ax^2 + bx + c,$$

the Nagell-Lutz theorem tells us that points of finite order have integer coordinates; so it is natural to ask whether the converse is true. However, a little experimentation shows that points with integer coordinates need not have finite order. We saw one example in Chapter IV, Section 3, where we showed that the curve $y^2 = x^3 + 3$ has no points of finite order, but it clearly has the integer point $(1, 2)$. Similarly, it is easy to show that the curve $y^2 = x^3 + 17$ has no points of finite order, yet it has lots of integer points, including

$$(-2, \pm 3), \quad (-1, \pm 4), \quad (2, \pm 5), \quad (4, \pm 9), \quad (8, \pm 23),$$

and six other points which we will leave as an exercise for you to discover.

Let's think a little bit about how many integer points we would expect. If the rank of C is zero, then $C(\mathbb{Q})$ is finite, and the Nagell-Lutz theorem says that those finitely many points are integer points. This is the trivial case because if there are only finitely many rational points, then there are certainly only finitely many integer points.

The situation becomes much more interesting when the rank is greater than zero. Suppose, for example, that the rank is 1 and that there are no non-trivial points of finite order. Then we can choose a generator P for $C(\mathbb{Q})$, and every point in $C(\mathbb{Q})$ has the form nP for some integer n. Suppose that we look at the sequence of points $P, 2P, 3P, \ldots$ If we write $nP = (x_n, y_n)$, then for $n \geq 3$ our explicit formula for the group law says that x_n is given by the formula

$$x_n = \left(\frac{y_{n-1} - y_1}{x_{n-1} - x_1} \right)^2 - a - x_{n-1} - x_1.$$

So even if P and $(n-1)P$ have integer coordinates, there is no reason to expect that nP should have integer coordinates. In fact, looking at this formula, it seems quite unlikely that very many of the nP's will have integer coordinates.

This intuition turns out to be correct, although it is by no means easy to prove. The general result, which is due to Siegel [1,2], goes as follows.

Siegel's Theorem. *Let C be a non-singular cubic curve given by an equation $F(x, y) = 0$ with integer coefficients. Then C has only finitely many points with integer coordinates.*

One warning is in order. The curve C consists of the points satisfying $F(x, y) = 0$ together with one or more points at infinity. C must be non-singular at every point, including the points at infinity.

By way of contrast, we can compare Siegel's theorem with the situation for linear and quadratic equations. If a linear equation

$$ax + by = c, \qquad \text{with } a, b, c \in \mathbb{Z},$$

has a solution in integers (x_0, y_0), then it has infinitely many solutions given by the recipe

$$(x_0 + bn, y_0 - an), \qquad n \in \mathbb{Z}.$$

Further, every integer solution has this form.

Similarly, quadratic equations can have infinitely many solutions. For example, consider the equation

$$x^2 - 2y^2 = 1.$$

This clearly has the solution $(3, 2)$. But now suppose we have a solution (x, y). Then it is easy to check that $(3x + 4y, 2x + 3y)$ is also a solution. If we start with $(3, 2)$ and apply this procedure, we get a sequence of solutions

$$(3, 2), \quad (17, 12), \quad (99, 70), \quad (577, 408), \ldots.$$

Since the x and y coordinates are strictly increasing, there are infinitely many distinct solutions; and it is possible to show that this gives all solutions with $x, y > 0$. Those of you who have studied Pell's equation

$$x^2 - Dy^2 = 1$$

will recognize this special case and will know how to generate all solutions for $D = 2$ using the powers of $3 + 2\sqrt{2}$.

There are several different proofs of Siegel's theorem, none of them easy. In the next section we will consider one special case where the proof is very easy and will discuss some further interesting questions that arise. The remainder of the chapter will be devoted to another case of Siegel's theorem, due to Axel Thue in 1909, whose proof uses many of the tools needed to do the general case. The proof of Thue's theorem is complicated, but just as in the proof of Mordell's theorem, the proof can be broken down into several manageable steps.

The proofs of Siegel's theorem and Thue's theorem have one other thing in common with the proof of Mordell's theorem. Recall that although Mordell's theorem tells us that the group of rational points on C is finitely generated, it does not provide us with a sure-fire method of finding generators. Similarly, Siegel's and Thue's theorems tell us that the set of points with integer coordinates is finite, but they do not provide a way for us to find all of the integer points. In the 1930's, Skolem [1] came up with a new proof of Siegel's theorem which in practice often allows one to find all solutions; but it, too, is not guaranteed to work for every curve. Finally, in 1966, Baker [2] gave an effective method to find all solutions.

2. Taxicabs and Sums of Two Cubes

The title of this section may provoke some curiosity because this is the first time in this book that we have referred to methods of conveyance. The reference has to do with a famous mathematical story. When the brilliant Indian mathematician Ramanujan was in the hospital in London, his colleague G.H. Hardy came to visit. Hardy remarked that he had come in taxicab number 1729, and surely that was a rather dull number. Ramanujan instantly replied that, to the contrary, 1729 is a very interesting number. It is the smallest number expressible as the sum of two cubes in two different ways. Thus,

$$1729 = 9^3 + 10^3 = 1^3 + 12^3.$$

So the taxicab number 1729 gives a cubic curve

$$x^3 + y^3 = 1729$$

which has two integer points. Of course, we can switch x and y, so we end up with the four points

$$(9, 10), \quad (10, 9), \quad (1, 12), \quad (12, 1).$$

We claim that there are no other integer points. This is a special case of Siegel's theorem, but in this case the proof is easy because the cubic $x^3 + y^3$ factors.

Suppose that (x, y) satisfies $x^3 + y^3 = 1729$. Then

$$(x + y)(x^2 - xy + y^2) = 1729 = 7 \cdot 13 \cdot 19.$$

So we have just to consider all possible factorizations $1729 = AB$ and solve

$$x + y = A, \qquad x^2 - xy + y^2 = B.$$

Substituting $y = A - x$ into the second equation, we find

$$3x^2 - 3Ax + A^2 - B = 0;$$

so for each factorization $1729 = AB$ we need to check if

$$\frac{3A \pm \sqrt{12B - 3A^2}}{6}$$

is an integer. Doing this, we find that we only get integer solutions for the two pairs $(A, B) = (13, 133)$ and $(A, B) = (19, 91)$, and these lead to the four solutions listed above.

More generally, if we have any cubic equation which factors as

$$(ax + by + c)(dx^2 + exy + fy^2 + gx + hy + i) = j$$

with $j \neq 0$, then we see immediately that there are only finitely many integer solutions.[†] Merely look at all possible factorizations $j = AB$, solve the pair of equations

$$ax + by + c = A, \qquad dx^2 + exy + fy^2 + gx + hy + i = B,$$

and see what integer solutions arise. This might be called the trivial case of Siegel's theorem because it can be solved by an elementary factorization argument.

But there are still many interesting questions that we can ask even about an equation, such as $x^3 + y^3 = m$, for which Siegel's theorem is trivial. For example, we know that there are finitely many solutions, but

---

[†] Well, not quite. For example, a silly equation like $x^3 = 1$ will have infinitely many solutions because y is arbitrary. We need to assume that the equation is not of the form $n(ax + by + c)^3 = m$.

can we bound how large they are? Well, yes, we can do that rather easily. We know that the solutions satisfy

$$x + y = A \quad \text{and} \quad x^2 - xy + y^2 = B$$

for some factorization $AB = m$. Hence

$$m \geq |B| = |x^2 - xy + y^2| = \frac{3}{4}x^2 + \left(\frac{1}{2}x - y\right)^2 \geq \frac{3}{4}x^2,$$

so we see that $|x| \leq 2\sqrt{m/3}$. The same argument gives the same bound for $|y|$, which proves the following theorem for the "taxicab equation."

Proposition. *Let $m \geq 1$ be an integer. Then every solution to the equation*

$$x^3 + y^3 = m$$

in integers $x, y \in \mathbb{Z}$ satisfies

$$\max\{|x|, |y|\} \leq 2\sqrt{\frac{m}{3}}.$$

Next we might ask for the number of solutions. Ramanujan's observation was that for $1 \leq m \leq 1728$, the equation $x^3 + y^3 = m$ has at most one solution in positive integers. And for $m = 1729$, there are two solutions. Here we are counting (x, y) and (y, x) as the same solution. Now one might ask whether there is a value for m so that there are three solutions, and four solutions, and so on? The answer is that for any $N \geq 1$ we can find an m so that the equation $x^3 + y^3 = m$ has at least N solutions.

To do this, we begin with the observation that there are equations

$$x^3 + y^3 = m$$

which have infinitely many *rational* solutions. For example, consider the curve $x^3 + y^3 = 9$, which has the rational point $(2, 1)$. As we saw in Chapter I, Section 3, there is essentially a one-to-one correspondence between the rational points on $x^3 + y^3 = 9$ and the rational points on the curve $Y^2 = X^3 - 48$ given by the formulas

$$X = \frac{12}{x+y}, \quad Y = 12\frac{x-y}{x+y}.$$

The point $(1, 2)$ on $x^3 + y^3 = 9$ corresponds to the point $Q = (4, 4)$ on $Y^2 = X^3 - 48$. If we compute $2Q = (28, -148)$ and $3Q = \left(\frac{73}{9}, \frac{595}{27}\right)$, we see that Q has infinite order because points of finite order have integer coordinates. (That's part of the Nagell-Lutz theorem, which we proved in Chapter II.) Hence, both $Y^2 = X^3 - 48$ and $x^3 + y^3 = 9$ have infinitely many rational points.

Since there are infinitely many rational points on $x^3 + y^3 = 9$, we can certainly find N such points, say $P_1, \ldots, P_N$. If $P = \left(\dfrac{a}{b}, \dfrac{c}{d} \right)$ is any rational point written in lowest terms, then substituting into the equation and clearing denominators gives

$$a^3 d^3 + c^3 b^3 = 9 b^3 d^3.$$

Thus, b^3 divides $a^3 d^3$, and similarly d^3 divides $c^3 b^3$. But $\gcd(a, b) = 1$ and $\gcd(c, d) = 1$, so we find that $b^3 = \pm d^3$. If we write our fractions with positive denominators, we have $b = d$. Thus, we can write the coordinates of $P_1, \ldots, P_N$ as

$$P_1 = \left(\frac{a_1}{d_1}, \frac{c_1}{d_1} \right), \ldots, P_N = \left(\frac{a_N}{d_N}, \frac{c_N}{d_N} \right).$$

Note that $P_1, \ldots, P_N$ are rational points on the curve $x^3 + y^3 = 9$.

Now for the main idea. We choose an m which in essence clears the denominators of the P_i's, thereby making them into integer points. The m that we choose is

$$m = 9 (d_1 d_2 \cdots d_N)^3.$$

Then if we multiply the x and y coordinates of any P_i by $d_1 d_2 \cdots d_N$, we will get an integer point on $x^3 + y^3 = m$. In other words, let

$$P_i' = (d_1 \cdots d_{i-1} a_i d_{i+1} \cdots d_N, \ d_1 \cdots d_{i-1} c_i d_{i+1} \cdots d_N).$$

Then $P_1', \ldots, P_N'$ are integer points on the curve

$$x^3 + y^3 = 9 (d_1 d_2 \cdots d_N)^3.$$

This proves our assertion, which we restate as a formal proposition.

Proposition. *For every integer $N \geq 1$, there is an integer $m \geq 1$ so that the cubic curve*
$$x^3 + y^3 = m$$
has at least N points with integer coordinates.

Of course, this does not strictly generalize Ramanujan's observation because he referred only to sums of two positive cubes. However, it is not hard to prove that if the curve $x^3 + y^3 = m$ has infinitely many rational solutions and if $m > 0$, then there are infinitely many rational solutions with x and y both positive. The idea is that the set of real points on this curve looks like the circle group, so the subgroup generated by a point of infinite order must be dense in the set of real points. Since there are real points with $x, y > 0$, an open set of such points will contain infinitely many rational points with $x, y > 0$. So if you want, you can add the words "and with x and y both positive" on to the end of our proposition.

In some sense, this proposition is a satisfactory answer to our question of how many integer points a cubic curve can have. Nonetheless, it probably leaves you a bit uneasy because we really haven't found a lot of intrinsically integer points. Instead, we have found lots of rational points and cleared their denominators. So possibly a more natural question would be the following:

> Given an integer N, is it possible to find an integer $m \geq 1$ so that the equation $x^3 + y^3 = m$ has at least N integer solutions with $\gcd(x, y) = 1$ and $x > y$?

Amazingly enough, the answer to this question is not known, even for quite small values of N. For $N = 3$, the smallest value of m is 3242197, which has the representations

$$3242197 = 141^3 + 76^3 = 138^3 + 85^3 = 202^3 + (-171)^3.$$

And if we also require that the solution be in positive integers, as Ramanujan did, then the answer for $N = 3$ was discovered by Paul Vojta in 1983:

$$15170835645 = 2468^3 + 517^3 = 2456^3 + 709^3 = 2152^3 + 1733^3.$$

For $N = 4$, it is still not known if there is an m with four positive solutions!

To conclude this chapter, we want to discuss an interesting relationship between the number of integer points and the rank of the group of rational points on elliptic curves. Serge Lang has given a general conjecture which has been proven for certain special sorts of curves, including the curves we have looked at in this section, as described in the following theorem.

Theorem. (Silverman [1]) Let $m \geq 1$ be an integer, and let C_m be the cubic curve

$$C_m : x^3 + y^3 = m.$$

There is a constant $\kappa > 1$, <u>independent of m</u>, so that

$$\#\{(x, y) \in C_m(\mathbb{Q}) : x, y \in \mathbb{Z}, \gcd(x, y) = 1\} \leq \kappa^{1 + \operatorname{rank} C_m(\mathbb{Q})}.$$

This theorem says that the integer points with $\gcd(x, y) = 1$ tend to be somewhat linearly independent in the group of rational points. In particular, if one could find enough such points, then one would know that the rank was large.

You may recall we asked earlier whether there are cubic curves of arbitrarily high rank, and noted that this is still an open question. So one way to answer this question about large ranks would be to find a sequence of m's so that the number of integer points in $C_m(\mathbb{Q})$ with $\gcd(x, y) = 1$ goes to infinity.

3. Thue's Theorem and Diophantine Approximation

In the last section we saw that it is easy to find all integer solutions to the
equation $x^3 + y^3 = m$. The reason it is easy is that the polynomial $x^3 + y^3$
factors as $(x+y)(x^2 - xy + y^2)$; and so by taking the various factorizations
of m, we end up with two equations for the two unknowns x and y.

Suppose, instead, we take a polynomial that does not factor; for example, suppose we look at the equation

$$x^3 + 2y^3 = m.$$

It is not clear whether an equation of this sort might have infinitely many
solutions. Notice that for equations of degree 2, the equation $x^2 - y^2 = m$
has only finitely many solutions, whereas $x^2 - 2y^2 = m$ often has infinitely
many. So our experience with polynomials that factor is not necessarily a
good guide.

More generally, we will look at cubic equations of the form

$$ax^3 + by^3 = c.$$

It turns out that such an equation has only finitely many solutions in integers, regardless of whether or not it factors. In this section we will show
how to reduce this problem to a question of approximating a certain real
number by rational numbers. We will outline how to prove the approximation theorem that we need. The remainder of the chapter will then be
devoted to giving the details of the proof.

The theorem of Thue that we will prove is the following:

Theorem.   (Thue [1]) *Let a, b, c be non-zero integers. Then the equation*

$$ax^3 + by^3 = c$$

has only finitely many solutions in integers x, y.

One trivial observation is that if (x, y) is a solution to $ax^3 + by^3 = c$,
then (ax, y) is a solution to $X^3 + a^2bY^3 = a^2c$. So it is enough to prove
Thue's theorem with $a = 1$. A second observation is that by replacing y
by $-y$ and/or b by $-b$ if necessary, it is enough to look at the equation

$$x^3 - by^3 = c, \qquad \text{with } b, c \in \mathbb{Z}, \ b > 0, \ c > 0.$$

This is the equation that we will prove has only finitely many integer solutions.

The factorization techniques we used in the last section worked extremely well, so suppose we try to use them again. Of course, unless b is a
perfect cube, we cannot factor $x^3 - by^3$ over the rational numbers; we will
need to use a cube root of b. So we let

$$\beta = \sqrt[3]{b};$$

and then we can factor

$$x^3 - by^3 = (x - \beta y)(x^2 + \beta xy + \beta^2 y^2).$$

It is important to note that this is NOT a factorization into integers; so we cannot just factor c and get two equations for x and y.

However, what we observe is that if (x, y) is a solution to $x^3 - by^3 = c$ with x and y large, then the difference $|x - \beta y|$ must be quite small. This is true because

$$x^2 + \beta xy + \beta^2 y^2 = \left(x + \tfrac{1}{2}\beta y\right)^2 + \tfrac{3}{4}\beta^2 y^2 \geq \tfrac{3}{4}\beta^2 y^2.$$

Hence

$$|c| = |x^3 - by^3| = |x - \beta y| \cdot |x^2 + \beta xy + \beta^2 y^2| \geq |x - \beta y| \cdot \tfrac{3}{4}\beta^2 y^2.$$

So if we divide by $\tfrac{3}{4}\beta^2 |y^3|$, we obtain the important inequality

$$\left|\frac{x}{y} - \beta\right| \leq \frac{4|c|}{3\beta^2} \cdot \frac{1}{|y|^3}.$$

This inequality says that if (x, y) is an integer solution to the equation $x^3 - by^3 = c$ with $|y|$ large, then the rational number $\dfrac{x}{y}$ is extremely close to the irrational number $\beta = \sqrt[3]{b}$. In order to prove that there are only finitely many integer solutions, we will prove that there are only finitely many rational numbers with this approximation property. The study of estimates of this sort is called the *Theory of Diophantine Approximation*. We will prove the following Diophantine Approximation Theorem, which is due to Thue.

Diophantine Approximation Theorem.   (Thue [1]) *Let b be a positive integer which is not a perfect cube, and let $\beta = \sqrt[3]{b}$. Let C be a fixed positive constant. Then there are only finitely many pairs of integers (p, q) with $q > 0$ which satisfy the inequality*

$$\left|\frac{p}{q} - \beta\right| \leq \frac{C}{q^3}. \tag{$*$}$$

Assuming the truth of the Diophantine Approximation Theorem, how can we finish the proof that $x^3 - by^3 = c$ has only finitely many integer solutions? First, if b is a perfect cube, then $x^3 - by^3$ factors, so we can use the elementary argument from Section 2. Second, if we have a solution with $y = 0$, then $x^3 = c$, so there is at most one possible value for x. Third, for all solutions (x, y) with $y \neq 0$, we saw above that

$$\left|\frac{x}{y} - \beta\right| \leq \frac{C}{|y|^3} \qquad \text{with } C = \frac{4|c|}{3\beta^2}.$$

So from the Diophantine Approximation Theorem there are only finitely many possible pairs (x, y) with $y > 0$. To deal with solutions having $y < 0$, we rewrite our inequality as

$$\left| \frac{-x}{-y} - \beta \right| \le \frac{C}{|y|^3},$$

and then the Diophantine Approximation Theorem again shows that there are only finitely many pairs (x, y).

Now "all" that is left is to prove the Diophantine Approximation Theorem. To motivate the argument used in the actual proof, we are first going to describe an idea for the proof which does not quite work.

As above, we consider the factorization

$$x^3 - by^3 = (x - \beta y)(x^2 + \beta xy + \beta^2 y^2).$$

Suppose that $\frac{p}{q}$ satisfies the estimate $(*)$ in the Diophantine Approximation Theorem. Substituting $x = p$ and $y = q$ into our identity and dividing by q^3 yields

$$\frac{p^3 - bq^3}{q^3} = \left(\frac{p}{q} - \beta \right) \left(\frac{p^2}{q^2} + \beta \frac{p}{q} + \beta^2 \right).$$

We make two observations concerning this last equation. First, since b is not a perfect cube, the integer $p^3 - bq^3$ is not zero; hence,

$$\left| \frac{p^3 - bq^3}{q^3} \right| \ge \frac{1}{q^3}.$$

Second, from $(*)$ we have

$$\left| \frac{p}{q} \right| \le \beta + \frac{C}{q^3} \le \beta + C,$$

so

$$\left| \frac{p^2}{q^2} + \beta \frac{p}{q} + \beta^2 \right| \le (\beta + C)^2 + \beta(\beta + C) + \beta^2 \le C',$$

where we have written C' for the constant $3\beta^2 + 3\beta C + C^2$. The crucial fact is that C' depends only on β and C; it is the same for every choice of $\frac{p}{q}$.

Substituting these two inequalities into the equation given above, we have shown that there is a constant C' so that for every rational number $\frac{p}{q}$,

$$\left| \frac{p}{q} - \beta \right| \ge \frac{1}{C'q^3}. \qquad (**)$$

Recall we are trying to prove that for every constant C, there are only finitely many rational numbers $\dfrac{p}{q}$ which satisfy the inequality

$$\left| \frac{p}{q} - \beta \right| \leq \frac{C}{q^3}. \tag{$*$}$$

Comparing $(*)$ and $(**)$, we do not seem to have learned anything. (We can conclude that $C' \geq 1/C$, but in fact we already know that C' is fairly large.) The problem is that the bounds in both $(*)$ and $(**)$ involve a multiple of $\dfrac{1}{q^3}$.

Now there is nothing we can do about $(*)$; that's what we are trying to prove. But suppose that we could prove a stronger version of $(**)$ with some exponent less than 3, say for the sake of illustration that we could prove

$$\left| \frac{p}{q} - \beta \right| \geq \frac{1}{C'q^{2.9}}. \tag{$**)'$}$$

Then comparing $(*)$ with $(**)'$, we would find

$$\frac{1}{C'q^{2.9}} \leq \left| \frac{p}{q} - \beta \right| \leq \frac{C}{q^3},$$

and so

$$q \leq (CC')^{10}.$$

In other words, we would have shown that every solution to $(*)$ has its denominator bounded by a number $(CC')^{10}$ which depends only on C and b. Then $(*)$ would imply that the numerator is also bounded, so we could conclude that $(*)$ has only finitely many solutions.

How might we improve on $(**)$? Let's summarize how we proved $(**)$. We took the polynomial $f(X) = X^3 - b$ which has integer coefficients and β as a root. Evaluating $f(X)$ at p/q, we noted that $|f(p/q)|$ is no smaller than $1/q^3$. On the other hand, factoring $f(X)$, we saw that $|f(p/q)|$ equals $|p/q - \beta|$ times something that is bounded. Comparing the upper and lower bounds for $|f(p/q)|$ yielded $(**)$.

One way to improve $(**)$ might be to use some other polynomial in place of $X^3 - b$. More precisely, suppose that we find a polynomial $F(X)$ with integer coefficients which is divisible by $(X - \beta)^n$, where $n \geq 1$ is some (presumably large) integer. Then $F(X)$ factors as

$$F(X) = (X - \beta)^n G(X)$$

for some polynomial $G(X) \in \mathbb{R}[X]$, so just as above we can show that

$$\left| F\left(\frac{p}{q} \right) \right| \leq C'' \left| \frac{p}{q} - \beta \right|^n.$$

Here, C'' depends on C and on the polynomial $F(X)$, but it is the same for all $\frac{p}{q}$'s.

On the other hand, if $F(p/q) \neq 0$, then we immediately derive the lower bound

$$\left| F\left(\frac{p}{q}\right) \right| = \frac{|\text{non-zero integer}|}{q^d} \geq \frac{1}{q^d},$$

where d is the degree of F. Comparing the upper and lower bounds and taking n^{th} roots, we find

$$\left| \frac{p}{q} - \beta \right| \geq \frac{1}{\sqrt[n]{C''}} \cdot \frac{1}{q^{d/n}}.$$

So if $d < 3n$ (strict inequality), we will be done.

Unfortunately, we can show that d is always greater than or equal to $3n$. Here's why. We have assumed that $F(X)$ is divisible by $(X - \beta)^n$, where $\beta = \sqrt[3]{b}$. We have further assumed that $F(X)$ has integer coefficients. Since the minimal polynomial of β over $\mathbb{Q}$ is $X^3 - b$, it follows that $F(X)$ must be divisible by $(X^3 - b)^n$, from which we see that $\deg(F) \geq 3n$. So our first attempt to improve $(**)$ meets with failure.

Thue's brilliant idea, which enabled him to improve $(**)$, was to take a polynomial of two variables $F(X, Y) \in \mathbb{Z}[X, Y]$. He chose a polynomial which vanishes to high order at the point (β, β), and then he compared upper and lower bounds for the value $\left| F\left(\frac{p_1}{q_1}, \frac{p_2}{q_2}\right) \right|$, where $\frac{p_1}{q_1}$ and $\frac{p_2}{q_2}$ are certain rational numbers which satisfy $(*)$. So Thue's proof breaks down into three parts:

(1) Find a good polynomial $F(X, Y)$.

(2) Derive an upper bound for $\left| F\left(\frac{p_1}{q_1}, \frac{p_2}{q_2}\right) \right|$ in terms of $\left| \frac{p_1}{q_1} - \beta \right|$ and $\left| \frac{p_2}{q_2} - \beta \right|$.

(3) Derive a lower bound for $\left| F\left(\frac{p_1}{q_1}, \frac{p_2}{q_2}\right) \right|$. In particular, show that this value is not zero. This is technically the hardest part of the proof!

This description of the proof is certainly very sketchy. We will now describe each of the steps in more detail, leaving the actual proofs for the subsequent sections of this chapter. It is important to understand this outline of the proof before proceeding; otherwise, it is easy to get bogged down in the numerous details.

| Step I: Construction of an Auxiliary Polynomial |

We begin by constructing a polynomial $F(X, Y)$ with integer coefficients such that $F(X, Y)$ vanishes to very high order at the point (β, β). We will need to find such an F so that the coefficients of F are not too large.

Step II: The Auxiliary Polynomial Is Small

We assume that there are infinitely many pairs of integers (p, q) that satisfy the Diophantine inequality $(*)$, and we aim to derive a contradiction. Under this assumption, we can find a rational number p_1/q_1 satisfying $(*)$ and with q_1 quite large. Then we can find a second rational number p_2/q_2 satisfying $(*)$ with q_2 much larger than q_1. Having done this, we consider the value of the polynomial $F(X, Y)$ at the point $(p_1/q_1, p_2/q_2)$. Since $F(X, Y)$ vanishes to high order at (β, β), and since $(*)$ says that each p_i/q_i is close to β, we find that $F(p_1/q_1, p_2/q_2)$ is quite small.

Step III: The Auxiliary Polynomial Does Not Vanish

This is the subtlest part of the proof. We want to show that $F(p_1/q_1, p_2/q_2)$ is not zero. For the sake of this outline of the proof, we will suppose that we can show this. Then by writing the non-zero rational number

$$F\left(\frac{p_1}{q_1}, \frac{p_2}{q_2}\right) \quad \text{as} \quad \frac{\text{non-zero integer}}{q_1^d q_2^e},$$

we get the lower bound

$$\left| F\left(\frac{p_1}{q_1}, \frac{p_2}{q_2}\right) \right| \geq \frac{1}{q_1^d q_2^e}.$$

This lower bound will contradict the upper bound from Step II, thereby completing the proof of the theorem.

Unfortunately, there is one additional complication to the proof. In Step III, we will not actually be able to show that $F(p_1/q_1, p_2/q_2)$ is not zero. Instead, we will show that some derivative of F does not vanish at $(p_1/q_1, p_2/q_2)$. This means that in Step II we will need to give an upper bound for the values of derivatives of F. It is not hard to do this, so we hope that you will not be deterred by the small notational inconveniences this entails.

4. Construction of an Auxiliary Polynomial

In this section we are going to construct a polynomial $F(X, Y)$ with reasonably small integer coefficients and the property that F vanishes to high order at (β, β). The way that we will build F is by solving a system of linear equations with integer coefficients. Results which describe integer solutions to systems of linear equations are often named after Siegel because he was the first to formalize this procedure.

Siegel's Lemma. *Let $N > M$ be positive integers, and let*

$$a_{11}T_1 + \cdots + a_{1N}T_N = 0$$
$$\vdots \qquad \ddots \qquad \vdots \qquad \vdots$$
$$a_{M1}T_1 + \cdots + a_{MN}T_N = 0$$

be a system of linear equations with integer coefficients. Then there is a solution $(t_1, \ldots, t_N)$ to this system of equations with $t_1, \ldots, t_N$ integers, not all zero, and satisfying

$$\max_{1 \leq i \leq N} |t_i| < 2\left(4N \max_{\substack{1 \leq i \leq M \\ 1 \leq j \leq N}} |a_{ij}|\right)^{\frac{M}{N-M}}.$$

The statement of this lemma looks complicated, but it is really saying something very easy. The system of homogeneous linear equations has more variables than equations, so we know it has solutions. Since the coefficients are integers, there will be rational solutions; and by clearing the denominators of the rational solutions, we can find integer solutions. So it is obvious that there are non-zero integer solutions. The last part of the lemma then says that we can find some solution which is not too large; precisely, we can find a solution whose size is bounded in terms of the number of equations M, the number of variables N, and the size of the coefficients a_{ij}. This, too, is not surprising; so the real content of the lemma is the precise form of the bound.

PROOF. (Of Siegel's Lemma) For any vector $\mathbf{t} = (t_1, \ldots, t_N)$ with integer coordinates, we let

$$\|\mathbf{t}\| = \max_{1 \leq i \leq N} |t_i|$$

be the largest of the absolute values of its coordinates. Similarly, we will let A be the matrix

$$A = \begin{pmatrix} a_{11} & \cdots & a_{1N} \\ \vdots & \ddots & \vdots \\ a_{M1} & \cdots & a_{MN} \end{pmatrix} \quad \text{and} \quad \|A\| = \max_{\substack{1 \leq i \leq M \\ 1 \leq j \leq N}} |a_{ij}|.$$

So the statement of Siegel's lemma is that the equation $A\mathbf{t} = \mathbf{0}$ has a vector solution $\mathbf{t} \neq \mathbf{0}$ with integer coordinates satisfying

$$\|\mathbf{t}\| < 2\left(4N\|A\|\right)^{M/(N-M)}.$$

If $\mathbf{t} = (t_1, \ldots, t_N)$ is any vector, we can estimate the size of the vector

$$A\mathbf{t} = \left(\sum_{j=1}^{N} a_{1j}t_j, \ldots, \sum_{j=1}^{N} a_{Mj}t_j\right)$$

as follows. The ith coordinate of At satisfies the inequality

$$\left| \sum_{j=1}^{N} a_{ij} t_j \right| \leq \sum_{j=1}^{N} |a_{ij} t_j|$$

$$\leq N \left(\max_{1 \leq j \leq N} |a_{ij}| \right) \left(\max_{1 \leq j \leq N} |t_j| \right)$$

$$\leq N \|A\| \cdot \|t\|.$$

Hence

$$\|At\| \leq N \|A\| \cdot \|t\|.$$

Thus if t is a vector with size $\|t\| \leq H$, then its image At has size $\|At\| \leq N\|A\|H$. In particular, multiplication by the matrix A maps the set of integer vectors

$$T_H = \big\{ t = (t_1, \ldots, t_N) : t_i \in \mathbb{Z}, \ \|t\| \leq H \big\}$$

into the set of integer vectors

$$U_H = \big\{ u = (u_1, \ldots, u_M) : u_i \in \mathbb{Z}, \ \|u\| \leq N\|A\|H \big\}.$$

We claim that if H is large enough, then T_H will have more elements than U_H, and so there will be two elements of T_H with the same image in U_H. (This is the famous "pigeonhole principle." Here T_H is the set of pigeons, U_H is the set of pigeonholes, and multiplication by the matrix A is the rule we use to assign each pigeon to a pigeonhole.)

How many elements are there in T_H and U_H? Each vector in T_H has N coordinates, and each coordinate is an integer satisfying $-H \leq t_i \leq H$. So

$$\#T_H = \big(2[H] + 1 \big)^N,$$

where $[H]$ is the greatest integer in H. Similarly,

$$\#U_H = \big(2[N\|A\|H] + 1 \big)^M.$$

Since $N > M$, we see that $\#T_H$ will be larger than $\#U_H$ provided H is large enough; but we need to be more precise. If we assume that $H \geq 1$, then

$$\#T_H \geq \big(2(H-1) + 1 \big)^N = (2H - 1)^N \geq H^N;$$

and similarly

$$\#U_H \leq (2N\|A\|H + 1)^M \leq (3N\|A\|H)^M.$$

Combining these two estimates, we find that:

$$\text{If} \quad H > (3N\|A\|)^{\frac{M}{N-M}}, \quad \text{then} \quad \#T_H > \#U_H.$$

Now we can finish the proof of Siegel's lemma. Let

$$H = \left(4N\|A\|\right)^{\frac{M}{N-M}}.$$

Then T_H has more elements than U_H. Since multiplication by the matrix A sends elements of T_H to elements of U_H, there must be two distinct vectors $\mathbf{t}', \mathbf{t}'' \in T_H$ with the same image $A\mathbf{t}' = A\mathbf{t}''$. Then $\mathbf{t} = \mathbf{t}' - \mathbf{t}''$ satisfies

$$A\mathbf{t} = \mathbf{0}, \text{ and}$$

$$\|\mathbf{t}\| = \|\mathbf{t}' - \mathbf{t}''\| \leq \|\mathbf{t}'\| + \|\mathbf{t}''\| \leq 2H = 2\left(4N\|A\|\right)^{\frac{M}{N-M}}.$$

Further, $\mathbf{t} \neq \mathbf{0}$ because $\mathbf{t}'$ and $\mathbf{t}''$ are distinct. Hence, the vector $\mathbf{t}$ has all of the desired properties, which completes the proof of Siegel's lemma. $\square$

Now we are ready to construct our auxiliary polynomial. We recall that $b > 0$ is a fixed integer, and $\beta = \sqrt[3]{b}$. We let n be a large positive integer that we will specify later. (For those who are curious, the value that we eventually choose for n will depend on b and on the two rational numbers p_1/q_1 and p_2/q_2 which are close to β.) Then we let m be the integer satisfying

$$m \leq \frac{2}{3}n < m+1.$$

(That is, m is the greatest integer in $\frac{2}{3}n$.) We are going to construct a non-zero polynomial

$$F(X,Y) = P(X) + YQ(X)$$

with integer coefficients so that $P(X)$ and $Q(X)$ have degree at most $m+n$ and so that $F(X,\beta)$ is divisible by $(X - \beta)^n$. We will also need to keep track of the size of the coefficients of F.

It is convenient to use superscripts to denote differentiation with respect to X. However, if the integer coefficients of a polynomial have a common factor, we will want to cancel the common factor out. For example, let's look at the simple polynomial $f(x) = x^n$ and differentiate it k times. We obtain $n(n-1)\cdots(n-k+1)x^{n-k}$. If we multiply and divide by $(n-k)!$, then we see that

$$\frac{d^k(x^n)}{dx^k} = \frac{n!}{(n-k)!}x^{n-k}.$$

You have probably already noticed that the ratio $\frac{n!}{(n-k)!}$ is always divisible by $k!$; in fact, the quantity $\frac{n!}{(n-k)!k!}$ is exactly the binomial coefficient $\binom{n}{k}$ giving the number of ways of choosing k objects out of a set of n objects.

This shows that when we take the kth derivative of any polynomial, we can divide every coefficient by $k!$ and still have integer coefficients. So we define $F^{(k)}(X, Y)$ to be $1/k!$ times the kth derivative of $F(X, Y)$ with respect to X:

$$F^{(k)}(X, Y) = \frac{1}{k!} \frac{\partial^k}{\partial X^k} F(X, Y) = \frac{1}{k!} \left(\frac{d^k P(X)}{dX^k} + Y \frac{d^k Q(X)}{dX^k} \right).$$

Then $F^{(k)}(X, Y)$ will have integer coefficients if $F(X, Y)$ does.

The condition that $F(X, \beta)$ be divisible by $(X - \beta)^n$ is equivalent to its first $n - 1$ derivatives vanishing at $X = \beta$, so we will choose the coefficients of $F(X, Y)$ so as to force

$$F(\beta, \beta) = F^{(1)}(\beta, \beta) = \cdots = F^{(n-1)}(\beta, \beta) = 0.$$

We will write

$$P(X) = \sum_{i=0}^{m+n} u_i X^i \quad \text{and} \quad Q(X) = \sum_{i=0}^{m+n} v_i X^i.$$

Then

$$F^{(k)}(X, Y) = \sum_{i=k}^{m+n} \binom{i}{k} (u_i X^{i-k} + v_i X^{i-k} Y);$$

so

$$F^{(k)}(\beta, \beta) = \sum_{i=k}^{m+n} \binom{i}{k} (u_i \beta^{i-k} + v_i \beta^{i-k+1})$$

$$= \sum_{i=0}^{m+n-k} \binom{i+k}{k} \beta^i u_{i+k} + \sum_{i=1}^{m+n-k+1} \binom{i+k-1}{k} \beta^i v_{i+k-1}$$

$$= \sum_{i=0}^{m+n-k+1} \left\{ \binom{i+k}{k} \beta^i u_{i+k} + \binom{i+k-1}{k} \beta^i v_{i+k-1} \right\}.$$

For this last equation we make the convention that $u_i = v_i = 0$ if either $i < 0$ or $i > m + n$.

Our goal is to choose the u_i's and v_i's so that this last quantity vanishes for all $0 \leq k < n$. We can simplify matters a bit by recalling that $\beta^3 = b$, so every power β^i is an integer times one of 1, β, or β^2. We write $i = 3j + \ell$, and break the last sum up into a double sum over j and ℓ.

$$F^{(k)}(\beta, \beta) = \sum_{\ell=0}^{2} \left\{ \sum_{j} \binom{3j + \ell + k}{k} b^j u_{3j+\ell+k} \right.$$

$$\left. + \binom{3j + \ell + k - 1}{k} b^j v_{3j+\ell+k-1} \right\} \beta^\ell.$$

The quantity in braces will be always be an integer. On the other hand, 1, β, and β^2 are linearly independent over $\mathbb{Q}$. (That is, if we have a sum $A + B\beta + C\beta^2 = 0$ with $A, B, C \in \mathbb{Q}$, then necessarily $A = B = C = 0$.) So we are forced to choose the u_i's and v_i's so that

$$\sum_j \binom{3j + \ell + k}{k} b^j u_{3j+\ell+k} + \binom{3j + \ell + k - 1}{k} b^j v_{3j+\ell+k-1} = 0$$

for every $\ell \in \{0, 1, 2\}$ and every $k \in \{0, 1, \ldots, n-1\}$.

Although our equations are rather messy, the astute reader will see that we are in exactly the right situation to apply Siegel's lemma. We have $3n$ homogeneous linear equations (one for each $0 \le \ell \le 2$ and $0 \le k < n$) in the $2(m + n + 1)$ variables $\{u_0, \ldots, u_{m+n}, v_0, \ldots, v_{m+n}\}$. Further, these equations have integer coefficients. So Siegel's lemma tells us that there is a non-zero solution in integers satisfying

$$\max_{0 \le i \le m+n} \{|u_i|, |v_i|\} \le 2\big(4 \cdot 2(m + n + 1) \cdot \mu\big)^{\frac{3n}{2(m+n+1)-3n}}.$$

Here we have let μ denote the largest coefficient in the equations, which we now need to estimate.

First we observe that

$$\binom{N}{M} \le 2^N \qquad \text{for any integers } N, M \ge 0.$$

[This is immediate once we notice that the binomial coefficient $\binom{N}{M}$ is one of the terms in the binomial expansion of $(1+1)^N$.] Hence,

$$\max_{\substack{j,\ell,k \\ 0 \le 3j+\ell \le m+n \\ 0 \le k < n}} \binom{3j + \ell + k}{k} b^j \le \max_{\substack{0 \le i \le m+n \\ 0 \le k < n}} 2^{i+k} b^{i/3}$$

$$= 2^{m+2n-1} b^{(m+n)/3}$$

$$< (4b)^{m+n}.$$

For the other part of our upper bound for $\max\{|u_i|, |v_i|\}$ we can use the very coarse estimate

$$4 \cdot 2(m + n + 1) \le 2^{m+n+3} \le 4^{m+n}.$$

(We will assume $m \ge 3$.) Putting all of this together gives

$$\max_{0 \le i \le m+n} \{|u_i|, |v_i|\} \le 2 \cdot \big((16b)^{m+n}\big)^{\frac{3n}{2(m+n+1)-3n}}.$$

We can also simplify the exponent. Since m satisfies $m + 1 > \frac{2}{3}n$, we find that

$$\frac{3n}{2(m+n+1) - 3n} = \frac{3}{2\frac{m+1}{n} - 1} \le 9.$$

Substituting this into the above estimate for $\max\{|u_i|, |v_i|\}$, we find we have proven the following result, which was the main goal of this section.

Auxiliary Polynomial Theorem.   *Let b be an integer, let $\beta = \sqrt[3]{b}$, and let m, n be integers satisfying*

$$m + 1 > \frac{2}{3}n \geq m \geq 3.$$

Then there is a non-zero polynomial with integer coefficients

$$F(X, Y) = P(X) + Q(X)Y = \sum_{i=0}^{m+n} u_i X^i + v_i X^i Y$$

having the following two properties:

$$F^{(k)}(\beta, \beta) = 0 \qquad \text{for all } 0 \leq k < n. \tag{i}$$

$$\max_{0 \leq i \leq m+n} \left\{ |u_i|, |v_i| \right\} \leq 2 \cdot (16b)^{9(m+n)}. \tag{ii}$$

Example. Although the computations in this section have been somewhat complicated, we would like to demonstrate with an example that the entire argument is really just an easy exercise in linear algebra. We will take

$$n = 5 \qquad \text{and} \qquad m = 3,$$

and we will construct an auxiliary polynomial for the irrational number

$$\beta = \sqrt[3]{2}.$$

The first thing we do is write out

$$F(X, Y) = \sum_{i=0}^{8} u_i X^i + v_i X^i Y.$$

Next we take the kth derivative with respect to X, divide by $k!$, and evaluate at $(X, Y) = (\sqrt[3]{2}, \sqrt[3]{2})$. We need to do this for each $k = 0, 1, 2, 3, 4$ and set the resulting quantities equal to 0. This gives equations (remember that $\beta^3 = 2$)

$$0 = \ F(\beta, \beta) \ = u_0 + v_0\beta + \ u_1\beta + \ v_1\beta^2 + \cdots + \ 4u_8\beta^2 + \ \ 8v_8$$

$$0 = F^{(1)}(\beta, \beta) = u_1 + v_1\beta + 2u_2\beta + 2v_2\beta^2 + \cdots + \ 32u_8\beta + \ 32v_8\beta^2$$

$$\vdots \qquad \qquad \vdots$$

$$0 = F^{(4)}(\beta, \beta) = u_4 + v_4\beta + 5u_5\beta + 5v_5\beta^2 + \cdots + 140u_8\beta + 140v_8\beta^2$$

Now in each of these equations we group the terms according to the power of β they contain. For example, the last equation is

$$0 = F^{(4)}(\beta, \beta) = \{u_4 + 70u_7 + 30v_6\} + \{5u_5 + 140u_8 + v_4 + 70v_7\}\beta$$
$$+ \{15u_6 + 5v_5 + 140v_8\}\beta^2.$$

Since we want the u_i's and the v_i's to be integers, the only way to force this quantity to be zero is to make each of the expressions in brackets equal to zero. Doing this for each of the $F^{(k)}(\beta,\beta)$'s gives us the following 15 linear equations in the 18 variables $\{u_0,\ldots,u_8,v_0,\ldots,v_8\}$, which we write in matrix form.

$$
\begin{pmatrix}
1 & 0 & 0 & 2 & 0 & 0 & 4 & 0 & 0 & 0 & 0 & 2 & 0 & 0 & 4 & 0 & 0 & 8 \\
0 & 1 & 0 & 0 & 2 & 0 & 0 & 4 & 0 & 1 & 0 & 0 & 2 & 0 & 0 & 4 & 0 & 0 \\
0 & 0 & 1 & 0 & 0 & 2 & 0 & 0 & 4 & 0 & 1 & 0 & 0 & 2 & 0 & 0 & 4 & 0 \\
0 & 1 & 0 & 0 & 8 & 0 & 0 & 28 & 0 & 0 & 0 & 0 & 6 & 0 & 0 & 24 & 0 & 0 \\
0 & 0 & 2 & 0 & 0 & 10 & 0 & 0 & 32 & 0 & 1 & 0 & 0 & 8 & 0 & 0 & 28 & 0 \\
0 & 0 & 0 & 3 & 0 & 0 & 12 & 0 & 0 & 0 & 0 & 2 & 0 & 0 & 10 & 0 & 0 & 32 \\
0 & 0 & 1 & 0 & 0 & 20 & 0 & 0 & 112 & 0 & 0 & 0 & 0 & 12 & 0 & 0 & 84 & 0 \\
0 & 0 & 0 & 3 & 0 & 0 & 30 & 0 & 0 & 0 & 0 & 1 & 0 & 0 & 20 & 0 & 0 & 112 \\
0 & 0 & 0 & 0 & 6 & 0 & 0 & 42 & 0 & 0 & 0 & 0 & 3 & 0 & 0 & 30 & 0 & 0 \\
0 & 0 & 0 & 1 & 0 & 0 & 40 & 0 & 0 & 0 & 0 & 0 & 0 & 0 & 20 & 0 & 0 & 224 \\
0 & 0 & 0 & 0 & 4 & 0 & 0 & 70 & 0 & 0 & 0 & 1 & 0 & 0 & 40 & 0 & 0 & 0 \\
0 & 0 & 0 & 0 & 0 & 10 & 0 & 0 & 112 & 0 & 0 & 0 & 0 & 4 & 0 & 0 & 70 & 0 \\
0 & 0 & 0 & 0 & 1 & 0 & 0 & 70 & 0 & 0 & 0 & 0 & 0 & 0 & 30 & 0 & 0 & 0 \\
0 & 0 & 0 & 0 & 0 & 5 & 0 & 0 & 140 & 0 & 0 & 0 & 0 & 1 & 0 & 0 & 70 & 0 \\
0 & 0 & 0 & 0 & 0 & 0 & 15 & 0 & 0 & 0 & 0 & 0 & 0 & 0 & 5 & 0 & 0 & 140
\end{pmatrix}
\begin{pmatrix} u_0 \\ u_1 \\ u_2 \\ u_3 \\ u_4 \\ u_5 \\ u_6 \\ u_7 \\ u_8 \\ v_0 \\ v_1 \\ v_2 \\ v_3 \\ v_4 \\ v_5 \\ v_6 \\ v_7 \\ v_8 \end{pmatrix}
=
\begin{pmatrix} 0 \\ 0 \\ 0 \\ 0 \\ 0 \\ 0 \\ 0 \\ 0 \\ 0 \\ 0 \\ 0 \\ 0 \\ 0 \\ 0 \\ 0 \end{pmatrix}
$$

This set of equations has three free variables. Using standard row reduction, we can solve for all of the variables in terms of v_6, v_7, and v_8.

$$
\begin{aligned}
& u_0 = -8v_8 && u_1 = -4v_6 && u_2 = -4v_7 \\
& u_3 = -64v_8 && u_4 = -5v_6 && u_5 = -5v_7 \\
& u_6 = -20v_8 && u_7 = -\tfrac{5}{14}v_6 && u_8 = -\tfrac{5}{14}v_7 \\
& v_0 = \tfrac{10}{7}v_6 && v_1 = \tfrac{10}{7}v_7 && v_2 = 40v_8 \\
& v_3 = 5v_6 && v_4 = 5v_7 && v_5 = 32v_8.
\end{aligned}
$$

To find an auxiliary polynomial (for $n = 5$), it only remains to choose particular values for v_6, v_7, v_8. We want the coefficients of $F(X,Y)$ to be integers, so we will need to make sure that v_6 and v_7 are divisible by 14. Also, we want the coefficients of $F(X,Y)$ to be as small as possible. We list some natural choices for v_6, v_7, v_8 and the corresponding auxiliary polynomial $F(X,Y)$:

v_6	v_7	v_8	$F(X,Y)$
14	0	0	$-56X - 70X^4 - 5X^7 + 20Y + 70X^3Y + 14X^6Y$
0	14	0	$-56X^2 - 70X^5 - 5X^8 + 20XY + 70X^4Y + 14X^7Y$
0	0	1	$-8 - 64X^3 - 20X^6 + 40X^2Y + 32X^5Y + X^8Y$

It is now easy take derivatives and directly verify that each of these polynomials satisfies

$$
\frac{\partial^k F}{\partial X^k}\left(\sqrt[3]{2}, \sqrt[3]{2}\right) = 0 \qquad \text{for all } k = 0, 1, 2, 3, 4.
$$

Taking the last polynomial in our table, we find that there is an auxiliary polynomial for $b = 2$, $n = 5$, and $m = 3$ whose largest coefficient has magnitude 64. It is instructive to compare this with our theorem, which tells us that there is an auxiliary polynomial whose coefficients are no larger than

$$2 \cdot (16b)^{9(m+n)} = 2 \cdot 32^{72} \approx 4.7 \cdot 10^{108}.$$

It almost seems superfluous to point out that the estimate provided by our Auxiliary Polynomial Theorem is far from the best possible such result!

Later we are going to use the auxiliary polynomial $F(X, Y)$ by evaluating it at rational numbers which are close to $\sqrt[3]{2}$. On the one hand, we expect such a value to be very small because $F(X, Y)$ vanishes to very high order at $(\sqrt[3]{2}, \sqrt[3]{2})$; whereas on the other hand we will be able to show that the value cannot be too small. For example, suppose we take the auxiliary polynomial

$$F(X, Y) = -8 - 64X^3 - 20X^6 + 40X^2 Y + 32X^5 Y + X^8 Y$$

found above. The rational numbers

$$\frac{29}{23} = 1.2608\ldots \qquad \text{and} \qquad \frac{635}{504} = 1.2599206\ldots$$

are quite close to

$$\sqrt[3]{2} = 1.2599210\ldots.$$

And sure enough, we find that

$$F\left(\frac{29}{23}, \frac{635}{504}\right) = -\frac{2816387629}{23^8 \cdot 504} \approx -0.0000714$$

is quite small. This last calculation serves to illustrate the Smallness Theorem which we will prove in the next section.

5. The Auxiliary Polynomial Is Small

The auxiliary polynomial $F(X, Y)$ we constructed in the last section vanishes to very high order at the point (β, β). So if p_1/q_1 and p_2/q_2 are two rational numbers which are close to β, then we would expect $F(p_1/q_1, p_2/q_2)$ to be quite small. This is indeed the case, as we will now prove.

Smallness Theorem. *Let $F(X, Y)$ be an auxiliary polynomial as described in the Auxiliary Polynomial Theorem. Then there is a constant $c_1 > 0$, depending only on b, so that for any real numbers x, y such that $|x - \beta| \le 1$ and any integer $0 \le t < n$, we have*

$$\left| F^{(t)}(x, y) \right| \le c_1^n \left\{ |x - \beta|^{n-t} + |y - \beta| \right\}.$$

(N.B. It is essential that c_1 depend only on b, and not on n or t or F.)

PROOF. We know that many of the partial derivatives of $F(X,Y)$ vanish at (β, β). We can exploit this fact by writing out the Taylor series expansion of F around the point (β, β). Since Y appears only to the first power in $F(X,Y) = P(X) + YQ(X)$, we find

$$F(X,Y) = \sum_{k,j} \frac{1}{k!j!} \cdot \frac{\partial^{k+j}F}{\partial X^k \partial Y^j}(\beta, \beta) \cdot (X-\beta)^k(Y-\beta)^j$$

$$= \sum_{k=0}^{m+n} F^{(k)}(\beta, \beta)(X-\beta)^k + \sum_{k=0}^{m+n} Q^{(k)}(\beta)(X-\beta)^k(Y-\beta).$$

Now we use the fact that $F^{(k)}(\beta, \beta) = 0$ for all $0 \le k < n$, which means that the first sum starts with the $k = n$ term. Thus,

$$F(X,Y) = \sum_{k=n}^{m+n} F^{(k)}(\beta, \beta)(X-\beta)^k + \sum_{k=0}^{m+n} Q^{(k)}(\beta)(X-\beta)^k(Y-\beta).$$

We really want to estimate $F^{(t)}(x,y)$, so we differentiate t times with respect to X and divide by $t!$. This yields

$$F^{(t)}(X,Y) = \sum_{k=n}^{m+n} F^{(k)}(\beta, \beta)\binom{k}{t}(X-\beta)^{k-t}$$

$$+ \sum_{k=0}^{m+n} Q^{(k)}(\beta)\binom{k}{t}(X-\beta)^{k-t}(Y-\beta)$$

$$= \left\{ \sum_{k=n}^{m+n} F^{(k)}(\beta, \beta)\binom{k}{t}(X-\beta)^{k-n} \right\} \cdot (X-\beta)^{n-t}$$

$$+ \left\{ \sum_{k=0}^{m+n} Q^{(k)}(\beta)\binom{k}{t}(X-\beta)^{k-t} \right\} \cdot (Y-\beta).$$

This last formula reveals the reason we have done this computation. If we substitute values for X and Y that are close to β, then the last expression will be small, due to the presence of the factors $(X-\beta)^{n-t}$ and $(Y-\beta)$.

So now we put $X = x$ and $Y = y$, take the absolute value of both sides, and use the triangle inequality. We find

$$\left| F^{(t)}(x,y) \right| \le \left\{ \sum_{k=n}^{m+n} \left| F^{(k)}(\beta, \beta) \right| \binom{k}{t} |x-\beta|^{k-n} \right\} \cdot |x-\beta|^{n-t}$$

$$+ \left\{ \sum_{k=0}^{m+n} \left| Q^{(k)}(\beta) \right| \binom{k}{t} |x-\beta|^{k-t} \right\} \cdot |y-\beta|.$$

$$(**)$$

Compare this estimate with the estimate we are trying to prove. All that remains is to show that the quantities in braces are bounded by c_1^n for some constant c_1 depending only on b.

First, we observe that for any integer $k \leq m+n$ and any exponent $e \geq 0$,

$$\binom{k}{t}|x - \beta|^e \leq 2^{m+n} \qquad \text{because } |x - \beta| \leq 1.$$

Next, we write $F(X,Y) = P(X) + Q(X)Y = \sum u_i X^i + v_i X^i Y$ as usual. Then we can estimate

$$\left|F^{(k)}(\beta,\beta)\right| = \left|\sum_{i=k}^{m+n} \binom{i}{k}(u_i\beta^{i-k} + v_i\beta^{i-k+1})\right|$$

$$\leq (m+n+1) \cdot \max_{0 \leq i \leq m+n}\binom{i}{k} \cdot 2\max_{0 \leq i \leq m+n}\{|u_i|,|v_i|\} \cdot \beta^{m+n}$$

$$\leq 2^{m+n} \cdot 2^{m+n} \cdot 2 \cdot 2(16b)^{9(m+n)} \cdot b^{(m+n)/3}$$

$$= 4(2^{38}b^{28/3})^{m+n}.$$

Notice we have made use of the upper bound for the coefficients of F provided by the Auxiliary Polynomial Theorem.

This allows us to bound the first sum in braces in (**) by

$$\sum_{k=n}^{m+n}\left|F^{(k)}(\beta,\beta)\right|\binom{k}{t}|x-\beta|^{k-n} \leq (m+1)\cdot 4(2^{38}b^{28/3})^{m+n}\cdot 2^{m+n}$$

$$\leq (2^{42}b^{28/3})^{m+n}$$

$$\leq (2^{70}b^{140/9})^n \qquad \text{because } m \leq \tfrac{2}{3}n.$$

A similar calculation gives a bound for $Q^{(k)}(\beta)$:

$$\left|Q^{(k)}(\beta)\right| = \left|\sum_{i=k}^{m+n}\binom{i}{k}v_i\beta^{i-k}\right|$$

$$\leq (m+n+1)\cdot 2^{m+n}\cdot \max_{0 \leq i \leq m+n}|v_i|\cdot\beta^{m+n}$$

$$\leq 2^{2(m+n)}\cdot 2(16b)^{9(m+n)}\cdot b^{(m+n)/3}$$

$$= 2(2^{38}b^{28/3})^{m+n}.$$

And then the second sum in braces in (**) is bounded by

$$\sum_{k=0}^{m+n}\left|Q^{(k)}(\beta)\right|\binom{k}{t}|x-\beta|^{k-t} \leq (m+n+1)\cdot 2(2^{38}b^{28/3})^{m+n}\cdot 2^{m+n}$$

$$\leq (2^{41}b^{28/3})^{m+n}$$

$$\leq (2^{205/3}b^{140/9})^n.$$

We now have upper bounds for both of the bracketed expressions in (∗∗). Substituting these bounds into (∗∗) gives

$$\left|F^{(t)}(x,y)\right| \le (2^{70}b^{140/9})^n|x-\beta|^{n-t} + (2^{205/3}b^{140/9})^n|y-\beta|$$
$$\le c_1^n\{|x-\beta|^{n-t} + |y-\beta|\},$$

where we may take $c_1 = 2^{70}b^{140/9}$. This is precisely the estimate we have been aiming to prove. $\qquad\square$

6. The Auxiliary Polynomial Does Not Vanish

In the last section we showed that an auxiliary polynomial $F(X,Y)$ will be small if it is evaluated at a point close to (β,β). In this section we would like to show that if the point has rational coordinates, then the auxiliary polynomial does not vanish. Unfortunately, we are not able to prove such a strong result. Instead, we will show that some derivative $F^{(t)}(X,Y)$ does not vanish with t not too large.

Non-Vanishing Theorem. *Let $F(X,Y)$ be an auxiliary polynomial as described in the Auxiliary Polynomial Theorem. Let p_1/q_1 and p_2/q_2 be rational numbers in lowest terms. There is a constant $c_2 > 0$, depending only on b, and an integer*

$$0 \le t \le 1 + \frac{c_2 n}{\log q_1},$$

so that

$$F^{(t)}\left(\frac{p_1}{q_1}, \frac{p_2}{q_2}\right) \ne 0.$$

[N.B. As always, it is crucial that the constant c_2 depends only on b.]

PROOF. We write $F(X,Y) = P(X) + YQ(X)$ as usual. We are going to look at the *Wronskian polynomial $W(X)$* defined by

$$W(X) = \det\begin{vmatrix} P(X) & Q(X) \\ P'(X) & Q'(X) \end{vmatrix} = P(X)Q'(X) - Q(X)P'(X).$$

Why is the Wronskian a natural object to look at?

We are looking for some derivative of $F(X,Y)$ which does not vanish at $(p_1/q_1, p_2/q_2)$. Suppose, for example, that we are unlucky and find that F and its first derivative both vanish:

$$F\left(\frac{p_1}{q_1}, \frac{p_2}{q_2}\right) = 0 \quad \text{and} \quad F^{(1)}\left(\frac{p_1}{q_1}, \frac{p_2}{q_2}\right) = 0.$$

This means that

$$P\left(\frac{p_1}{q_1}\right) + Q\left(\frac{p_1}{q_1}\right)\frac{p_2}{q_2} = 0,$$

$$P'\left(\frac{p_1}{q_1}\right) + Q'\left(\frac{p_1}{q_1}\right)\frac{p_2}{q_2} = 0.$$

Eliminating p_2/q_2 from these two equations, we obtain

$$W\left(\frac{p_1}{q_1}\right) = P\left(\frac{p_1}{q_1}\right)Q'\left(\frac{p_1}{q_1}\right) - Q\left(\frac{p_1}{q_1}\right)P'\left(\frac{p_1}{q_1}\right) = 0.$$

So the Wronskian appears naturally when we assume that some derivatives $F^{(t)}(p_1/q_1, p_2/q_2)$ vanish and try to draw conclusions.

Now we will work more generally. Suppose that we let T be the largest integer such that

$$F^{(t)}\left(\frac{p_1}{q_1},\frac{p_2}{q_2}\right) = P^{(t)}\left(\frac{p_1}{q_1}\right) + Q^{(t)}\left(\frac{p_1}{q_1}\right)\frac{p_2}{q_2} = 0 \qquad \text{for all } 0 \le t < T.$$

Our goal is to show that T cannot be too large.

If we take pairs of these equations and eliminate p_2/q_2 from them, we get relations

$$P^{(t)}\left(\frac{p_1}{q_1}\right)Q^{(s)}\left(\frac{p_1}{q_1}\right) - P^{(s)}\left(\frac{p_1}{q_1}\right)Q^{(t)}\left(\frac{p_1}{q_1}\right) = 0 \qquad \text{for all } 0 \le s,t < T.$$

We can relate this to the Wronskian by differentiating $W(X)$. Thus

$$W^{(r)}(X) = \sum_{i+j=r} \frac{i!(j+1)!}{r!}\{P^{(i)}(X)Q^{(j+1)}(X) - Q^{(i)}(X)P^{(j+1)}(X)\}.$$

Substituting $X = p_1/q_1$, we find from the above relations that every term in the sum vanishes (provided $r < T - 1$.) Hence,

$$W^{(r)}\left(\frac{p_1}{q_1}\right) = 0 \qquad \text{for all } 0 \le r < T - 1.$$

This means that $\dfrac{p_1}{q_1}$ is a $(T-1)$-fold root of $W(X)$, so

$$\left(X - \frac{p_1}{q_1}\right)^{T-1}\bigg|\, W(X).$$

But $W(X)$ has integer coefficients, so Gauss' lemma says that $W(X)$ is divisible by $(q_1 X - p_1)^{T-1}$ in the polynomial ring $\mathbb{Z}[X]$. (Recall that Gauss' lemma says that if a polynomial with integer coefficients factors in $\mathbb{Q}[X]$, then it factors in $\mathbb{Z}[X]$.) In other words, there is a polynomial $V(X)$ with integer coefficients such that

$$W(X) = (q_1 X - p_1)^{T-1} V(X).$$

In order to exploit this factorization, we need to estimate the size of the coefficients of $W(X)$. This is not difficult because the Auxiliary Polynomial Theorem gives us a bound for the coefficients of $P(X)$ and $Q(X)$. If we write, as usual, $P(X) = \sum u_i X^i$ and $Q(X) = \sum v_i X^i$, then

$$W(X) = P(X)Q'(X) - Q(X)P'(X) = \sum_{i,j} j(u_i v_j - v_i u_j) X^{i+j-1}.$$

So the largest coefficient of $W(X)$ is bounded by

$$\max_{i,j \leq m+n} \left| j(u_i v_j - v_i u_j) \right| \leq 2(m+n) \left(\max_{i \leq m+n} \{u_i, v_i\} \right)^2$$

$$\leq 2(m+n) \cdot \left(2 \cdot (16b)^{9(m+n)} \right)^2 \leq c_3^n,$$

where c_3 is a constant depending only on b. (Note we always assume that $m \leq \frac{2}{3}n$, as specified in the Auxiliary Polynomial Theorem.)

On the other hand, since $V(X)$ has integer coefficients, the leading coefficient[†] of the product $(q_1 X - p_1)^{T-1} V(X)$ is at least q_1^{T-1}. Thus, $W(X)$ has a coefficient which is at least as large as q_1^{T-1}. So we have shown that

$$q_1^{T-1} \leq (\text{largest coefficient of } W(X)) \leq c_3^n.$$

Taking logarithms and defining a new constant $c_2 = \log c_3$, we find that

$$T \leq 1 + \frac{c_2 n}{\log q_1}.$$

It only remains to recall that we chose T as the largest integer for which the derivatives $F^{(t)}(p_1/q_1, p_2/q_2)$ vanish for all $0 \leq t < T$. We have just found an upper bound for T. It follows that there is some integer

$$0 \leq t \leq 1 + \frac{c_2 n}{\log q_1} \quad \text{such that} \quad F^{(t)}\left(\frac{p_1}{q_1}, \frac{p_2}{q_2} \right) \neq 0.$$

This (almost) concludes the proof of the Non-Vanishing Theorem.

What's left is that we must show that the Wronskian polynomial $W(X)$ is not identically zero. Suppose to the contrary that $W(X) = 0$. This means that $P'(X)Q(X) = Q'(X)P(X)$, so (by the quotient rule!)

$$\frac{d}{dX}\left(\frac{P(X)}{Q(X)} \right) = 0.$$

[†] Actually, we need to check that $W(X)$ is not the zero polynomial. We will verify this at the end.

Thus, the ratio $\dfrac{P(X)}{Q(X)}$ is constant, say $P(X) = aQ(X)$. Note that $a \in \mathbb{Q}$.

Now we have $F(X, Y) = P(X) + YQ(X) = (a + Y)Q(X)$. From the Auxiliary Polyomial Theorem we know that

$$F^{(k)}(\beta, \beta) = (a + \beta)Q^{(k)}(\beta) = 0 \qquad \text{for all } 0 \le k < n.$$

Since a is rational, $a + \beta \neq 0$, so $Q^{(k)}(\beta) = 0$. In other words, β is an n-fold root of $Q(X)$, so $(X - \beta)^n \mid Q(X)$.

But $\beta = \sqrt[3]{b}$, and $Q(X)$ has rational coefficients, so $Q(X)$ must be divisible by the nth power of the minimal polynomial of β,

$$(X^3 - b)^n \mid Q(X).$$

In particular, $Q(X)$ has degree at least $3n$.

However, we know that the degree of $Q(X)$ is at most $m + n$, where m satisfies $m \le \frac{2}{3}n$; so the degree of $Q(X)$ is at most $\frac{5}{3}n$. This contradiction shows that $W(X)$ is not the zero polynomial, which completes the proof of the Non-Vanishing Theorem.   □

7. Proof of the Diophantine Approximation Theorem

We have now assembled all of the tools needed to prove the Diophantine Approximation Theorem.

Diophantine Approximation Theorem.   (Thue) *Let b be a positive integer which is not a perfect cube, and let $\beta = \sqrt[3]{b}$. Let C be a fixed positive constant. Then there are only finitely many pairs of integers (p, q) with $q > 0$ which satisfy the inequality*

$$\left| \frac{p}{q} - \beta \right| \le \frac{C}{q^3}. \tag{$*$}$$

PROOF. We give a proof by contradiction. So we suppose that there are infinitely many pairs (p, q) that satisfy the inequality $(*)$. Let c_1 and c_2 be the two constants appearing in the Smallness Theorem (Section 5) and the Non-Vanishing Theorem (Section 6), respectively. We emphasize again that these constants depend only on the integer b, which is fixed throughout our discussion.

The inequality $(*)$ implies in particular (since $q \ge 1$) that

$$|p - \beta q| \le C.$$

Since we are assuming infinitely many solutions (p, q) to $(*)$, we see that the values of q must tend toward infinity. (Otherwise, both p and q would be bounded, which means there would be only finitely many pairs.) Hence we can find a solution (p_1, q_1) to $(*)$ whose second coordinate satisfies

$$q_1 > e^{9c_2} \qquad \text{and} \qquad q_1 > (2c_1 C)^{18}. \tag{1}$$

Then, since our assumption is that there are still infinitely many more solutions, we can find one (p_2, q_2) whose second coordinate is even larger, say satisfying[†]

$$q_2 > q_1^{65}. \tag{2}$$

Next, we let n be the integer satisfying

$$n \le \frac{9}{8} \frac{\log q_2}{\log q_1} < n + 1.$$

Exponentiating, we can also write this as

$$q_1^{\frac{8}{9}n} \le q_2 \le q_1^{\frac{8}{9}(n+1)}. \tag{3}$$

Notice that inequality (2) implies that $\dfrac{\log q_2}{\log q_1} > 65$, so

$$n \ge \frac{9}{8} \cdot 65 - 1 > 72. \tag{4}$$

Now we start to make use of our theorems. We use the Auxiliary Polynomial Theorem (Section 4) for our chosen value of n to find a polynomial $F(X, Y)$. Then we apply the Non-Vanishing Theorem (Section 6) to find an integer t such that

$$t \le 1 + \frac{c_2 n}{\log q_1} \qquad \text{and} \qquad F^{(t)}\left(\frac{p_1}{q_1}, \frac{p_2}{q_2}\right) \ne 0. \tag{5}$$

Notice that from (1) we get the estimate

$$t \le 1 + \frac{1}{9}n. \tag{6}$$

We are going to look at the rational number $F^{(t)}\left(\frac{p_1}{q_1}, \frac{p_2}{q_2}\right)$. More precisely, we are going to derive contradictory upper and lower bounds for this

---

[†] How do we know to choose the exponents 9 and 18 and 65 in (1) and (2)? The answer is that initially, we do not know. What we did was to write down the proof leaving the exponents as variables. Then, at the end, we could see what values would work. But there is nothing magical about 9, 18, and 65; any larger values will also work. And if you redo our calculation being a little more careful, you will find that smaller values will work, too.

number, which will finish the proof of the theorem. We start with the lower bound.

The auxiliary polynomial $F^{(t)}(X,Y)$ has integer coefficients and degree at most $m+n$ in X and degree 1 in Y. So multiplying everything out and putting it over a common denominator, we find

$$F^{(t)}\left(\frac{p_1}{q_1}, \frac{p_2}{q_2}\right) = \frac{\text{integer}}{q_1^{m+n}q_2}.$$

Further, we know from (5) that the integer in the numerator is not zero. There being no integers strictly between 0 and 1, we deduce that the absolute value of the numerator is at least 1. Hence,

$$\left|F^{(t)}\left(\frac{p_1}{q_1}, \frac{p_2}{q_2}\right)\right| \geq \frac{1}{q_1^{m+n}q_2}.$$

Now the Auxiliary Polynomial Theorem says $m \leq \frac{2}{3}n$, while (3) gives $q_2 < q_1^{(8/9)(n+1)}$, so we obtain the fundamental lower bound

$$\left|F^{(t)}\left(\frac{p_1}{q_1}, \frac{p_2}{q_2}\right)\right| \geq \frac{1}{q_1^{\frac{23}{9}n+\frac{8}{9}}}. \tag{7}$$

To find a complementary upper bound, we turn to the Smallness Theorem (Section 5), which says

$$\left|F^{(t)}\left(\frac{p_1}{q_1}, \frac{p_2}{q_2}\right)\right| \leq c_1^n \left\{\left|\frac{p_1}{q_1} - \beta\right|^{n-t} + \left|\frac{p_2}{q_2} - \beta\right|\right\}$$

$$\leq c_1^n \left\{\left(\frac{C}{q_1^3}\right)^{n-t} + \frac{C}{q_2^3}\right\} \qquad \text{from } (*),$$

$$\leq c_1^n \left\{\left(\frac{C}{q_1^3}\right)^{\frac{8}{9}n-1} + \frac{C}{q_1^{\frac{8}{3}n}}\right\} \qquad \text{from (6) and (3),}$$

$$\leq \frac{(2c_1C)^n}{q_1^{\frac{8}{3}n-3}}$$

$$\leq \frac{1}{q_1^{\frac{47}{18}n-3}} \qquad \text{from (1).} \tag{8}$$

Combining our lower bound (7) with our upper bound (8), we discover that

$$\frac{1}{q_1^{\frac{23}{9}n+\frac{8}{9}}} \leq \left|F^{(t)}\left(\frac{p_1}{q_1}, \frac{p_2}{q_2}\right)\right| \leq \frac{1}{q_1^{\frac{47}{18}n-3}},$$

so

$$q_1^{\frac{1}{18}n - \frac{35}{9}} \leq 1.$$

On the other hand, (4) says that $n \geq 72$. So we find

$$q_1^{\frac{1}{9}} \leq 1.$$

This is an obvious absurdity because the integer q_1 is certainly larger than 2 [e.g., from (1)]. We have arrived at the desired contradiction, which completes the proof that there are only finitely many pairs of integers (p, q) with $q > 0$ satisfying the inequality (∗). □

8. Further Developments

In this chapter we have proven that an equation

$$ax^3 + by^3 = c$$

has only finitely many solutions in integers x, y. The proof depends on a Diophantine Approximation Theorem which says roughly that it is not possible to use rational numbers p/q to very closely approximate a cube root $\sqrt[3]{b}$. With a small amount of modification, the proof that we gave can be adapted to prove the following stronger result, which Thue published in 1909.

Theorem. (Thue [1]) *Let $\beta \in \mathbb{R}$ be the root of an irreducible polynomial $f(X) \in \mathbb{Q}[X]$ having degree $d = \deg(f) \geq 3$. Let $\varepsilon > 0$ and $C > 0$ be any positive numbers. Then there are only finitely many pairs of integers (p, q) with $q > 0$ which satisfy the inequality*

$$\left| \frac{p}{q} - \beta \right| \leq \frac{C}{q^{\frac{1}{2}d + 1 + \varepsilon}}.$$

We proved the particular version of this theorem with $f(X) = X^3 - b$, $d = 3$, and $\varepsilon = \frac{1}{2}$.

A number of mathematicians have strengthened Thue's result. Notice that one way to make it stronger would be to decrease the exponent of q appearing on the right-hand side. So we might ask for what value of $\tau(d)$ is it true that there are only finitely many rational numbers satisfying

$$\left| \frac{p}{q} - \beta \right| \leq \frac{C}{q^{\tau(d) + \varepsilon}}.$$

Thue's result says that we may take $\tau(d) = \frac{1}{2}d + 1$. The following list illustrates some of the progress made on this problem.

Liouville	1851	$\tau(d) = d$
Thue	1909	$\tau(d) = \frac{1}{2}d + 1$
Siegel	1921	$\tau(d) = 2\sqrt{d}$
Gelfond, Dyson	1947	$\tau(d) = \sqrt{2d}$
Roth	1955	$\tau(d) = 2.$

Roth's theorem, which somewhat surprisingly says that for every d we may take $\tau(d) = 2$, is the stongest theorem of this form that is possible. That is, if we tried to take $\tau(d)$ to be any constant value less than 2, then the theorem would not be true. However, even Roth's theorem is not the end of the story. There are higher dimensional generalizations due to Schmidt [1], Vojta [1], and Faltings [1].

The proof that we gave for our special case of Thue's theorem contains all of the ingredients which appear in general. One constructs an auxiliary polynomial, evaluates it at some rational numbers, shows that it (or a small derivative) does not vanish, and derives a contradiction by giving upper and lower bounds for its magnitude. Siegel, Gelfond, and Dyson obtain their stronger results by taking a general polynomial $F(X, Y)$, rather than the special sort of polynomial $P(X) + YQ(X)$ used by Thue. Roth improves this by using an auxiliary polynomial $F(X_1, \ldots, X_r)$ of many variables. However, when working with a polynomial in many variables, it is extremely difficult to prove the analogue of what we called the Non-Vanishing Theorem. The major new technique developed by Roth was an intricate inductive procedure designed to show that some fairly small partial derivative of his auxiliary polynomial $F(X_1, \ldots, X_r)$ does not vanish when evaluated at $(p_1/q_1, \ldots, p_r/q_r)$.

In our concentration on proving the Diophantine Approximation Theorem, we ignored the problem of *effectivity*. That is, we proved that there are only finitely many pairs of integers (p, q) satisfying the inequality

$$\left| \frac{p}{q} - \sqrt[3]{b} \right| \le \frac{1}{q^3}; \tag{*}$$

but if we are given a particular value for b, such as $b = 2$, does our proof give us a way of actually finding all such pairs?

The answer is NO! If you look at our proof, you will find it says the following. If we can find a solution (p_1, q_1) to (*) with q_1 very large (how large depends on b,) then we can bound the coordinates of every other solution in terms of b and q_1. So if we can find that first large solution, then we can find them all. But suppose that there are no large solutions. "Ah," you may say, "then we just take all small solutions and we're done." However, nothing we have proven gives us any way of verifying that there are no large solutions. So if we find one large solution, then we can find all solutions; but if we cannot find a large solution, then we have no way of proving that the solution set we have found is complete. (This is a little subtle. You should stop and think about it for a minute.)

This is not a good state of affairs. In 1966, Baker devised a new method to prove a Diophantine Approximation Theorem which is effective. Unfortunately, Baker's theorem is not even as strong as Thue's result; but it is strong enough to deduce effective bounds for the integer solutions to cubic equations. The bounds tend to be quite large, as the following result illustrates.

Theorem. (Baker [2, page 45]) *Let* $a, b, c \in \mathbf{Z}$ *be integers, and let*

$$H = \max\{|a|, |b|, |c|\}.$$

Then every point (x, y) *on the elliptic curve*

$$y^2 = x^3 + ax^2 + bx + c$$

with integer coordinates $x, y \in \mathbf{Z}$ *satisfies*

$$\max\{|x|, |y|\} \le \exp\big((10^6 H)^{10^6}\big).$$

For special curves such as the ones we considered in this chapter, Baker's method yields somewhat better estimates. For example, Baker [1] gives the estimate

$$\left|\frac{p}{q} - \sqrt[3]{2}\right| \ge \frac{10^{-6}}{q^{2.9955}},$$

valid for all rational numbers p/q. In Section 3 we showed that any solution to

$$x^3 - 2y^3 = c$$

in integers $x, y \in \mathbf{Z}$ satisfies

$$\left|\frac{x}{y} - \sqrt[3]{2}\right| \le \frac{4|c|}{3\sqrt[3]{4}} \cdot \frac{1}{y^3}.$$

Combining this inequality with Baker's result, we find that

$$|y| \le 10^{1317} \cdot |c|^{2000/9}.$$

So again the bound is large, but at least it only grows like a power of $|c|$ rather than an exponential of a power of $|c|$.

EXERCISES

5.1. Define a sequence of pairs of integers by the following rule:

$$(x_0, y_0) = (1, 0);$$
$$(x_{i+1}, y_{i+1}) = (3x_i + 4y_i, 2x_i + 3y_i) \qquad \text{for } i \geq 0.$$

(a) Prove that every (x_i, y_i) is a solution to the equation

$$x^2 - 2y^2 = 1.$$

(b) Suppose that (x, y) is a solution to the equation in (a) with x and y positive integers. Prove that there is some index $i \geq 0$ such that $(x, y) = (x_i, y_i)$. *Hint.* If $y > 0$, prove that

$$(x, y) = (3u + 4v, 2u + 3v)$$

for positive integers u, v satisfying $u^2 - 2v^2 = 1$ and $v < y$.)

5.2. Let a, b, c be non-zero integers, and suppose that (x, y) is a solution in integers to the equation

$$ax^3 + bxy^2 = c.$$

Prove that

$$\max\{|ax^2|, |by^2|\} \leq 1 + \max\{|a|, |b|\}c^2.$$

5.3. Find all integer solutions to the following equations.
 (a) $x^2y + xy^2 = 240$.
 (b) $(x - 2y + 1)(79x^2 + 4xy - 34y^2) = 98$.
 (c) $(x - 2y + 1)(403x^2 - 388xy + 394y^2 + 1412x - 1612y) = 1218$.

5.4. Let $m \geq 2$ be an integer. Prove that the equation $x^3 + y^3 = m$ has no more than $d(m)$ solutions in pairs of integers (x, y), where $d(m)$ is the number of distinct (positive) divisors of m. [N.B. For this exercise we are counting (u, v) and (v, u) as distinct solutions if $u \neq v$.]

5.5. Let m be a *prime*.
 (a) Prove that the equation $x^3 + y^3 = m$ has no solutions in integers unless either $m = 2$, or else $m = 3u^2 + 3u + 1$ for some integer u.
 (b) Find all primes m less than 300 for which the equation in (a) has an integer solution.

5.6. For this exercise we will look at the curves

$$C_d : y^2 = x^3 + d.$$

We let $C_d(\mathbb{Z})$ denote the set of integer points:

$$C_d(\mathbb{Z}) = \{(x,y) : x, y \in \mathbb{Z},\ y^2 = x^3 + d\}.$$

(a) Prove that for every integer $N \geq 1$ there is an integer $d \geq 1$ so that $C_d(\mathbb{Z})$ contains at least N points.

(b) More precisely, prove that there is a constant $\kappa > 0$ and a sequence of integers $1 \leq d_1 < d_2 < d_3 < \cdots$ so that

$$\#C_{d_i}(\mathbb{Z}) \geq \kappa \log \log d_i.$$

(*Hint.* Take a rational point P of infinite order on some C_d, look at the rational points $P, 2P, 4P, 8P, \ldots$, and clear the denominators. Use the height formula $h(2P) \leq 4h(P) + \kappa$ to keep track of the size of the denominators.)

(c) * Same as in (b), but prove the better lower bound

$$\#C_d(\mathbb{Z}) \geq \kappa(\log d_i)^{1/3}.$$

(d) Show that C_{17} has at least 16 integer points. How many integer points can you find on C_{2089}?

(e) ** Call an integer point *primitive* if $\gcd(x,y) = 1$. Either prove that the number of primitive integer points in $C_d(\mathbb{Z})$ is bounded independently of d, or else find a sequence of d's so that the number of primitive integer points in $C_d(\mathbb{Z})$ goes to infinity.

5.7. Let $\beta \in \mathbb{R}$ be any real number.

(a) Prove that there are infinitely many rational numbers p/q satisfying

$$\left| \frac{p}{q} - \beta \right| \leq \frac{1}{2q}.$$

(b) * Prove Dirichlet's result that there are infinitely many rational numbers p/q satisfying

$$\left| \frac{p}{q} - \beta \right| \leq \frac{1}{q^2}.$$

(This shows that the exponent 2 in Roth's theorem cannot be improved.)

5.8. Let $\beta \in \mathbb{R}$ be any real number. In this exercise we will consider solutions to the inequality

$$\left| \frac{p}{q} - \beta \right| \le \frac{1}{q^3}.$$

(a) Suppose that p/q and p'/q' are distinct solutions with $q' \ge q$. Prove that $q' \ge \frac{1}{2}q^2$.

(b) Suppose we have a sequence of distinct solutions $\dfrac{p_0}{q_0}, \dfrac{p_1}{q_1}, \dots, \dfrac{p_r}{q_r}$, labeled so that $4 \le q_0 \le q_1 \le \cdots \le q_r$. Prove that

$$q_r \ge 2^{2^r}.$$

5.9. Let $d \ge 3$ be an integer, and let b be an integer that is not a perfect d^{th} power.
(a) For any constant C, prove that there are only finitely many rational numbers p/q satisfying the inequality

$$\left| \frac{p}{q} - \sqrt[d]{b} \right| \le \frac{C}{q^3}.$$

(b) Let a, b, and c be non-zero integers. Prove that the equation

$$ax^d + by^d = c$$

has only finitely many solutions $x, y \in \mathbb{Z}$.

5.10. (a) * Prove the general version of Thue's Diophantine Approximation Theorem stated at the beginning of Section 8.
(b) * Let $a_0 t^d + a_1 t^{d-1} + \cdots + a_d$ be a polynomial with integer coefficients which is irreducible in $\mathbb{Q}[t]$. Prove that for any non-zero integer c, the equation

$$a_0 x^d + a_1 x^{d-1}y + a_2 x^{d-2}y^2 + \cdots + a_{d-1}xy^{d-1} + a_d y^d = c$$

has only finitely many solutions $x, y \in \mathbb{Z}$.

5.11. Let C be a non-singular cubic curve given by a Weierstrass equation

$$y^2 = x^3 + ax^2 + bx + c$$

with integer coefficients. Let $P \in C(\mathbb{Q})$, and suppose that there is an integer $n \ge 1$ such that nP has integer coordinates. Prove that P has integer coordinates. (*Hint.* Consider the subgroups $C(p)$ defined in Chapter II, Section 4.)

CHAPTER VI

Complex Multiplication

1. Abelian Extensions of $\mathbb{Q}$

In this chapter we want to describe how points of finite order on certain elliptic curves can be used to generate interesting extension fields of $\mathbb{Q}$. Here we mean points of finite order with arbitrary complex coordinates, not just the ones with rational coordinates that we studied in Chapter II. So we will need to use some basic theorems about extension fields and Galois theory, but nothing very fancy. We will start by reminding you of most of the facts we will be using, and you can look at any basic algebra text (such as Herstein [1] or Jacobson [1]) for the proofs and additional background material.

We will be interested in subfields of the complex numbers, $\mathbb{Q} \subset K \subset \mathbb{C}$. Then K is a $\mathbb{Q}$ vector space, and the *degree of K over $\mathbb{Q}$* is defined to be

$$[K : \mathbb{Q}] = \text{dimension of } K \text{ as a } \mathbb{Q} \text{ vector space.}$$

If $[K : \mathbb{Q}]$ is finite, then we say that K is a *number field*.

An important technique for studying number fields is to look at the set of field homomorphisms

$$\sigma : K \hookrightarrow \mathbb{C}.$$

Recall that a homomorphism of fields is always one-to-one because a field has no non-trivial ideals. Also note that since by definition $\sigma(1) = 1$, any σ automatically satisfies $\sigma(a) = a$ for all $a \in \mathbb{Q}$. One proves that the number of homomorphisms $K \hookrightarrow \mathbb{C}$ is always exactly equal to the degree $[K : \mathbb{Q}]$.

Now it sometimes happens that the image $\sigma(K)$ is equal to the original field K. Then σ is actually an isomorphism from K to itself, in which case we call σ an *automorphism* of K. Note this does not mean $\sigma(\alpha) = \alpha$ for every $\alpha \in K$, but merely that $\sigma(\alpha) \in K$. If this happens for every σ, then we call K a *Galois extension of* $\mathbb{Q}$. More generally, define

$$\text{Aut}(K) = \{\text{automorphisms } \sigma : K \to K\}.$$

Clearly, $\mathrm{Aut}(K)$ is a group, since if $\sigma, \tau \in \mathrm{Aut}(K)$, then we can form an element $\sigma\tau \in \mathrm{Aut}(K)$ by taking the composition $(\sigma\tau)(\alpha) = \sigma(\tau(\alpha))$. The number field K is a Galois extension of $\mathbb{Q}$ if and only if

$$\# \mathrm{Aut}(K) = [K : \mathbb{Q}].$$

In this case we write $\mathrm{Gal}(K/\mathbb{Q})$ instead of $\mathrm{Aut}(K)$ and call $\mathrm{Gal}(K/\mathbb{Q})$ the *Galois group of $K/\mathbb{Q}$*.

This is all fairly abstract. How does one actually find number fields that are Galois over $\mathbb{Q}$? The answer is simple. Take any polynomial with rational coefficients $f(X) \in \mathbb{Q}[X]$. Factor $f(X)$ over the complex numbers

$$f(X) = a(X - \alpha_1)(X - \alpha_2) \cdots (X - \alpha_n),$$

and let

$$K = \mathbb{Q}(\alpha_1, \alpha_2, \ldots, \alpha_n)$$

be the the smallest subfield of $\mathbb{C}$ containing all of the α_i's. Then any homomorphism $\sigma : K \to \mathbb{C}$ is determined by the values of $\sigma(\alpha_1), \ldots, \sigma(\alpha_n)$; and each $\sigma(\alpha_i)$ has to be a root of $f(X)$, so equals some α_j. In particular, $\sigma(\alpha_i) \in K$, so $\sigma(K) = K$. [The inclusion $\sigma(K) \subset K$ is clear, and then equality follows by comparing the degrees of $\sigma(K)$ and K over $\mathbb{Q}$.] The field K is called the *splitting field of $f(X)$* over $\mathbb{Q}$, and we have just seen that such a splitting field is a Galois extension. Conversely, one can prove that if a number field K is a Galois extension of $\mathbb{Q}$, then it is the splitting field of some polynomial $f(X) \in \mathbb{Q}[X]$.

This fact helps to explain why Galois extensions are both useful and important. The study of roots of polynomials lies at the classical base of much of algebra and number theory. In order to study those roots, one can instead study the fields that the roots generate. And if one takes the field generated by all of the roots, then one gets a Galois extension, which has attached to it a certain finite group. So by using basic facts from group theory, one can often make interesting deductions about the roots of the original polynomial. Schematically, one might imagine this process as follows:

$$\boxed{\begin{array}{c}\text{Roots of} \\ \text{Polynomials}\end{array}} \xleftrightarrow{\text{Field Theory}} \boxed{\begin{array}{c}\text{Extension} \\ \text{Fields}\end{array}} \xleftrightarrow{\text{Galois Theory}} \boxed{\begin{array}{c}\text{Group} \\ \text{Theory}\end{array}}$$

The easiest sorts of groups are abelian groups, so it is natural to begin by looking at Galois extensions $K/\mathbb{Q}$ whose Galois groups are abelian. One way such extensions arise is in the study of Fermat's equation

$$x^n + y^n = 1.$$

If we try to apply the factorization techniques used throughout this book, we might move the x^n to the other side of the equation and factor

$$y^n = 1 - x^n = (1 - x)(1 - \zeta x)(1 - \zeta^2 x) \cdots (1 - \zeta^{n-1} x).$$

Here $\zeta \in \mathbb{C}$ is a *primitive n^{th} root of unity* (i.e., $\zeta^n = 1$, and $\zeta^j \neq 1$ for all $1 \leq j < n$). For example, we could take $\zeta = e^{2\pi i/n}$. In order to study Fermat's equation, we are led (following Kummer) to look at the field $\mathbb{Q}(\zeta)$.

A field generated by roots of unity, such as the field $\mathbb{Q}(\zeta)$, is called a *cyclotomic field* (from the Greek *kyklos*, circle, because roots of unity lie on the unit circle $|z| = 1$ in the complex plane). Note $\mathbb{Q}(\zeta)$ will contain all of the powers of ζ, so it is the splitting field of the polynomial $X^n - 1$ over $\mathbb{Q}$. Thus, $\mathbb{Q}(\zeta)$ is a Galois extension of $\mathbb{Q}$. We next describe its Galois group.

An automorphism $\sigma : \mathbb{Q}(\zeta) \to \mathbb{Q}(\zeta)$ is determined by the value of $\sigma(\zeta)$; and that value will also be a primitive n^{th} root of unity, because σ will preserve the order of an element. Every primitive n^{th} root of unity is a power of ζ, more precisely a power ζ^t for some integer t relatively prime to n. Thus, we get a one-to-one map of sets

$$t : \mathrm{Gal}(\mathbb{Q}(\zeta)/\mathbb{Q}) \longrightarrow \left(\frac{\mathbb{Z}}{n\mathbb{Z}}\right)^*$$

completely determined by the property

$$\sigma(\zeta) = \zeta^{t(\sigma)} \qquad \text{for } \sigma \in \mathrm{Gal}(\mathbb{Q}(\zeta)/\mathbb{Q}).$$

Here $(\mathbb{Z}/n\mathbb{Z})^*$ is the group of units in $\mathbb{Z}/n\mathbb{Z}$,

$$\left(\frac{\mathbb{Z}}{n\mathbb{Z}}\right)^* = \{m \,(\mathrm{mod}\ n) \,:\, \gcd(m, n) = 1\}.$$

We claim that the map t is actually a homomorphism. The proof is easy. If $\sigma, \tau \in \mathrm{Gal}(\mathbb{Q}(\zeta)/\mathbb{Q})$, then

$$\zeta^{t(\sigma\tau)} = (\sigma\tau)(\zeta) = \sigma(\tau(\zeta)) = \sigma(\zeta^{t(\tau)})$$
$$= (\sigma(\zeta))^{t(\tau)} = (\zeta^{t(\sigma)})^{t(\tau)} = \zeta^{t(\sigma)t(\tau)}.$$

Hence, $t(\sigma\tau) \equiv t(\sigma)t(\tau) \,(\mathrm{mod}\ n)$, which proves our assertion.

We have just proven that there is a one-to-one homomorphism

$$t : \mathrm{Gal}(\mathbb{Q}(\zeta)/\mathbb{Q}) \longrightarrow \left(\frac{\mathbb{Z}}{n\mathbb{Z}}\right)^*.$$

Since $(\mathbb{Z}/n\mathbb{Z})^*$ is an abelian group, the same is true of $\mathrm{Gal}(\mathbb{Q}(\zeta)/\mathbb{Q})$. This completes the proof of the following proposition.

Proposition. *The Galois group of a cyclotomic extension is abelian. More precisely, if ζ is a primitive n^{th} root of unity, then there is a one-to-one homomorphism*

$$t : \mathrm{Gal}(\mathbb{Q}(\zeta)/\mathbb{Q}) \longrightarrow \left(\frac{\mathbb{Z}}{n\mathbb{Z}}\right)^*$$

determined by the property $\sigma(\zeta) = \zeta^{t(\sigma)}$.

In fact, the map t is an isomorphism; but the proof is not easy unless $n = p$ is prime, in which case it can be proven by checking that $[\mathbb{Q}(\zeta) : \mathbb{Q}] = p - 1$.

We now want to talk more generally about field extensions $F \subset K$ with F not necessarily equal to $\mathbb{Q}$. Recall that for such an extension field, one defines

$$\text{Aut}_F(K) = \left\{ \begin{array}{c} \text{automorphisms } \sigma : K \to K \text{ such} \\ \text{that } \sigma(a) = a \text{ for all } a \in F \end{array} \right\}.$$

If $[K : F] = \# \text{Aut}_F(K)$, then one says says that K/F is a Galois extension and writes $\text{Gal}(K/F)$ instead of $\text{Aut}_F(K)$.

Now suppose that we have a subextension of a cyclotomic field,

$$\mathbb{Q} \subset F \subset \mathbb{Q}(\zeta).$$

The Fundamental Theorem of Galois Theory tells us that $F/\mathbb{Q}$ will itself be a Galois extension if and only if $\text{Gal}(\mathbb{Q}(\zeta)/F)$ is a normal subgroup of $\text{Gal}(\mathbb{Q}(\zeta)/\mathbb{Q})$. But we just proved that $\text{Gal}(\mathbb{Q}(\zeta)/\mathbb{Q})$ is an abelian group, so all of its subgroups are normal. Hence, $F/\mathbb{Q}$ is Galois, and Galois Theory tells us that there is an isomorphism

$$\frac{\text{Gal}(\mathbb{Q}(\zeta)/\mathbb{Q})}{\text{Gal}(\mathbb{Q}(\zeta)/F)} \xrightarrow{\sim} \text{Gal}(F/\mathbb{Q}).$$

Hence, every subfield of a cyclotomic field is a Galois extension of $\mathbb{Q}$ with abelian Galois group. Amazingly enough, the converse is also true.

Theorem. (Kronecker-Weber Theorem) *Let F be a Galois number field whose Galois group $\text{Gal}(F/\mathbb{Q})$ is abelian. Then there exists a cyclotomic extension $\mathbb{Q}(\zeta)$ of $\mathbb{Q}$ such that*

$$F \subset \mathbb{Q}(\zeta).$$

Hence, the Galois extensions of $\mathbb{Q}$ with abelian Galois group are precisely the subfields of cyclotomic fields.

The proof of the Kronecker-Weber theorem is quite difficult, although nowadays people would say that it is an immediate corollary of class field theory (which is too complicated for us to even describe). But we can prove a special case for you.

Suppose we take F to be a quadratic extension $F = \mathbb{Q}(\sqrt{p})$, where p is a prime. Then $F/\mathbb{Q}$ is a Galois extension whose Galois group is a cyclic group of order two; in particular, the Galois group is abelian, so by the Kronecker-Weber theorem, F should be contained in some cyclotomic extension.

To prove this for odd p, we let $\zeta \in \mathbb{C}$ be a primitive p^{th} root of unity, and let γ be the quadratic Gauss sum

$$\gamma = \sum_{a=0}^{p-1} \zeta^{a^2}.$$

Then one can check that

$$\gamma^2 = \begin{cases} p & \text{if } p \equiv 1 \,(\text{mod } 4), \\ -p & \text{if } p \equiv -1 \,(\text{mod } 4), \end{cases}$$

(cf. Exercise 4.4, where $\gamma = 2\alpha + 1$). Hence, if $p \equiv 1 \,(\text{mod } 4)$, then $\mathbb{Q}(\sqrt{p}) = \mathbb{Q}(\gamma) \subset \mathbb{Q}(\zeta)$. Similarly, if $p \equiv -1 \,(\text{mod } 4)$, then we can use the inclusions

$$\mathbb{Q}(\sqrt{p}) \subset \mathbb{Q}(i, \sqrt{-p}) = \mathbb{Q}(i, \gamma) \subset \mathbb{Q}(i, \zeta) = \mathbb{Q}(\zeta'),$$

where ζ' is a primitive $4p^{\text{th}}$ root of unity. So this proves the Kronecker-Weber theorem for the quadratic extension $\mathbb{Q}(\sqrt{p})$ if p is an odd prime. For $p = 2$ we leave it to the reader to check that $\sqrt{2} = \zeta + \zeta^{-1}$ for $\zeta = e^{2\pi i/8}$.

If we use a little complex analysis, the Kronecker-Weber theorem becomes even more astonishing. To calculate an nth root of unity, we can use the exponential function

$$f(z) = e^{2\pi i z} = \sum_{k=0}^{\infty} \frac{(2\pi i z)^k}{k!}.$$

This is an entire (i.e., everywhere holomorphic) function on $\mathbb{C}$. If we evaluate this function at the rational number $1/n$, we get a complex number

$$f\left(\frac{1}{n}\right) = \sum_{k=0}^{\infty} \frac{(2\pi i)^k}{n^k k!}$$

given by a convergent series. Now we have three amazing facts:

(i) This series converges to a number which is a root of a polynomial equation having rational coefficients (viz., it is a root of $X^n - 1$).
(ii) The field extension of $\mathbb{Q}$ generated by this number is a Galois extension with abelian Galois group.
(iii) Every Galois extension of $\mathbb{Q}$ with abelian Galois group is contained in one of these extensions.

So the abelian extensions of $\mathbb{Q}$ can be described in terms of certain special values of the holomorphic function $f(z) = e^{2\pi i z}$. Further, recall our homomorphism (really an isomorphism)

$$t : \text{Gal}\big(\mathbb{Q}(\zeta)/\mathbb{Q}\big) \longrightarrow \left(\frac{\mathbb{Z}}{n\mathbb{Z}}\right)^*,$$

where we can take $\zeta = f(1/n) = e^{2\pi i/n}$. Then we can describe the effect of an element $\sigma \in \mathrm{Gal}\big(\mathbb{Q}(\zeta)/\mathbb{Q}\big)$ on ζ very easily in terms of the function f:

$$\sigma\left(f\left(\frac{1}{n}\right)\right) = f\left(\frac{t(\sigma)}{n}\right).$$

The question now arises whether a similar theory exists for other fields. Kronecker's Jugendtraum (literally, "dream of youth") was to construct such a theory for quadratic imaginary fields, that is, quadratic extensions F of $\mathbb{Q}$ such that F is not contained in $\mathbb{R}$. Kronecker's hope was to find a holomorphic (or meromorphic) function $f(z)$ with the following property: for every Galois extension K/F with abelian Galois group, there should be special values $f(a_1), \ldots, f(a_n)$ of $f(z)$ so that the field $F\big(f(a_1), \ldots, f(a_n)\big)$ generated by these values is Galois with abelian Galois group, and so that

$$K \subset F\big(f(a_1), \ldots, f(a_n)\big).$$

He further hoped that if $\sigma \in \mathrm{Gal}\big(F\big(f(a_1), \ldots, f(a_n)\big)/F\big)$, then one would be able to describe $\sigma\big(f(a_i)\big)$ in some simple manner as another special value of the function $f(z)$. Thus, Kronecker's Jugendtraum is true for $F = \mathbb{Q}$, since we can take $f(z) = e^{2\pi i z}$ and the special values $f(j/n)$, $j \in (\mathbb{Z}/n\mathbb{Z})^*$.

Kronecker and his contemporaries were largely able to construct such a theory for quadratic imaginary fields using a construction that is intimately tied up with the theory of elliptic curves. This is the theory we will be discussing in the remainder of this chapter.

More generally, one can start with any number field F and ask for a description of all Galois extensions K/F such that $\mathrm{Gal}(K/F)$ is abelian. The class field theory alluded to above gives such a description, but it does so in a somewhat indirect manner. Except in certain special cases, the extension of Kronecker's Jugendtraum to number fields is still very much an open question.

2. Algebraic Points on Cubic Curves

As usual, let C be an elliptic curve given by a Weierstrass equation

$$C : y^2 = x^3 + ax^2 + bx + c$$

with rational coefficients $a, b, c \in \mathbb{Q}$. Up to now we have been mainly concerned with points on this curve having either rational or integer coordinates, although we have occasionally talked about the real points $C(\mathbb{R})$ and the complex points $C(\mathbb{C})$. More generally, if $K \subset \mathbb{C}$ is any subfield of the complex numbers, then we can look at the set of K-rational points on C:

$$C(K) = \big\{(x,y) : x, y \in K \quad \text{and} \quad y^2 = x^3 + ax^2 + bx + c\big\} \cup \{\mathcal{O}\}.$$

It is clear from the formulas for the addition law on C that $C(K)$ is closed under addition, so it is a subgroup of $C(\mathbb{C})$.

For example, consider the curve

$$y^2 = x^3 - 4x^2 + 16.$$

The discriminant of the cubic polynomial is $D = 45056 = 2^{12} \cdot 11$, and one can easily check (e.g., using the Nagell-Lutz theorem) that the rational points of finite order on C form a group of order five,

$$\{\mathcal{O}, (0, \pm 4), (4, \pm 4)\}.$$

With somewhat more effort, one can in fact prove that $C(\mathbb{Q})$ consists of only these five points; there are no points of infinite order.

However, if we replace $\mathbb{Q}$ by an extension field, matters may drastically change. For example, if we take the field $\mathbb{Q}(\sqrt{-2})$, then C contains the point

$$P = (8 + 4\sqrt{-2}, 12 + 16\sqrt{-2}) \in C(\mathbb{Q}(\sqrt{-2})).$$

We can use the duplication formula to compute $2P$,

$$2P = \left(\frac{-124 + 56\sqrt{-2}}{(3 + 4\sqrt{-2})^2}, \frac{-276 - 448\sqrt{-2}}{(3 + 4\sqrt{-2})^3}\right).$$

The point P has infinite order, so $C(\mathbb{Q}(\sqrt{-2}))$ contains infinitely many points.

Suppose now that K is a Galois extension of $\mathbb{Q}$. Then for any point $P = (x, y) \in C(K)$ and any element $\sigma \in \mathrm{Gal}(K/\mathbb{Q})$, we can define a point

$$\sigma(P) = (\sigma(x), \sigma(y)).$$

We will also set $\sigma(\mathcal{O}) = \mathcal{O}$. (This makes sense because in homogeneous coordinates, $\mathcal{O} = [0, 1, 0]$, so we should have $\sigma(\mathcal{O}) = [\sigma(0), \sigma(1), \sigma(0)] = [0, 1, 0] = \mathcal{O}$.) Of course, we should check that $\sigma(P)$ is on the curve C; but even more importantly, we must investigate how the map $P \to \sigma(P)$ interacts with the group law on C. All of this (and more) is contained in the following elementary proposition.

Proposition. *Let C be an elliptic curve defined by an equation with coefficients in $\mathbb{Q}$, and let K be a Galois extension of $\mathbb{Q}$.*
(a) $C(K)$ is a subgroup of $C(\mathbb{C})$.
(b) For $P \in C(K)$ and $\sigma \in \mathrm{Gal}(K/\mathbb{Q})$, define $\sigma(P)$ by

$$\sigma(P) = \begin{cases} (\sigma(x), \sigma(y)) & \text{if } P = (x, y), \\ \mathcal{O} & \text{if } P = \mathcal{O}. \end{cases}$$

Then $\sigma(P) \in C(K)$.

(c) For all $P \in C(K)$ and all $\sigma, \tau \in \mathrm{Gal}(K/\mathbb{Q})$,

$$(\sigma\tau)(P) = \sigma\big(\tau(P)\big).$$

Further, the identity element $e \in \mathrm{Gal}(K/\mathbb{Q})$ acts trivially, $e(P) = P$.
(d) For all $P, Q \in C(K)$ and all $\sigma \in \mathrm{Gal}(K/\mathbb{Q})$,

$$\sigma(P + Q) = \sigma(P) + \sigma(Q) \qquad \text{and} \qquad \sigma(-P) = -\sigma(P).$$

Hence, $\sigma(nP) = n(\sigma(P))$ for all integers n.
(e) If $P \in C(K)$ has order n and if $\sigma \in \mathrm{Gal}(K/\mathbb{Q})$, then $\sigma(P)$ also has order n.

PROOF. (a) If P_1 and P_2 are in $C(K)$, so their x and y coordinates are in K, then it is clear from the explicit formulas for the addition law on C that $P_1 \pm P_2$ also have coordinates in K. Hence, $C(K)$ is closed under addition and subtraction, so it is a subgroup of $C(\mathbb{C})$.
(b) Let $P = (x, y) \in C(K)$. The coordinates of $\sigma(P)$ are in K, so we just need to check that $\sigma(P)$ is a point of C. We use the fact that P is on C and that $\sigma : K \to K$ is a homomorphism which fixes $\mathbb{Q}$. Thus,

$$
\begin{aligned}
P \in C &\Longrightarrow y^2 - x^3 - ax^2 - bx - c = 0 \\
&\Longrightarrow \sigma(y^2 - x^3 - ax^2 - bx - c) = 0 \\
&\Longrightarrow \sigma(y)^2 - \sigma(x)^3 - \sigma(a)\sigma(x)^2 - \sigma(b)\sigma(x) - \sigma(c) = 0 \\
&\qquad\qquad\qquad\qquad\qquad\qquad \text{because } \sigma \text{ is a homomorphism,} \\
&\Longrightarrow \sigma(y)^2 - \sigma(x)^3 - a\sigma(x)^2 - b\sigma(x) - c = 0 \quad \text{because } \sigma \text{ fixes } \mathbb{Q}, \\
&\Longrightarrow \sigma(P) = \big(\sigma(x), \sigma(y)\big) \in C.
\end{aligned}
$$

(c) We will leave this as an exercise.
(d) As in (b), this part follows from the fact that the addition law is given by rational functions with coefficients in $\mathbb{Q}$. There are several cases to check; we will do one, and leave the others for you.

Write $P = (x_1, y_1)$, $Q = (x_2, y_2)$, and $P + Q = (x_3, y_3)$. We will assume that $P \neq \pm Q$. Then the formulas we derived in Chapter I, Section 4 say that

$$x_3 = \left(\frac{y_2 - y_1}{x_2 - x_1}\right)^2 - a - x_1 - x_2, \qquad y_3 = \left(\frac{y_2 - y_1}{x_2 - x_1}\right)(x_1 - x_3) - y_1.$$

Using the fact that σ is a homomorphism which fixes $\mathbb{Q}$, we compute

$$\sigma(x_3) = \left(\frac{\sigma(y_2) - \sigma(y_1)}{\sigma(x_2) - \sigma(x_1)}\right)^2 - a - \sigma(x_1) - \sigma(x_2),$$

$$\sigma(y_3) = \left(\frac{\sigma(y_2) - \sigma(y_1)}{\sigma(x_2) - \sigma(x_1)}\right)(\sigma(x_1) - \sigma(x_3)) - \sigma(y_1).$$

Hence,

$$\sigma(P+Q) = \big(\sigma(x_3), \sigma(y_3)\big) = \big(\sigma(x_1), \sigma(y_1)\big) + \big(\sigma(x_2), \sigma(y_2)\big) = \sigma(P) + \sigma(Q).$$

The fact that $\sigma(-P) = -\sigma(P)$ is even easier because if $P = (x, y)$, then

$$\sigma(-P) = \sigma(x, -y) = \big(\sigma(x), \sigma(-y)\big) = \big(\sigma(x), -\sigma(y)\big) = -\sigma(P).$$

Finally, by repeatedly applying the formula $\sigma(P + Q) = \sigma(P) + \sigma(Q)$, we easily find that $\sigma(nP) = n\sigma(P)$ for $n \geq 0$; and then $\sigma(-P) = -\sigma(P)$ shows that it is also true for $n < 0$.

(e) Suppose that $P \in C(K)$ has order n. Then using (d) we find

$$n\sigma(P) = \sigma(nP) = \sigma(\mathcal{O}) = \mathcal{O},$$

so $\sigma(P)$ has order dividing n. To see that the order is exactly n, we suppose that $m\sigma(P) = \mathcal{O}$. Again using (d), this implies that $\sigma(mP) = \mathcal{O}$. Now we apply σ^{-1} to both sides to deduce that

$$\mathcal{O} = \sigma^{-1}(\mathcal{O}) = \sigma^{-1}\big(\sigma(mP)\big) = mP.$$

Since P has order exactly n, this implies that $m \geq n$. Hence, $\sigma(P)$ also has order exactly n. $\square$

In the last section we defined a cyclotomic field as the splitting field (over $\mathbb{Q}$) of a polynomial $X^n - 1$. To clarify the analogy with elliptic curves, we want to reformulate this as follows.

Consider the group $\mathbb{C}^*$ of non-zero complex numbers with the group law being multiplication. For any integer n, raising to the nth power gives a homomorphism from $\mathbb{C}^*$ to itself:

$$\lambda_n : \mathbb{C}^* \longrightarrow \mathbb{C}^*, \qquad \lambda_n(z) = z^n.$$

The kernel of the homomorphism λ_n consists of precisely the set of nth roots of unity in $\mathbb{C}$. So a cyclotomic field is the field generated over $\mathbb{Q}$ by the elements in the kernel of some n^{th}-power homomorphism $\lambda_n : \mathbb{C}^* \to \mathbb{C}^*$.

Now we are going to do the same thing, replacing the group $\mathbb{C}^*$ with the elliptic curve $C(\mathbb{C})$ and replacing the nth-power homomorphism with the *multiplication-by-n map*

$$\lambda_n : C(\mathbb{C}) \longrightarrow C(\mathbb{C}), \qquad \lambda_n(P) = nP.$$

The kernel of λ_n is a subgroup of $C(\mathbb{C})$ which we will denote by

$$C[n] = \ker(\lambda_n) = \{P \in C(\mathbb{C}) : nP = \mathcal{O}\}.$$

It is the set of points of order dividing n. It is easy to describe $C[n]$ as an abstract group, at least if you believe the analytic description of $C(\mathbb{C})$ that we discussed in Chapter II, Section 2. (We will sketch a more algebraic proof for you in the exercises.)

Proposition. *As an abstract group,*

$$C[n] \cong \frac{\mathbb{Z}}{n\mathbb{Z}} \oplus \frac{\mathbb{Z}}{n\mathbb{Z}};$$

that is, $C[n]$ is a direct sum of two cyclic groups of order n.

PROOF. Recall from Chapter II, Section 2 that $C(\mathbb{C})$ is isomorphic (as a group) to $\mathbb{C}/L$, where

$$L = \mathbb{Z}\omega_1 + \mathbb{Z}\omega_2 = \{m_1\omega_1 + m_2\omega_2 : m_1, m_2 \in \mathbb{Z}\}$$

is a lattice in $\mathbb{C}$. With this description, it is easy to see what $C[n]$ looks like; in fact, we can give an explicit isomorphism

$$
\begin{array}{ccc}
\dfrac{\mathbb{Z}}{n\mathbb{Z}} \oplus \dfrac{\mathbb{Z}}{n\mathbb{Z}} & \longrightarrow & C[n] \subset \dfrac{\mathbb{C}}{L}, \\[2mm]
(a_1, a_2) & \longmapsto & \dfrac{a_1}{n}\omega_1 + \dfrac{a_2}{n}\omega_2.
\end{array}
$$

$\square$

As we have seen, cyclotomic extensions are generated by the elements in the kernel of the nth power map $\mathbb{C}^* \to \mathbb{C}^*$. In a similar manner, we want to look at the field extensions generated by the points in $C[n]$. Now a point $P = (x, y) \in C[n]$ has two coordinates, so we might consider the field generated by all of the coordinates of all of the points in $C[n]$. The following proposition suggests that this is a good field to look at.

Proposition. *Let C be an elliptic curve given by a Weierstrass equation*

$$C : y^2 = x^3 + ax^2 + bx + c$$

with rational coefficients $a, b, c \in \mathbb{Q}$.
(a) Let $P = (x_1, y_1) \in C$ be a point of order n. Then x_1 and y_1 are algebraic over $\mathbb{Q}$ (i.e., x_1 and y_1 are the roots of a polynomial with rational coefficients).
(b) Let

$$\{(x_1, y_1), \ldots, (x_m, y_m), \mathcal{O}\} = C[n]$$

be the complete set of points of $C(\mathbb{C})$ of order dividing n. From above, m is equal to $n^2 - 1$. Let

$$K = \mathbb{Q}(x_1, y_1, \ldots, x_m, y_m)$$

be the field generated by the coordinates of all of the points in $C[n]$. Then K is a Galois extension of $\mathbb{Q}$. [N.B. In general, Gal$(K/\mathbb{Q})$ will not be abelian.]

PROOF. (a) We will give a computational proof, although in truth it is not difficult to adapt the proof of (b) to simultaneously prove (a).

If we are given a point $P = (x, y)$ and an integer $n \geq 2$, how can we tell if $nP = \mathcal{O}$? For $n = 2$, we have seen that

$$2P = \mathcal{O} \iff x^3 + ax^2 + bx + c = 0;$$

so the x coordinate of a point of order two is clearly algebraic. In general, if we repeatedly use the addition formula, we can find a multiplication-by-n formula similar to the duplication formula. For large values of n the formula would be extremely complicated, but the fact that the addition law is given by rational functions means that if $P = (x, y)$, then the

$$(x \text{ coordinate of } nP) = \frac{\text{polynomial in } x \text{ and } y}{\text{polynomial in } x \text{ and } y}.$$

In fact, since the x coordinates of nP and $-nP = n(x, -y)$ are the same, it is not hard to see (e.g., by induction) that we can take the polynomials to depend only on x:

$$(x \text{ coordinate of } nP) = \frac{\phi_n(x)}{\psi_n(x)},$$

where $\phi_n(x)$ and $\psi_n(x)$ are relatively prime polynomials in $\mathbb{Q}[x]$. Then a point $P = (x_1, y_1)$ will have order dividing n if and only if $\psi_n(x_1) = 0$.

This proves that the x coordinate of a point of order n is algebraic because it is a root of the polynomial $\psi_n(x)$; and since $y^2 = x^3 + ax^2 + bx + c$, the y coordinate will be algebraic, too.

(b) Let $\sigma : K \to \mathbb{C}$ be a field homomorphism. In order to prove that K is Galois over $\mathbb{Q}$, we must verify that $\sigma(K) = K$.

The map σ is completely determined by where it sends the x_i's and y_i's. What are the allowable possibilities for these values? By assumption, each point P_i is in $C[n]$; and from the proposition we proved,

$$\mathcal{O} = \sigma(\mathcal{O}) = \sigma(nP_i) = n\sigma(P_i),$$

so $\sigma(P_i)$ is also in $C[n]$. This means that $\sigma(P_i)$ is one of the P_j's (with $i = j$ being allowed). Hence, the x and y coordinates of $\sigma(P_i)$ are already in K; that is, $\sigma(x_i), \sigma(y_i) \in K$. This is true for each $1 \leq i \leq m$, and so $\sigma(K) \subset K$, which completes the proof that K is a Galois extension of $\mathbb{Q}$.

Addendum: Here is the alternative (albeit somewhat fancier) proof of (a) that we mentioned. We have just seen that every field homomorphism $\sigma : K \to \mathbb{C}$ is determined by specifying some permutation of the points $P_1, \ldots, P_m$. In particular, this means that there are only finitely many such homomorphisms. But, if some x_i or y_i were not algebraic over $\mathbb{Q}$, then the field K would have infinite degree over $\mathbb{Q}$, so there would be infinitely many distinct homomorphisms $K \to \mathbb{C}$. Therefore, all of the x_i's and y_i's must be algebraic over $\mathbb{Q}$. $\square$

Example. Let's see how this proposition works in practice. We will look at the elliptic curve

$$C : y^2 = x^3 + x.$$

If $P = (x, y)$ is a point on this curve, then it is easy enough to compute $2P$.

$$2P = \left(\frac{x^4 - 2x^2 + 1}{4y^2}, \frac{x^6 + 5x^4 - 5x^2 - 1}{8y^3} \right).$$

First we will look at the points of order three. We observe that

$$P = (x, y) \text{ has order three} \Longleftrightarrow \left(\begin{array}{c} \text{the } x \text{ coordinate of } 2P \text{ equals} \\ \text{the } x \text{ coordinate of } P \end{array} \right)$$

$$\Longleftrightarrow \frac{x^4 - 2x^2 + 1}{4y^2} = x$$

$$\Longleftrightarrow 3x^4 + 6x^2 - 1 = 0 \quad \text{because } y^2 = x^3 + x.$$

So the points of order three in $C(\mathbb{C})$ are the points whose x coordinates satisfy the polynomial equation

$$3x^4 + 6x^2 - 1 = 0.$$

In particular, the coordinates of the points of order three on C are algebraic numbers.

Now each x gives two possible values for y (note points with $y = 0$ have order two, not order three). This gives eight points of order three, and together with $\mathcal{O}$ they form the group $C[3] \cong \mathbb{Z}/3\mathbb{Z} \oplus \mathbb{Z}/3\mathbb{Z}$.

Since our equation is so simple, we can solve it explicitly. Thus,

$$\alpha = \sqrt{\frac{2\sqrt{3} - 3}{3}} \quad \text{satisfies} \quad 3\alpha^4 + 6\alpha^2 - 1 = 0;$$

and the other three roots are $-\alpha$, $\left(i\sqrt{3}\,\alpha\right)^{-1}$, and $-\left(i\sqrt{3}\,\alpha\right)^{-1}$. Substituting into $y^2 = x^3 + x$, we can then find the y coordinates. Thus, if we let

$$\beta = \sqrt[4]{\frac{8\sqrt{3} - 12}{9}} = \sqrt{\frac{2\alpha}{\sqrt{3}}},$$

then the nine points in $C[3]$ are

$$C[3] = \left\{ \mathcal{O}, (\alpha, \pm\beta), (-\alpha, \pm i\beta), \left(\frac{i}{\sqrt{3}\alpha}, \pm\frac{2\sqrt{-i}}{\sqrt[4]{27}\beta} \right), \left(\frac{-i}{\sqrt{3}\alpha}, \pm\frac{2\sqrt{i}}{\sqrt[4]{27}\beta} \right) \right\}.$$

It is a nice exercise to check that the field generated by the coordinates of these points is $\mathbb{Q}(\beta, i)$, and that $\mathrm{Gal}\big(\mathbb{Q}(\beta, i)/\mathbb{Q}\big)$ is a non-abelian group of order 16. Recall that we never claimed that elliptic curves would give abelian Galois groups over $\mathbb{Q}$; instead, we said they would give abelian Galois groups over certain quadratic fields. For this elliptic curve, we will prove in Section 5 (as a special case of our main theorem) that $\mathrm{Gal}\big(\mathbb{Q}(\beta,i)/\mathbb{Q}(i)\big)$ is an abelian group. You might try to prove this directly, without any reference to elliptic curves.

Next we will look at the points on C having order four. Since a point has order two if and only if its y coordinate is 0, we find that

$$P = (x, y) \text{ has order four} \iff 2P \text{ has order two}$$
$$\iff \text{the } y \text{ coordinate of } 2P \text{ is } 0$$
$$\iff x^6 + 5x^4 - 5x^2 - 1 = 0.$$

So the points of order four in $C(\mathbb{C})$ are the 12 points whose x coordinates satisfy the polynomial equation

$$x^6 + 5x^4 - 5x^2 - 1 = 0.$$

Of course, there are also three points of order two and one point of order one, which altogether give the 16 points in $C[4]$.

The polynomial factors as

$$x^6 + 5x^4 - 5x^2 - 1 = (x - 1)(x + 1)(x^4 + 6x^2 + 1).$$

Further, if we let $\alpha = (\sqrt{2} - 1)\,i$, then

$$x^4 + 6x^2 + 1 = (x - \alpha)(x + \alpha)(x - \alpha^{-1})(x + \alpha^{-1}).$$

And if we let $\beta = (1 + i)\,(\sqrt{2} - 1)$, then $\beta^2 = \alpha^3 + \alpha$. So a little algebra gives us the complete set of points of order four on C:

$$\Big\{(1, \pm\sqrt{2}), (-1, \pm i\sqrt{2}), (\alpha, \pm\beta), (-\alpha, \pm i\beta),$$
$$(\alpha^{-1}, \pm\alpha^{-2}\beta), (-\alpha^{-1}, \pm i\alpha^{-2}\beta)\Big\}.$$

Hence, the points of order four generate the field $\mathbb{Q}\left(i, \sqrt{2}\right)$.

3. A Galois Representation

In the last section we considered the field

$$\mathbb{Q}\left(x_1, y_1, \ldots, x_m, y_m\right\},$$

where $\left\{\mathcal{O}, (x_1, y_1), \ldots, (x_m, y_m)\right\}$ is the set $C[n]$ of points having order dividing n. This field will be our main object of study in the remainder of this chapter, so it is convenient to give it a special name. We will call it the *field of definition of $C[n]$* (*over* $\mathbb{Q}$) and will denote it $\mathbb{Q}(C[n])$. Thus,

$$\mathbb{Q}(C[n]) = \begin{pmatrix} \text{field generated over } \mathbb{Q} \text{ by the } x \text{ and} \\ y \text{ coordinates of all points in } C[n] \end{pmatrix}.$$

Later, if we need to replace $\mathbb{Q}$ by another field F, we will write $F(C[n])$.

We proved in the last section that $\mathbb{Q}(C[n])$ is a Galois extension of $\mathbb{Q}$; we now begin to describe its Galois group. If $\sigma \in \mathrm{Gal}(\mathbb{Q}(C[n])/\mathbb{Q})$ and if $P \in C[n]$, then we know from Section 2 that $\sigma(P) \in C[n]$. Thus, each σ induces a permutation of the set $C[n]$. But this permutation is not completely arbitrary. For example, we also proved in Section 2 that

$$\sigma(P + Q) = \sigma(P) + \sigma(Q), \quad \sigma(-P) = -\sigma(P), \quad \text{and} \quad \sigma(\mathcal{O}) = \mathcal{O}.$$

In other words, each $\sigma \in \mathrm{Gal}(\mathbb{Q}(C[n])/\mathbb{Q})$ gives a homomorphism from $C[n]$ to itself:

$$C[n] \longrightarrow C[n], \qquad P \longmapsto \sigma(P).$$

Further, this homomorphism has an inverse, namely the homomorphism corresponding to σ^{-1}. Thus, each $\sigma \in \mathrm{Gal}(\mathbb{Q}(C[n])/\mathbb{Q})$ gives a group isomorphism from $C[n]$ to itself.

Using the description of $C[n]$ proven in the last section, we can describe these isomorphisms in a very explicit manner. Recall we proved that $C[n]$ is a direct sum of two cyclic groups of order n,

$$C[n] \cong \frac{\mathbb{Z}}{n\mathbb{Z}} \oplus \frac{\mathbb{Z}}{n\mathbb{Z}}.$$

So $C[n]$ can be generated by two "basis" elements, say P_1 and P_2. (There are lots of possible choices for P_1 and P_2, just as in a vector space there are many different bases. It will not matter which basis we choose.) Then the n^2 elements of $C[n]$ are given by the set

$$\left\{ a_1 P_1 + a_2 P_2 : a_1, a_2 \in \frac{\mathbb{Z}}{n\mathbb{Z}} \right\}.$$

In other words, every element of $C[n]$ can be written as $a_1 P_1 + a_2 P_2$ for a *unique* pair of elements $a_1, a_2 \in \mathbb{Z}/n\mathbb{Z}$.

Now suppose that $h : C[n] \to C[n]$ is any homomorphism from $C[n]$ to itself. Then

$$h(a_1 P_1 + a_2 P_2) = a_1 h(P_1) + a_2 h(P_2),$$

so h is completely determined once we know the value of $h(P_1)$ and $h(P_2)$. Conversely, if we take any two points $Q_1, Q_2 \in C[n]$, then we get a homomorphism from $C[n]$ to itself by the rule

$$a_1 P_1 + a_2 P_2 \longmapsto a_1 Q_1 + a_2 Q_2.$$

Note the analogy with linear algebra. A linear transformation between vector spaces can be given by specifying the image of each element in a basis. So we are really just doing linear algebra, except that the scalars for our vector space are in the ring $\mathbb{Z}/n\mathbb{Z}$, rather than in a field. (A vector space with scalars in a ring R is called an R-module. It turns out that not every R-module actually has a basis; but luckily for us, $C[n]$ does.)

So a homomorphism $h : C[n] \to C[n]$ is determined by the values of $h(P_1)$ and $h(P_2)$. Now each of $h(P_1)$ and $h(P_2)$ is itself a linear combination of P_1 and P_2, say

$$h(P_1) = \alpha_h P_1 + \gamma_h P_2,$$

$$h(P_2) = \beta_h P_1 + \delta_h P_2.$$

Here $\alpha_h, \beta_h, \gamma_h, \delta_h$ are elements of $\mathbb{Z}/n\mathbb{Z}$ that are uniquely determined by h. It is suggestive to write this is in matrix notation

$$\bigl(h(P_1), h(P_2)\bigr) = (P_1, P_2) \begin{pmatrix} \alpha_h & \beta_h \\ \gamma_h & \delta_h \end{pmatrix}.$$

Then if $g : C[n] \to C[n]$ is another homomorphism, it is easy to check that the composition $g \circ h$ is given by the usual matrix product:

$$\begin{pmatrix} \alpha_{goh} & \beta_{goh} \\ \gamma_{goh} & \delta_{goh} \end{pmatrix} = \begin{pmatrix} \alpha_g & \beta_g \\ \gamma_g & \delta_g \end{pmatrix} \begin{pmatrix} \alpha_h & \beta_h \\ \gamma_h & \delta_h \end{pmatrix}.$$

For example,

$$\begin{aligned}
\alpha_{goh} P_1 + \gamma_{goh} P_2 &= (g \circ h)(P_1) \\
&= g\bigl(h(P_1)\bigr) \\
&= g(\alpha_h P_1 + \gamma_h P_2) \\
&= \alpha_h g(P_1) + \gamma_h g(P_2) \\
&= \alpha_h(\alpha_g P_1 + \gamma_g P_2) + \gamma_h(\beta_g P_1 + \delta_g P_2) \\
&= (\alpha_h \alpha_g + \gamma_h \beta_g) P_1 + (\alpha_h \gamma_g + \gamma_h \delta_g) P_2.
\end{aligned}$$

This shows that one column is correct; we will leave you to check the other.

The homomorphisms $C[n] \to C[n]$, $P \mapsto \sigma(P)$, that we looked at above are actually isomorphisms; that is, they have inverses. How is the existence of an inverse to a homomorphism $h : C[n] \to C[n]$ reflected in the matrix for h? If we take $g = h^{-1}$, then the matrix for $g \circ h$ is the identity matrix, so we find that

$$\begin{pmatrix} 1 & 0 \\ 0 & 1 \end{pmatrix} = \begin{pmatrix} \alpha_{h^{-1}} & \beta_{h^{-1}} \\ \gamma_{h^{-1}} & \delta_{h^{-1}} \end{pmatrix} \begin{pmatrix} \alpha_h & \beta_h \\ \gamma_h & \delta_h \end{pmatrix}.$$

Thus the matrix associated to an isomorphism is invertible. And conversely, any invertible matrix can be used to define an isomorphism of $C[n]$ to itself.

This suggests that we should look at the set (actually group) of invertible 2×2 matrices with coefficients in $\mathbb{Z}/n\mathbb{Z}$. More generally, one can look at square matrices of any size with coefficients in any commutative ring R. The resulting group is called the *general linear group* and is denoted by

$$\mathrm{GL}_r(R) = \left\{ \begin{array}{l} r \times r \text{ matrices } A \text{ with coefficients} \\ \text{in } R \text{ and satisfying } \det(A) \in R^* \end{array} \right\}.$$

The condition that the determinant be a unit is equivalent to requiring that A^{-1} exists. (Note that A^{-1} is required to have coefficients in the ring R.) The proof of this fact for general rings is the same as the proof you saw in linear algebra when R is a field. Of course, for 2×2 matrices we can just write everything out. Thus, if $A = \begin{pmatrix} \alpha & \beta \\ \gamma & \delta \end{pmatrix}$ is a matrix with coefficients in R and if its determinant $\Delta = \alpha\delta - \beta\gamma$ is a unit in R, then the inverse matrix

$$\begin{pmatrix} \alpha & \beta \\ \gamma & \delta \end{pmatrix}^{-1} = \begin{pmatrix} \delta/\Delta & -\beta/\Delta \\ -\gamma/\Delta & \alpha/\Delta \end{pmatrix}$$

has coefficients in R. Conversely, if A has an inverse with coefficients in R, then $1 = (\det A)(\det A^{-1})$, so $\det A$ is a unit.

Let's look at an example, say the case of 2×2 matrices with coefficients in $\mathbb{Z}/2\mathbb{Z}$. It is easy to list all such matrices with non-zero determinant. There are six of them:

$$\begin{pmatrix} 1 & 0 \\ 0 & 1 \end{pmatrix}, \begin{pmatrix} 1 & 0 \\ 1 & 1 \end{pmatrix}, \begin{pmatrix} 1 & 1 \\ 0 & 1 \end{pmatrix}, \begin{pmatrix} 1 & 1 \\ 1 & 0 \end{pmatrix}, \begin{pmatrix} 0 & 1 \\ 1 & 0 \end{pmatrix}, \begin{pmatrix} 0 & 1 \\ 1 & 1 \end{pmatrix}.$$

The group $\mathrm{GL}_2(\mathbb{Z}/2\mathbb{Z})$ is isomorphic to the symmetric group on three letters; a quick way to get an isomorphism is to look at the way that the matrices permute the three non-zero vectors in the vector space $(\mathbb{Z}/2\mathbb{Z})^2$.

Let us briefly recapitulate. To each element $\sigma \in \mathrm{Gal}(\mathbb{Q}(C[n])/\mathbb{Q})$ we have associated an isomorphism from $C[n]$ to itself. And to each such isomorphism, we have associated a matrix in $\mathrm{GL}_2(\mathbb{Z}/n\mathbb{Z})$. So we get a map

$$\rho_n : \mathrm{Gal}(\mathbb{Q}(C[n])/\mathbb{Q}) \longrightarrow \mathrm{GL}_2\left(\frac{\mathbb{Z}}{n\mathbb{Z}}\right), \qquad \rho_n(\sigma) = \begin{pmatrix} \alpha_\sigma & \beta_\sigma \\ \gamma_\sigma & \delta_\sigma \end{pmatrix},$$

where $\alpha_\sigma, \beta_\sigma, \gamma_\sigma, \delta_\sigma$ are determined by the rule

$$\sigma(P_1) = \alpha_\sigma P_1 + \gamma_\sigma P_2,$$
$$\sigma(P_2) = \beta_\sigma P_1 + \delta_\sigma P_2.$$

Further, the matrix computation we did above shows that

$$\rho_n(\sigma\tau) = \rho_n(\sigma)\rho_n(\tau);$$

hence, ρ_n is a homomorphism. So we have constructed a homomorphism from the complicated group $\text{Gal}(\mathbb{Q}(C[n])/\mathbb{Q})$ that we are trying to study into a group of matrices $\text{GL}_2(\mathbb{Z}/n\mathbb{Z})$. Such a homomorphism is called a *representation*.[†] Since $\text{Gal}(\mathbb{Q}(C[n])/\mathbb{Q})$ is a Galois group, the representation ρ_n is often called a *Galois representation*.

We have now proven a lot of important facts, which we record in the following theorem.

Galois Representation Theorem.   *Let C be an elliptic curve given by a Weierstrass equation with rational coefficients, and let $n \geq 2$ be an integer. Fix generators P_1 and P_2 for $C[n]$. Then the map*

$$\rho_n : \text{Gal}\big(\mathbb{Q}(C[n])/\mathbb{Q}\big) \longrightarrow \text{GL}_2\left(\frac{\mathbb{Z}}{n\mathbb{Z}}\right)$$

described above is a one-to-one homomorphism of groups.

PROOF. We have proven everything except that ρ_n is one-to-one. Suppose that $\sigma \in \text{Gal}\big(\mathbb{Q}(C[n])/\mathbb{Q}\big)$ is in the kernel of ρ_n, so $\rho_n(\sigma) = \left(\begin{smallmatrix} 1 & 0 \\ 0 & 1 \end{smallmatrix}\right)$. This means that $\sigma(P_1) = P_1$ and $\sigma(P_2) = P_2$, from which it follows that $\sigma(P) = P$ for every $P \in C[n]$. Since by definition $\sigma(x, y) = \big(\sigma(x), \sigma(y)\big)$, this means that σ fixes the x and y coordinates of every point in $C[n]$. Now recall that $\mathbb{Q}(C[n])$ is generated over $\mathbb{Q}$ by the x and y coordinates of the points in $C[n]$. Hence, σ fixes the generators of $\mathbb{Q}(C[n])$, and so it fixes the entire field $\mathbb{Q}(C[n])$. This means that σ is the identity element of $\text{Gal}\big(\mathbb{Q}(C[n])/\mathbb{Q}\big)$, which proves that the kernel of ρ_n consists of only one element. Therefore, ρ_n is one-to-one.   □

Notice the analogy with the cyclotomic extensions studied in Section 1. If we choose a generator $\zeta \in \mathbb{C}^*$ for the group of nth roots of unity, then we get a homomorphism

$$t : \text{Gal}\big(\mathbb{Q}(\zeta)/\mathbb{Q}\big) \longrightarrow \text{GL}_1\left(\frac{\mathbb{Z}}{n\mathbb{Z}}\right) = \left(\frac{\mathbb{Z}}{n\mathbb{Z}}\right)^*$$

[†] The theory of group representations is an extremely powerful tool for studying groups. It is used extensively in mathematics, physics, and chemistry. We will not need any of the general theory, but for those who are interested, a very nice introduction to the representation theory of finite groups is given in Serre [2].

according to the rule $\sigma(\zeta) = \zeta^{t(\sigma)}$. The homomorphism t is called the nth cyclotomic representation of $\mathbb{Q}$. As we mentioned (but did not prove) in Section 1, the cyclotomic representation is not only one-to-one, it is actually an isomorphism; so the Galois group of $\mathbb{Q}(\zeta)$ over $\mathbb{Q}$ is isomorphic to the unit group of the ring $\mathbb{Z}/n\mathbb{Z}$.

We have now done a lot of abstract theory, so this might be a good time to look at some particular elliptic curves and explicitly determine the representation ρ_n for some small value of n, such as $n = 2$.

Example 1. Consider the elliptic curve given by the equation

$$C : y^2 = x(x-1)(x-2).$$

Then

$$C[2] = \{\mathcal{O}, (0,0), (1,0), (2,0)\}$$

consists entirely of rational points. So $\mathbb{Q}(C[2]) = \mathbb{Q}$, and hence the Galois group $\text{Gal}(\mathbb{Q}(C[2])/\mathbb{Q})$ is the trivial group $\{\sigma_0\}$. The representation

$$\rho_2 : \text{Gal}(\mathbb{Q}(C[2])/\mathbb{Q}) \longrightarrow \text{GL}_2\left(\frac{\mathbb{Z}}{2\mathbb{Z}}\right)$$

is given by $\rho_2(\sigma_0) = \left(\begin{smallmatrix} 1 & 0 \\ 0 & 1 \end{smallmatrix}\right)$. In particular, the image of ρ_2 is definitely not onto; so in contrast to the case of the cyclotomic representation, the Galois representation associated to an elliptic curve need not be an isomorphism.

Example 2. Next we look at the elliptic curve

$$C : y^2 = x^3 + x.$$

The points of order two are not all rational, but they are easy to write down:

$$C[2] = \{\mathcal{O}, (0,0), (i,0), (-i,0)\}.$$

(Here, as always, we will use i to represent $\sqrt{-1}$.) Thus, $\mathbb{Q}(C[2]) = \mathbb{Q}(i)$, and the Galois group $\text{Gal}(\mathbb{Q}(C[2])/\mathbb{Q}) = \{\sigma_0, \sigma_1\}$ contains two elements, the identity σ_0 and complex conjugation σ_1.

To find the representation ρ_2, we need to choose generators for $C[2]$, say we take $P_1 = (0,0)$ and $P_2 = (i,0)$. Then

$$\sigma_1(P_1) = \sigma_1(0,0) = (0,0) = P_1,$$
$$\sigma_1(P_2) = \sigma_1(i,0) = (-i,0) = P_1 + P_2.$$

So the matrix associated to σ_1 is $\left(\begin{smallmatrix} 1 & 1 \\ 0 & 1 \end{smallmatrix}\right)$, and the representation ρ_2 is given explicitly by

$$\rho_2(\sigma_0) = \begin{pmatrix} 1 & 0 \\ 0 & 1 \end{pmatrix}, \qquad \rho_2(\sigma_1) = \begin{pmatrix} 1 & 1 \\ 0 & 1 \end{pmatrix}.$$

Example 3. Finally we will examine the elliptic curve

$$C : y^2 = x^3 - 2.$$

The points of order two are

$$C[2] = \left\{ \mathcal{O}, \left(\sqrt[3]{2}, 0 \right), \left(\zeta \sqrt[3]{2}, 0 \right), \left(\zeta^2 \sqrt[3]{2}, 0 \right) \right\} = \{ \mathcal{O}, P_1, P_2, P_3 \},$$

where $\zeta = \frac{1}{2} \left(-1 + \sqrt{-3} \right)$ is a primitive 3rd root of unity. So the field generated by the points of order two is

$$\mathbb{Q}(C[2]) = \mathbb{Q} \left(\sqrt{-3}, \sqrt[3]{2} \right).$$

The Galois group $\text{Gal}(\mathbb{Q}(C[2])/\mathbb{Q})$ has order six; it is the full symmetric group on the set consisting of the three non-zero points in $C[2]$.

We will write

$$\text{Gal}(\mathbb{Q}(C[2])/\mathbb{Q}) = \{ e, \sigma, \sigma^2, \tau, \sigma\tau, \sigma^2\tau \},$$

where σ and τ are determined by the properties

$$\sigma \left(\sqrt[3]{2} \right) = \zeta \sqrt[3]{2}, \qquad \tau \left(\sqrt[3]{2} \right) = \sqrt[3]{2},$$

$$\sigma \left(\sqrt{-3} \right) = \sqrt{-3}, \qquad \tau \left(\sqrt{-3} \right) = -\sqrt{-3}.$$

Then σ and τ satisfy the relations $\sigma^3 = \tau^2 = e$ and $\sigma\tau = \tau\sigma^2$. We also observe that $\tau(\zeta) = \zeta^2$.

Next we need to choose generators for $C[2]$; we will take the points P_1 and P_2 described above. Then

$$\sigma(P_1) = \sigma \left(\sqrt[3]{2}, 0 \right) = \left(\zeta \sqrt[3]{2}, 0 \right) = P_2,$$

$$\sigma(P_2) = \sigma \left(\zeta \sqrt[3]{2}, 0 \right) = \left(\zeta^2 \sqrt[3]{2}, 0 \right) = P_3 = P_1 + P_2.$$

From this computation we can read off $\rho_2(\sigma) = \left(\begin{smallmatrix} 0 & 1 \\ 1 & 1 \end{smallmatrix} \right)$.

Similarly, we have

$$\tau(P_1) = \tau \left(\sqrt[3]{2}, 0 \right) = \left(\sqrt[3]{2}, 0 \right) = P_1,$$

$$\tau(P_2) = \tau \left(\zeta \sqrt[3]{2}, 0 \right) = \left(\zeta^2 \sqrt[3]{2}, 0 \right) = P_3 = P_1 + P_2.$$

So the matrix for τ is $\rho_2(\tau) = \left(\begin{smallmatrix} 1 & 1 \\ 0 & 1 \end{smallmatrix} \right)$.

Since the representation ρ_2 is a homomorphism, and since σ and τ generate $\text{Gal}(\mathbb{Q}(C[2])/\mathbb{Q})$, we can use the values of $\rho_2(\sigma)$ and $\rho_2(\tau)$ to compute ρ_2 for any element of $\text{Gal}(\mathbb{Q}(C[2])/\mathbb{Q})$. For example,

$$\rho_2(\sigma^2\tau) = \rho_2(\sigma)^2 \rho_2(\tau) = \begin{pmatrix} 0 & 1 \\ 1 & 1 \end{pmatrix}^2 \begin{pmatrix} 1 & 1 \\ 0 & 1 \end{pmatrix} = \begin{pmatrix} 1 & 0 \\ 1 & 1 \end{pmatrix}.$$

Of course, one can also compute directly that

$$(\sigma^2\tau)(P_1) = P_1 + P_2 \quad \text{and} \quad (\sigma^2\tau)(P_2) = P_2.$$

Recall that one of our goals in this chapter is to construct field extensions with abelian Galois groups. Naturally, we are going to use the fields $\mathbb{Q}(C[n])$ that we have been studying. We have proven that there is a one-to-one homomorphism

$$\rho_n : \text{Gal}\big(\mathbb{Q}(C[n])/\mathbb{Q}\big) \longrightarrow \text{GL}_2\left(\frac{\mathbb{Z}}{n\mathbb{Z}}\right).$$

We have also seen that ρ_n need not be onto. That's good because the group $\text{GL}_2(\mathbb{Z}/n\mathbb{Z})$ with $n \geq 2$ is never an abelian group! For example, the matrices $\left(\begin{smallmatrix} 1 & 1 \\ 0 & 1 \end{smallmatrix}\right)$ and $\left(\begin{smallmatrix} 0 & 1 \\ 1 & 0 \end{smallmatrix}\right)$ do not commute. (You should check this.)

It turns out that for most elliptic curves the representations ρ_n are "almost" onto. It is only for a very special class of elliptic curves, called *elliptic curves with complex multiplication*, that we will get abelian Galois groups. We will save the actual definition of complex multiplication for the next section, but to finish our general discussion of representations, we want to quote the following beautiful (and difficult) theorem of Serre which explains one sense in which the ρ_n's are "almost" onto.

Theorem. (Serre [3,4], 1972) *Let C be an elliptic curve given by a Weierstrass equation with rational coefficients. Assume that C does not have complex multiplication. There is an integer $N \geq 1$, depending on the curve C, so that if n is any integer relatively prime to N, then the Galois representation*

$$\rho_n : \text{Gal}\big(\mathbb{Q}(C[n])/\mathbb{Q}\big) \longrightarrow \text{GL}_2\left(\frac{\mathbb{Z}}{n\mathbb{Z}}\right)$$

is an isomorphism.

4. Complex Multiplication

The complex points on an elliptic curve $C(\mathbb{C})$ form an abelian group, and for any abelian group and any integer n, there is a *multiplication-by-n homomorphism*

$$C(\mathbb{C}) \xrightarrow[\text{by } n]{\text{multiplication}} C(\mathbb{C}), \quad P \longmapsto nP.$$

The kernel of this homomorphism is precisely $C[n]$, the set of points of order dividing n.

The multiplication by n homomorphism on $C(\mathbb{C})$ has the special property that it is defined by rational functions; that is, the x and y coordinates of nP are given by rational functions of the x and y coordinates of P. For example, if $P = (x, y)$ is a point on the elliptic curve

$$y^2 = x^3 + ax^2 + bx + c,$$

then after a moderate amount of computation one finds that

$$2P = \left(\frac{g(x)}{4y^2}, \frac{h(x)}{8y^3} \right), \quad \text{where}$$

$$g(x) = x^4 - 2bx^2 - 8cx + b^2 - 4ac,$$

$$h(x) = x^6 + 2ax^5 + 5bx^4 + 20cx^3 + 5(4ac - b^2)x^2$$
$$+ 2(4a^2c - ab^2 - 2bc)x + 4abc - b^3 - 8c^2.$$

In general, a homomorphism $\phi : C(\mathbb{C}) \to C(\mathbb{C})$ which is defined by rational functions is called an *isogeny*. That is, an isogeny is a homomorphism $\phi : C(\mathbb{C}) \to C(\mathbb{C})$ that has the form

$$\phi(x, y) = \left(\frac{\text{polynomial in } x \text{ and } y}{\text{polynomial in } x \text{ and } y}, \frac{\text{polynomial in } x \text{ and } y}{\text{polynomial in } x \text{ and } y} \right).$$

More generally, one can look at isogenies $\phi : C(\mathbb{C}) \to \overline{C}(\mathbb{C})$ between two different elliptic curves. For example, consider the two elliptic curves

$$C : y^2 = x^3 + ax^2 + bx \quad \text{and} \quad \overline{C} : \overline{y}^2 = \overline{x}^3 + \overline{a}\,\overline{x}^2 + \overline{b}\,\overline{x}$$

that we studied in Chapter III, Section 4, where $\overline{a} = -2a$ and $\overline{b} = a^2 - 4b$. We showed in Chapter III that the function

$$\phi : C(\mathbb{C}) \longrightarrow \overline{C}(\mathbb{C}), \qquad \phi(x, y) = \left(\frac{y^2}{x^2}, \frac{y(x^2 - b)}{x^2} \right)$$

is a homomorphism. Thus ϕ is an isogeny from the elliptic curve C to the elliptic curve $\overline{C}$.

We will be particularly interested in isogenies from an elliptic curve to itself; such isogenies are called *endomorphisms* (or sometimes *algebraic endomorphisms*, to emphasize the fact that they are defined by rational functions). We have just seen that every elliptic curve has the multiplication-by-n endomorphisms, one for each integer n. For most elliptic curves, that's the whole story; there will be no other endomorphisms. However, there are some elliptic curves with additional endomorphisms. We will be focusing our attention on these special elliptic curves, which provides some justification for giving them a special name.

Definition. Let C be an elliptic curve. We say that C has *complex multiplication* (or CM for short) if there is an endomorphism $\phi : C \to C$ which is not a multiplication by n map.

It might be helpful at this point to give a few examples of elliptic curves having complex multiplication.

Example 1. The elliptic curve

$$C : y^2 = x^3 + x$$

has the complex multiplication

$$\phi(x, y) = (-x, iy)$$

because if $y^2 = x^3 + x$, then

$$(iy)^2 = -y^2 = -x^3 - x = (-x)^3 + (-x).$$

(Here $i = \sqrt{-1}$ as usual.)

Example 2. The elliptic curve

$$C : y^2 = x^3 + 1$$

has the complex multiplication

$$\phi(x, y) = \left(\frac{-1 + \sqrt{-3}}{2} x, -y \right).$$

Since $\left(\dfrac{-1 + \sqrt{-3}}{2} \right)^3 = 1$, it is easy to see that $\phi(x, y)$ is indeed a point on C.

Example 3. We recalled above an isogeny ϕ between two different curves

$$C : y^2 = x^3 + ax^2 + bx \qquad \text{and} \qquad \overline{C} : \overline{y}^2 = \overline{x}^3 + \overline{a}\,\overline{x}^2 + \overline{b}\overline{x}.$$

Suppose that we choose a and b so that these two curves are isomorphic. Then composing the isogeny $\phi : C \to \overline{C}$ with the isomorphism $\overline{C} \xrightarrow{\sim} C$ will give an endomorphism of C. For example, if we take $a = 0$, then the curves $C : y^2 = x^3 + bx$ and $\overline{C} : \overline{y}^2 = \overline{x}^3 - 4b\overline{x}$ are isomorphic via the map

$$\overline{C} \longrightarrow C, \qquad (\overline{x}, \overline{y}) \longmapsto \left(\frac{i}{2}\overline{x}, \frac{i-1}{4}\overline{y} \right).$$

Composing this with the isogeny $\phi : C \to \overline{C}$ from above gives the endomorphism

$$\psi : C \longrightarrow C, \qquad (x, y) \longmapsto \left(\frac{iy^2}{2x^2}, \frac{(i-1)y(x^2 - b)}{4x^2} \right).$$

This endomorphism may look mysterious, but it really isn't. Notice that the curve $C : y^2 = x^3 + bx$ is essentially the same as the curve from Example 1; in particular, it has the obvious endomorphism defined by $(x, y) \to (-x, -iy)$. Then it is not hard to check that the complicated map ψ is really just

$$\psi(x, y) = (x, y) + (-x, -iy).$$

(N.B. The $+$ sign means addition on the elliptic curve C.)

More generally, if ϕ_1 and ϕ_2 are endomorphisms of C, then we can define a new endomorphism $\phi_1 + \phi_2$ by

$$(\phi_1 + \phi_2) : C \longrightarrow C, \qquad (\phi_1 + \phi_2)(P) = \phi_1(P) + \phi_2(P).$$

We can also get a new endomorphism by taking the composition,

$$(\phi_1 \phi_2) : C \longrightarrow C, \qquad (\phi_1 \phi_2)(P) = \phi_1\big(\phi_2(P)\big).$$

With this "addition" and "multiplication," the set of endomorphisms of C becomes a ring. If C does not have complex multiplication, then this ring is just isomorphic to $\mathbb{Z}$, the ordinary ring of integers. But, if C does have complex multiplication, then the endomorphism ring is strictly larger than $\mathbb{Z}$. It is an interesting question (which we will partially answer in the exercises) as to what sort of ring it can be.

You may have noticed that we did not completely verify that the maps in Examples 1, 2, and 3 are endomorphisms. We did show that they are maps from C to C given by rational functions, but we did not check that they are homomorphisms. Using the explicit formulas for the group law, it is tedious (but not difficult) to check this. However, as the following theorem shows, there is actually no need to do the work. Unfortunately, the proof is too complicated for us to give. (See Silverman [2] III.4.8.)

Theorem. Let C and $\overline{C}$ be elliptic curves, and let $\phi : C(\mathbb{C}) \to \overline{C}(\mathbb{C})$ be a map given by rational functions and satisfying $\phi(\mathcal{O}) = \overline{\mathcal{O}}$. Then ϕ is automatically a homomorphism.

Why is an elliptic curve with an extra endomorphism said to have "complex multiplication"? Recall from Chapter II, Section 2 that the complex points on an elliptic curve look like $\mathbb{C}/L$, where

$$L = \{a_1\omega_1 + a_2\omega_2 : a_1, a_2 \in \mathbb{Z}\}$$

is a lattice in $\mathbb{C}$. So an endomorphism $\phi : C(\mathbb{C}) \to C(\mathbb{C})$ gives a holomorphic map

$$f : \frac{\mathbb{C}}{L} \longrightarrow \frac{\mathbb{C}}{L}.$$

This means that f is given by a convergent power series (at least in a neighborhood of 0),

$$f(z) = c_0 + c_1 z + c_2 z^2 + c_3 z^3 + \cdots.$$

We also know that f is a homomorphism, so $f(z_1 + z_2) = f(z_1) + f(z_2)$ for all z_1, z_2 in a neighborhood of 0. Of course, this equality is in the quotient $\mathbb{C}/L$, so we should really say that

$$f(z_1 + z_2) - f(z_1) - f(z_2) \in L \quad \text{for all } z_1, z_2 \text{ close to } 0.$$

But L consists of a discrete set of points in $\mathbb{C}$, and therefore contains no non-empty open set; but the image of a non-constant holomorphic function is open, so the difference $f(z_1 + z_2) - f(z_1) - f(z_2)$ must be constant. Putting $z_1 = z_2 = 0$, we see that the constant is $-c_0$, so

$$f(z_1 + z_2) + c_0 = f(z_1) + f(z_2) \quad \text{for all } z_1, z_2 \text{ close to } 0.$$

Further, since $f(0) = 0$ in $\mathbb{C}/L$, we find $c_0 \in L$. Hence the maps $z \to f(z)$ and $z \to f(z) - c_0$ give the same endomorphism of $\mathbb{C}/L$, so we may as well take the latter in place of the former. This means we may assume that $c_0 = 0$, so

$$f(z_1 + z_2) = f(z_1) + f(z_2) \quad \text{for all } z_1, z_2 \text{ close to } 0.$$

As you might suspect, there are not very many power series with this property.

Proposition. *Let $f(z)$ be a holomorphic function at 0 with the property that*

$$f(z_1 + z_2) = f(z_1) + f(z_2)$$

for all z_1, z_2 in some neighborhood of 0. Then $f(z) = cz$ for some $c \in \mathbb{C}$.

PROOF. Putting $z_1 = z_2 = 0$, we see that $f(0) = 2f(0)$, so $f(0) = 0$. Next we compute the derivative of f:

$$f'(z) = \lim_{h \to 0} \frac{f(z+h) - f(z)}{h}$$

$$= \lim_{h \to 0} \frac{(f(z) + f(h)) - f(z)}{h} \quad \text{from given property of } f,$$

$$= \lim_{h \to 0} \frac{f(h) - f(0)}{h} \quad \text{because } f(0) = 0,$$

$$= f'(0).$$

In other words, the derivative of $f(z)$ is constant, which means that $f(z)$ is linear, $f(z) = c_0 + c_1 z$. Then $0 = f(0) = c_0$, so $f(z) = c_1 z$. $\qquad\square$

Now let $\phi : C(\mathbb{C}) \to C(\mathbb{C})$ be an endomorphism. From the proposition, there is some $c \in \mathbb{C}$ so that ϕ is given by

$$f : \frac{\mathbb{C}}{L} \longrightarrow \frac{\mathbb{C}}{L}, \qquad f(z) = cz.$$

But c cannot be chosen completely arbitrarily because f is a function on the quotient group $\mathbb{C}/L$. Thus if $z_1, z_2 \in \mathbb{C}$ differ by an element of L, which means that they give the same element of $\mathbb{C}/L$, then we must have $f(z_1) = f(z_2)$. In terms of c, we find that

$$z_1 - z_2 \in L \Longrightarrow f(z_1) = f(z_2) \Longrightarrow cz_1 = cz_2 \text{ in } \frac{\mathbb{C}}{L} \Longrightarrow c(z_1 - z_2) \in L.$$

Hence, c must satisfy the condition $cL \subset L$; and conversely, if $cL \subset L$, then $f(z) = cz$ gives an endomorphism of $\mathbb{C}/L$. (Note: cL means the set of all complex numbers $c\omega$ with $\omega \in L$.)

So finally we ask "what are the possible values for c?" Since L is an abelian group, we will certainly have $cL \subset L$ if c is an integer. These are just the multiplication-by-c maps. If the elliptic curve has complex multiplication, then by definition there is at least one more value of $c \in \mathbb{C}$ such that $cL \subset L$. We will now prove that this additional value is necessarily complex (i.e., it is not real). So it is natural to say that the lattice L has *complex multiplication* because there is a complex number c such that $cL \subset L$. This is the origin of the appellation "complex multiplication" for elliptic curves with an extra endomorphism. (For additional information about the complex number c, see Exercise 6.15.)

Proposition. *Let $\mathbb{C}/L$ be an elliptic curve with a complex multiplication*

$$f : \frac{\mathbb{C}}{L} \longrightarrow \frac{\mathbb{C}}{L}, \qquad f(z \bmod L) = cz \, (\bmod \, L)$$

(i.e., $c \notin \mathbb{Z}$). Then c is not a real number.

PROOF. Choose generators for L, say

$$L = \mathbb{Z}\omega_1 + \mathbb{Z}\omega_2 = \{a_1\omega_1 + a_2\omega_2 : a_1, a_2 \in \mathbb{Z}\}.$$

Note that ω_1 and ω_2 are linearly independent over $\mathbb{R}$ because otherwise L would not be a lattice. In other words, if $r_1, r_2 \in \mathbb{R}$ satisfy $r_1\omega_1 + r_2\omega_2 = 0$, then $r_1 = r_2 = 0$.

We know that $cL \subset L$. In particular, $c\omega_1$ is in L, so we can find integers $A, B \in \mathbb{Z}$ so that

$$c\omega_1 = A\omega_1 + B\omega_2.$$

Thus

$$(c - A)\omega_1 - B\omega_2 = 0.$$

If c were real, we could conclude that $c - A = B = 0$, so $c = A$. This would contradict our assumption that $c \notin \mathbb{Z}$. Therefore, c is not real. □

5. Abelian Extensions of $\mathbb{Q}(i)$

In this section we will look at the elliptic curve

$$C : y^2 = x^3 + x$$

and the fields generated by its points of finite order. We have seen (Section 4, Example 1) that C has complex multiplication,

$$\phi : C \longrightarrow C, \qquad \phi(x,y) = (-x, iy).$$

Since the endomorphism ϕ involves i, it is probably not too surprising that we will be looking at extensions of the field $\mathbb{Q}(i)$. But there is a more concrete reason why $\mathbb{Q}(i)$ is the "right" field to study.

Let $K/\mathbb{Q}$ be any Galois extension with $i \in K$, and let $\sigma \in \text{Gal}(K/\mathbb{Q})$. Then for any point $P \in C(K)$, we have two ways to get a new point in $C(K)$, namely we can apply the endomorphism ϕ to P or we can apply the Galois element σ to P. We ask whether σ and ϕ commute. In other words, is it true that

$$\sigma\big(\phi(P)\big) = \phi\big(\sigma(P)\big) \quad \text{for every point } P \in C(K)?$$

Using the definitions, we see that

$$\sigma\big(\phi(P)\big) = \sigma(-x, iy) = \big(\sigma(-x), \sigma(iy)\big) = \big(-\sigma(x), \sigma(i)\sigma(y)\big),$$

$$\phi\big(\sigma(P)\big) = \phi\big(\sigma(x), \sigma(y)\big) = \big(-\sigma(x), i\sigma(y)\big).$$

So σ and ϕ will commute provided that $\sigma(i) = i$; in other words, they will commute if $\sigma \in \text{Gal}\big(K/\mathbb{Q}(i)\big)$. So if we are going to be using the map ϕ to study Galois groups, it makes sense to look at Galois extensions of $\mathbb{Q}(i)$ rather than of $\mathbb{Q}$.

Our main theorem says that the points of finite order on C generate abelian extensions of $\mathbb{Q}(i)$.

Theorem. *Let C be the elliptic curve*

$$y^2 = x^3 + x.$$

For each integer $n \geq 1$, let

$$K_n = \mathbb{Q}(i)\big(C[n]\big)$$

be the field generated by i and the coordinates of the points in $C[n]$. Then K_n is a Galois extension of $\mathbb{Q}(i)$, and its Galois group is abelian.

PROOF. The fields $\mathbb{Q}(i)$ and $\mathbb{Q}\big(C[n]\big)$ are both Galois over $\mathbb{Q}$ (cf. Section 2), so their compositum K_n is actually Galois over $\mathbb{Q}$. Hence K_n is certainly

Galois over $\mathbb{Q}(i)$. [If you do not like this proof, you can just rework the proof from Section 2, replacing $\mathbb{Q}$ by $\mathbb{Q}(i)$.]

Now comes the interesting part of the theorem, namely the fact that the group $\mathrm{Gal}(K_n/\mathbb{Q}(i))$ is abelian. We will use the representation that we studied in Section 3. Thus, we fix generators $P_1, P_2 \in C[n]$, and then we get a one-to-one homomorphism

$$\rho_n : \mathrm{Gal}(K_n/\mathbb{Q}(i)) \longrightarrow \mathrm{GL}_2\left(\frac{\mathbb{Z}}{n\mathbb{Z}}\right), \qquad \rho_n(\sigma) = \begin{pmatrix} \alpha_\sigma & \beta_\sigma \\ \gamma_\sigma & \delta_\sigma \end{pmatrix},$$

where $\alpha_\sigma, \beta_\sigma, \gamma_\sigma, \delta_\sigma$ are determined by the formulas

$$\sigma(P_1) = \alpha_\sigma P_1 + \gamma_\sigma P_2,$$
$$\sigma(P_2) = \beta_\sigma P_1 + \delta_\sigma P_2.$$

In a similar manner, the endomorphism $\phi : C \to C$ gives a homomorphism $\phi : C[n] \to C[n]$. [That is, if $P \in C[n]$, then $n\phi(P) = \phi(nP) = \phi(\mathcal{O}) = \mathcal{O}$, so $\phi(P) \in C[n]$.] There are thus elements $a, b, c, d \in \mathbb{Z}/n\mathbb{Z}$ such that

$$\phi(P_1) = aP_1 + cP_2,$$
$$\phi(P_2) = bP_1 + dP_2.$$

So the homomorphism $\phi : C[n] \to C[n]$ corresponds to the matrix $\begin{pmatrix} a & b \\ c & d \end{pmatrix}$.

Further, and this is one of the crucial steps in the proof, we saw above that for all $\sigma \in \mathrm{Gal}(K_n/\mathbb{Q}(i))$ and all $P \in C(K_n)$,

$$\sigma(\phi(P)) = \phi(\sigma(P)).$$

If we apply this with $P = P_1$ and $P = P_2$, this means that the matrices associated to σ and ϕ commute:

$$\begin{pmatrix} \alpha_\sigma & \beta_\sigma \\ \gamma_\sigma & \delta_\sigma \end{pmatrix} \begin{pmatrix} a & b \\ c & d \end{pmatrix} = \begin{pmatrix} a & b \\ c & d \end{pmatrix} \begin{pmatrix} \alpha_\sigma & \beta_\sigma \\ \gamma_\sigma & \delta_\sigma \end{pmatrix}.$$

There are two more steps needed to complete the proof of the theorem. First, we will show that the matrix for ϕ is not a *scalar matrix*; that is, it is not a multiple of the identity matrix. Second, we will use a little linear algebra to show that if a 2×2 matrix A is not a scalar matrix, then any two matrices that commute with A also commute with one another. From this we will be able to conclude that the image of ρ_n is an abelian subgroup of $\mathrm{GL}_2(\mathbb{Z}/n\mathbb{Z})$; and then since ρ_n is one-to-one, we will deduce that $\mathrm{Gal}(K_n/\mathbb{Q}(i))$ is also abelian.

Lemma 1. Let $A = \begin{pmatrix} a & b \\ c & d \end{pmatrix}$ be the matrix corresponding to ϕ as above.
(a) $A \in \mathrm{GL}_2(\mathbb{Z}/n\mathbb{Z})$.
(b) A is not a scalar matrix modulo ℓ for all primes ℓ dividing n. In other words, for every prime ℓ dividing n, at least one of the following three conditions is true:

    (i) $b \not\equiv 0 \pmod{\ell}$;

    (ii) $c \not\equiv 0 \pmod{\ell}$;

    (iii) $a \not\equiv d \pmod{\ell}$.

PROOF. (a) We need to show that $\det(A)$ is a unit in $\mathbb{Z}/n\mathbb{Z}$. If we compose ϕ with itself, we find that

$$\phi(\phi(P)) = \phi(\phi(x,y)) = \phi(-x, iy) = (x, -y) = -P.$$

Thus, the matrix A corresponding to ϕ satisfies $A^2 = \begin{pmatrix} -1 & 0 \\ 0 & -1 \end{pmatrix}$, which implies that $1 = \det(A^2) = \det(A)^2$. Therefore, $\det(A)$ is a unit, and hence $A \in \mathrm{GL}_2(\mathbb{Z}/n\mathbb{Z})$.

(b) Suppose to the contrary that there is some prime ℓ dividing n such that

$$\begin{pmatrix} a & b \\ c & d \end{pmatrix} \equiv \begin{pmatrix} m & 0 \\ 0 & m \end{pmatrix} \pmod{\ell}.$$

This means that $\phi : C[\ell] \to C[\ell]$ is the same as the multiplication-by-m map,

$$\phi(P) = mP \qquad \text{for all } P \in C[\ell].$$

Let $\tau : \mathbb{C} \to \mathbb{C}$ be complex conjugation. Once we fix a particular inclusion $K_n \subset \mathbb{C}$, we can think of τ as an element of $\mathrm{Gal}(K_n/\mathbb{Q})$. From Section 2 we know that $\tau(mP) = m\tau(P)$. On the other hand, $\tau(i) = -i$ by definition, so we find that

$$\tau(\phi(P)) = \tau(-x, iy) = \big(\tau(-x), \tau(iy)\big) = \big(-\tau(x), -i\tau(y)\big) = -\phi(\tau(P)).$$

This is true for any point in $C(K_n)$; so, in particular, it is true for points in $C[\ell]$. We therefore find that for every point $P \in C[\ell]$,

$$\begin{aligned} m\tau(P) &= \tau(mP) \\ &= \tau(\phi(P)) \\ &= -\phi(\tau(P)) \\ &= -m\tau(P) \quad \text{because } \tau(P) \text{ is also in } C[\ell]. \end{aligned}$$

Hence $2m\tau(P) = \mathcal{O}$ for every $P \in C[\ell]$.

But τ just permutes the elements in $C[\ell]$, and so $2mP = \mathcal{O}$ for every $P \in C[\ell]$. There are two possibilities — either $\ell = 2$, or ℓ divides m. But if $\ell|m$, then $\phi(P) = \mathcal{O}$ for every $P \in C[\ell]$, which is absurd because for example $\phi(\phi(P)) = -P$. So we must have $\ell = 2$.

But for $\ell = 2$ we can explicitly compute the matrix for ϕ. As generators for $C[2]$, we take $P_1 = (0,0)$ and $P_2 = (i,0)$. Then

$$\phi(P_1) = (0,0) = P_1 \qquad \text{and} \qquad \phi(P_2) = (-i,0) = P_1 + P_2,$$

so the matrix for $\phi : C[2] \to C[2]$ is $\begin{pmatrix} 1 & 1 \\ 0 & 1 \end{pmatrix}$. Since this matrix is not diagonal, we have also eliminated $\ell = 2$, which completes the proof of Lemma 1.

$\square$

Lemma 2. *Let $A \in \mathrm{GL}_2(\mathbb{Z}/n\mathbb{Z})$ be a matrix which is not a scalar matrix modulo ℓ for all primes ℓ dividing n. Then*

$$\left\{ B \in \mathrm{GL}_2\left(\frac{\mathbb{Z}}{n\mathbb{Z}}\right) \; : \; AB = BA \right\}$$

is an abelian subgroup of $\mathrm{GL}_2(\mathbb{Z}/n\mathbb{Z})$. (In other words, the matrices that commute with A also commute with one another.)

PROOF. It is easy to check that the set is a subgroup of $\mathrm{GL}_2(\mathbb{Z}/n\mathbb{Z})$; we will leave the verification to you. The hard part is to show that it is abelian. We are going to prove Lemma 2 one prime at a time.[†] In order to show that two numbers (or matrices) are congruent modulo n, it suffices to show that they are congruent modulo ℓ^e for all prime powers ℓ^e dividing n. So it is enough to prove Lemma 2 in the case that $n = \ell^e$ is a prime power.

The idea of the proof is very easy. By making a change-of-basis, we will put A in rational normal form:

$$A = \begin{pmatrix} 0 & * \\ 1 & * \end{pmatrix}.$$

Then we will explicitly describe all matrices which commute with such an A and will show that they also commute with one another. More precisely, we will prove the following two sublemmas:

Sublemma 2′. *Let $A \in \mathrm{GL}_2(\mathbb{Z}/\ell^e\mathbb{Z})$ be a matrix which is not a scalar matrix modulo ℓ. Then there is a change-of-basis matrix $T \in \mathrm{GL}_2(\mathbb{Z}/\ell^e\mathbb{Z})$ which puts A into rational normal form:*

$$T^{-1}AT = \begin{pmatrix} 0 & * \\ 1 & * \end{pmatrix}.$$

Sublemma 2″. *Let $A = \begin{pmatrix} 0 & * \\ 1 & * \end{pmatrix} \in \mathrm{GL}_2(\mathbb{Z}/n\mathbb{Z})$. Then*

$$\left\{ B \in \mathrm{GL}_2\left(\frac{\mathbb{Z}}{n\mathbb{Z}}\right) \; : \; AB = BA \right\}$$

[†] This may remind you of our proof of the Nagell-Lutz theorem in Chapter II, Section 4. There we proved that a certain rational number a/d was an integer by checking, for each prime ℓ, that ℓ did not divide d. In fact, this idea of looking at one prime at a time, which in fancy language is called *localization*, is one of the most powerful tools available in number theory. It is the algebraic equivalent of looking at a neighborhood of a point when you are studying real or complex analysis.

is an abelian subgroup of $GL_2(\mathbb{Z}/n\mathbb{Z})$.

We will start by proving Sublemma 2″ because the proof is just a calculation. We assume that A has the form $A = \begin{pmatrix} 0 & b \\ 1 & d \end{pmatrix}$ and ask which $B \in GL_2(\mathbb{Z}/n\mathbb{Z})$ commute with A. If we write out the condition $AB = BA$, we find

$$AB = BA \iff \begin{pmatrix} 0 & b \\ 1 & d \end{pmatrix} \begin{pmatrix} \alpha & \beta \\ \gamma & \delta \end{pmatrix} = \begin{pmatrix} \alpha & \beta \\ \gamma & \delta \end{pmatrix} \begin{pmatrix} 0 & b \\ 1 & d \end{pmatrix}$$

$$\iff \begin{pmatrix} b\gamma & b\delta \\ \alpha + d\gamma & \beta + d\delta \end{pmatrix} = \begin{pmatrix} \beta & b\alpha + d\beta \\ \delta & b\gamma + d\delta \end{pmatrix}.$$

Treating b and d as fixed quantities, we get four equations for the four variables $\alpha, \beta, \gamma, \delta$; but the equations are not independent. A little algebra shows that the general solution is

$$\beta = b\gamma \quad \text{and} \quad \delta = \alpha + d\gamma.$$

Hence,

$$\left\{ B \in GL_2 \left(\frac{\mathbb{Z}}{n\mathbb{Z}} \right) : AB = BA \right\}$$

$$= \left\{ \begin{pmatrix} \alpha & b\gamma \\ \gamma & \alpha + d\gamma \end{pmatrix} \in GL_2 \left(\frac{\mathbb{Z}}{n\mathbb{Z}} \right) : \alpha, \gamma \in \frac{\mathbb{Z}}{n\mathbb{Z}} \right\}.$$

Now to check that the matrices in this set commute with one another, we just need to take two of them, multiply them together in both directions, and verify that the answers are the same:

$$\begin{pmatrix} \alpha & b\gamma \\ \gamma & \alpha + d\gamma \end{pmatrix} \begin{pmatrix} \alpha' & b\gamma' \\ \gamma' & \alpha' + d\gamma' \end{pmatrix} = \begin{pmatrix} \alpha' & b\gamma' \\ \gamma' & \alpha' + d\gamma' \end{pmatrix} \begin{pmatrix} \alpha & b\gamma \\ \gamma & \alpha + d\gamma \end{pmatrix}.$$

We will leave the actual multiplication to you. So we have completed the proof of Sublemma 2″.

Now we tackle Sublemma 2′. We will write

$$A = \begin{pmatrix} a & b \\ c & d \end{pmatrix} \in GL_2 \left(\frac{\mathbb{Z}}{\ell^e \mathbb{Z}} \right).$$

Recall from linear algebra that to put a 2×2 matrix A into rational normal form, one takes a basis of the form $\{\vec{v}, A\vec{v}\}$. Then the columns of the change-of-basis matrix T are the two column vectors $\vec{v}$ and $A\vec{v}$, and one sees immediately that

$$AT = T \begin{pmatrix} 0 & * \\ 1 & * \end{pmatrix}.$$

This is what we will do; the only difficulty is to ensure that the matrix T we choose has an inverse. There will be three cases to consider.

Our assumption that A is not a scalar matrix modulo ℓ means that at least one of the following three conditions is true:

(i) $b \not\equiv 0 \,(\mathrm{mod}\, \ell)$;

(ii) $c \not\equiv 0 \,(\mathrm{mod}\, \ell)$;

(iii) $a \not\equiv d \,(\mathrm{mod}\, \ell)$.

Corresponding to these three possibilities, we make the following choice for the matrix T:

(i)   If $b \not\equiv 0 \,(\mathrm{mod}\, \ell)$, then $T = \begin{pmatrix} 0 & b \\ 1 & d \end{pmatrix}$.

(ii) If $b \equiv 0 \,(\mathrm{mod}\, \ell)$ and $c \not\equiv 0 \,(\mathrm{mod}\, \ell)$, then $T = \begin{pmatrix} 1 & a \\ 0 & c \end{pmatrix}$.

(iii) If $b \equiv c \equiv 0 \,(\mathrm{mod}\, \ell)$ and $a \not\equiv d \,(\mathrm{mod}\, \ell)$, then $T = \begin{pmatrix} 1 & a+c \\ 1 & b+d \end{pmatrix}$.

Note that in all three cases we have $\det(T) \not\equiv 0 \,(\mathrm{mod}\, \ell)$, so in all three case we have $T \in \mathrm{GL}_2(\mathbb{Z}/\ell^e\mathbb{Z})$. For example, in case (iii),

$$\det(T) = b + d - a - c \equiv d - a \not\equiv 0 \,(\mathrm{mod}\, \ell).$$

Hence, T is invertible; and since it is obvious in each case that $AT = T\begin{pmatrix} 0 & * \\ 1 & * \end{pmatrix}$, we conclude that $T^{-1}AT = \begin{pmatrix} 0 & * \\ 1 & * \end{pmatrix}$. So that completes the proof of Sublemma 2′.

Now we can use the sublemmas to finish the proof of Lemma 2. Let $A \in \mathrm{GL}_2(\mathbb{Z}/\ell^e\mathbb{Z})$ be a matrix that is not a scalar matrix modulo ℓ. Using Sublemma 2′, we find a matrix $T \in \mathrm{GL}_2(\mathbb{Z}/\ell^e\mathbb{Z})$ so that $T^{-1}AT = \begin{pmatrix} 0 & * \\ 1 & * \end{pmatrix}$. Next let $B, B' \in \mathrm{GL}_2(\mathbb{Z}/\ell^e\mathbb{Z})$ be two matrices that commute with A,

$$AB = BA \qquad \text{and} \qquad AB' = B'A.$$

This implies that

$$(T^{-1}AT)(T^{-1}BT) = (T^{-1}BT)(T^{-1}AT), \quad \text{and}$$
$$(T^{-1}AT)(T^{-1}B'T) = (T^{-1}B'T)(T^{-1}AT).$$

From Sublemma 2″ we conclude that $T^{-1}BT$ and $T^{-1}B'T$ commute:

$$(T^{-1}BT)(T^{-1}B'T) = (T^{-1}B'T)(T^{-1}BT).$$

Since T is invertible, this is the same as $BB' = B'B$, which finishes the proof of Lemma 2. $\square$

Now we possess all the tools needed to prove our main theorem. We have the representation

$$\rho_n : \mathrm{Gal}\big(K_n/\mathbb{Q}(i)\big) \longrightarrow \mathrm{GL}_2\left(\frac{\mathbb{Z}}{n\mathbb{Z}}\right),$$

and we have the matrix

$$A = \begin{pmatrix} a & b \\ c & d \end{pmatrix} \in \mathrm{GL}_2\left(\frac{\mathbb{Z}}{n\mathbb{Z}}\right)$$

corresponding to the homomorphism $\phi : C[n] \to C[n]$. We showed that A is not equal to a scalar matrix modulo ℓ for all primes ℓ dividing n (Lemma 1). Let $\sigma \in \mathrm{Gal}(K_n/\mathbb{Q}(i))$ be any element of our Galois group. We verified that σ and ϕ commute, which implies that the matrices A and $\rho_n(\sigma)$ commute: $A\rho_n(\sigma) = \rho_n(\sigma)A$. Applying Lemma 2, we conclude that the matrices in the set $\{\rho_n(\sigma) : \sigma \in \mathrm{Gal}(K_n/\mathbb{Q}(i))\}$ all commute with one another. Since the representation ρ_n is a homomorphism, it follows that

$$\rho_n(\sigma_1\sigma_2) = \rho_n(\sigma_1)\rho_n(\sigma_2) = \rho_n(\sigma_2)\rho_n(\sigma_1) = \rho_n(\sigma_2\sigma_1)$$
$$\text{for all } \sigma_1, \sigma_2 \in \mathrm{Gal}(K_n/\mathbb{Q}(i)).$$

Finally, we use the fact that ρ_n is one-to-one (Galois Representation Theorem, Section 3) to conclude that

$$\sigma_1\sigma_2 = \sigma_2\sigma_1 \quad \text{for all } \sigma_1, \sigma_2 \in \mathrm{Gal}(K_n/\mathbb{Q}(i)).$$

In other words, $\mathrm{Gal}(K_n/\mathbb{Q}(i))$ is abelian, which completes the proof of our main theorem. $\square$

You may recall that in the case of abelian extensions of $\mathbb{Q}$, not only did all of the cyclotomic fields have abelian Galois groups, but it was also true that every extension with abelian Galois group was contained in a cyclotomic extension. A similar statement holds for abelian extensions of $\mathbb{Q}(i)$. The proof is too difficult for us to give, but we would be remiss if we failed to at least state this beautiful result.

Theorem. Let $C : y^2 = x^3 + x$ be the elliptic curve we have been studying in this section. Let $F/\mathbb{Q}(i)$ be any Galois extension of $\mathbb{Q}(i)$ of finite degree, and suppose that $\mathrm{Gal}(F/\mathbb{Q}(i))$ is abelian. Then there is an integer $n \geq 1$ such that

$$F \subset K_n = \mathbb{Q}(i)(C[n]).$$

Earlier we talked about Kronecker's dream of constructing extension fields with abelian Galois groups by using special values of complex analytic functions. We have now shown how to construct abelian extensions of $\mathbb{Q}(i)$ by taking the coordinates of points of finite order on the elliptic curve $y^2 = x^3 + x$. We want to conclude by briefly explaining how this construction is a realization of Kronecker's dream.

We begin by writing $C(\mathbb{C}) \cong \mathbb{C}/L$ and choosing generators $L = \mathbb{Z}\omega_1 + \mathbb{Z}\omega_2$ for the lattice L. (See Chapter II, Section 2.) Then, as generators for $C[n]$, we can take

$$P_1 = \frac{\omega_1}{n} \quad \text{and} \quad P_2 = \frac{\omega_2}{n}.$$

Using P_1 and P_2, we get a representation

$$\rho_n : \mathrm{Gal}\big(K_n/\mathbb{Q}(i)\big) \longrightarrow \mathrm{GL}_2\left(\frac{\mathbb{Z}}{n\mathbb{Z}}\right)$$

as usual.

The isomorphism $C(\mathbb{C}) \cong \mathbb{C}/L$ described in Chapter II uses the Weierstrass $\wp$ function:

$$\frac{\mathbb{C}}{L} \longrightarrow C(\mathbb{C}), \qquad z \longmapsto \big(\wp(z), \wp'(z)\big).$$

So the x and y coordinates of points on C are the values of $\wp$ and $\wp'$. In particular, a point of order dividing n in $\mathbb{C}/L$ looks like

$$\frac{a_1\omega_1 + a_2\omega_2}{n} \quad \text{for some } a_1, a_2 \in \mathbb{Z}.$$

So K_n is generated by i and the numbers

$$\wp\left(\frac{a_1\omega_1 + a_2\omega_2}{n}\right) \quad \text{and} \quad \wp'\left(\frac{a_1\omega_1 + a_2\omega_2}{n}\right) \quad \text{for } 0 \le a_1, a_2 < n.$$

Since the K_n's are abelian extensions of $\mathbb{Q}(i)$, we have realized one part of Kronecker's Jugendtraum; we have generated abelian extensions of $\mathbb{Q}(i)$ using special values of meromorphic functions.

But more is true. We can use the representation ρ_n to describe how elements of $\mathrm{Gal}\big(K_n/\mathbb{Q}(i)\big)$ act on these special values:

$$
\begin{aligned}
\sigma\left(\wp\left(\frac{a_1\omega_1 + a_2\omega_2}{n}\right)\right) &= \sigma\big(x(a_1 P_1 + a_2 P_2)\big) \\
&= x\big(a_1\sigma(P_1) + a_2\sigma(P_2)\big) \\
&= x\big((a_1\alpha_\sigma + a_2\beta_\sigma)P_1 + (a_1\gamma_\sigma + a_2\delta_\sigma)P_2\big) \\
&= \wp\left(\frac{(a_1\alpha_\sigma + a_2\beta_\sigma)\omega_1}{n} + \frac{(a_1\gamma_\sigma + a_2\delta_\sigma)\omega_2}{n}\right).
\end{aligned}
$$

Or, letting $t_1 = \dfrac{\omega_1}{n}$ and $t_2 = \dfrac{\omega_2}{n}$ be our generators for the points of order n, we can write this in the more suggestive matrix notation

$$\sigma\left(\wp\left((t_1 \quad t_2)\begin{pmatrix} a_1 \\ a_2 \end{pmatrix}\right)\right) = \wp\left((t_1 \quad t_2)\,\rho_n(\sigma)\begin{pmatrix} a_1 \\ a_2 \end{pmatrix}\right).$$

And the same formula holds for $\wp'$. These formulas provide a concrete realization of Kronecker's Jugendtraum for the field $\mathbb{Q}(i)$.

EXERCISES

6.1. Recall that the discriminant of a (monic) polynomial

$$f(X) = (X - \alpha_1)(X - \alpha_2) \cdots (X - \alpha_n)$$

is defined to be

$$\text{Disc}(f) = \prod_{1 \le j < i \le n} (\alpha_i - \alpha_j)^2.$$

(a) Prove that

$$\text{Disc}(f) = (-1)^{(n^2-n)/2} \prod_{i=1}^{n} f'(\alpha_i).$$

(b) Let $f(X) = X^n - 1$. Prove that

$$\text{Disc}(f) = (-1)^{(n-1)(n-2)/2} n^n.$$

(c) Let ζ be a primitive n^{th} root of unity. Prove that the cyclotomic field $\mathbb{Q}(\zeta)$ contains $\sqrt{\text{Disc}(f)}$.

(d) Prove that $\mathbb{Q}(\sqrt{n})$ is contained in the cyclotomic field $\mathbb{Q}(\zeta')$, where ζ' is a primitive $4n^{\text{th}}$ root of unity. (You might want to begin with the case that n is odd.)

This exercise shows that every quadratic extension of $\mathbb{Q}$ is contained in a cyclotomic extension, thereby proving the Kronecker-Weber theorem for extensions $K/\mathbb{Q}$ satisfying $\text{Gal}(K/\mathbb{Q}) \cong \mathbb{Z}/2\mathbb{Z}$.

6.2. (a) Suppose that $\lambda(z)$ is a polynomial

$$\lambda(z) = a_0 z^n + a_1 z^{n-1} + \cdots + a_{n-1} z + a_n$$

of degree n such that $\lambda : \mathbb{C}^* \to \mathbb{C}^*$ is a homomorphism. Prove that $\lambda(z) = z^n$.

(b) Suppose that $\lambda(z)$ is a meromorphic function such that the map $\lambda : \mathbb{C}^* \to \mathbb{C}^*$ is a homomorphism. Prove that $\lambda(z) = z^n$ for some integer n.

6.3. Let C be an elliptic curve defined by a Weierstrass equation with coefficients in $\mathbb{Q}$, and let K be a Galois extension of $\mathbb{Q}$.

(a) Prove that for all $P \in C(K)$ and all $\sigma, \tau \in \text{Gal}(K/\mathbb{Q})$,

$$\tau\big(\sigma(P)\big) = (\tau\sigma)(P).$$

(This is a sort of associative law. One says that there is an *action of the group* $\text{Gal}(K/\mathbb{Q})$ *on the abelian group* $C(K)$. Group actions are very important in many areas of mathematics.)

(b) Prove that for all $P \in C(K)$ and all $\sigma \in \text{Gal}(K/\mathbb{Q})$,

$$\sigma(2P) = 2\sigma(P).$$

(Do not just quote the proposition in Section 2. When we proved that proposition, this is one of the cases that we left for you to do.)

6.4. We define a sequence of *division polynomials* $\psi_n \in \mathbf{Z}[x,y]$ for the elliptic curve

$$C : y^2 = x^3 + x$$

inductively by the following rules:

$$\psi_1 = 1, \qquad \psi_2 = 2y,$$
$$\psi_3 = 3x^4 + 6x^2 - 1,$$
$$\psi_4 = 4y(x^6 + 5x^4 - 5x^2 - 1),$$
$$\psi_{2n+1} = \psi_{n+2}\psi_n^3 - \psi_{n-1}\psi_{n+1}^3 \quad \text{for } n \geq 2,$$
$$2y\psi_{2n} = \psi_n(\psi_{n+2}\psi_{n-1}^2 - \psi_{n-2}\psi_{n+1}^2) \quad \text{for } n \geq 3.$$

Further, define ϕ_n and ω_n by

$$\phi_n = x\psi_n^2 - \psi_{n+1}\psi_{n-1},$$
$$4y\omega_n = \psi_{n+2}\psi_{n-1}^2 - \psi_{n-2}\psi_{n+1}^2.$$

(a) Prove that all of the ψ_n, ϕ_n, and ω_n are in $\mathbf{Z}[x,y]$.

(b) If n is odd, prove that ψ_n, ϕ_n, and $y^{-1}\omega_n$ are actually polynomials in $\mathbf{Z}[x,y^2]$. If n is even, prove that $(2y)^{-1}\psi_n$, ϕ_n, and ω_n are polynomials in $\mathbf{Z}[x,y^2]$. Hence, by replacing y^2 with $x^3 + x$, we can treat them as polynomials in $\mathbf{Z}[x]$.

(c) As polynomials in x, show that

$$\phi_n(x) = x^{n^2} + \text{lower order terms},$$
$$\psi_n(x)^2 = n^2 x^{n^2-1} + \text{lower order terms}.$$

(d) Prove that for any point $P = (x,y) \in C$,

$$nP = \left(\frac{\phi_n(P)}{\psi_n(P)^2}, \frac{\omega_n(P)}{\psi_n(P)^3} \right).$$

(e) Prove that $\psi_n(x)^2$ has no double roots. Prove that $\psi_n(x)^2$ and $\phi_n(x)$ have no common roots. (Here we are talking about roots in $\mathbf{C}$.)

(f) Let $P = (x,y) \in C(\mathbf{C})$ be a point on C. Prove that $nP = \mathcal{O}$ if and only if $\psi_n(x)^2 = 0$.

(g) Prove that for every n, the group $C[n]$ contains n^2 points. Deduce that

$$C[n] \cong \frac{\mathbf{Z}}{n\mathbf{Z}} \oplus \frac{\mathbf{Z}}{n\mathbf{Z}}.$$

6.5. * Redo the previous exercise for the elliptic curve

$$C : y^2 = x^3 + bx + c.$$

Everything will be the same except that ψ_3 and ψ_4 are given by the formulas

$$\psi_3 = 3x^4 + 6bx^2 + 12cx - b^2,$$
$$\psi_4 = 4y(x^6 + 5bx^4 + 20cx^3 - 5b^2x^2 - 4bcx - 8c^2 - b^3).$$

6.6. Let C be an elliptic curve given by a Weierstrass equation with rational coefficients. Let $m, n \geq 1$ be integers.
(a) If $\gcd(m, n) = 1$, prove that $\mathbb{Q}(C[mn])$ is the compositum of
the fields $\mathbb{Q}(C[m])$ and $\mathbb{Q}(C[n])$. (The compositum of two fields K_1
and K_2 is the smallest field containing both K_1 and K_2.)
(b) More generally, let $\ell = \mathrm{LCM}[m, n]$, and prove that $\mathbb{Q}(C[\ell])$ is the
compositum of the fields $\mathbb{Q}(C[m])$ and $\mathbb{Q}(C[n])$.

6.7. Let R be a commutative ring with multiplicative identity. Let A be
an $r \times r$ matrix with coefficients in R.
(a) If $\det A$ is a unit in R, prove that there is a matrix B with coefficients in R such that $AB = I$. (Here I is the $r \times r$ identity matrix.)
(b) If there is a matrix B with coefficients in R such that $AB = I$,
prove that $\det A$ is a unit in R.

6.8. Let A be an abelian group, and define $\mathrm{End}(A)$ to be the set of homomorphisms from A to itself,

$$\mathrm{End}(A) = \{\text{homomorphisms } A \to A\}.$$

Define an addition and a multiplication on $\mathrm{End}(A)$ by the rules

$$(g + h)(\alpha) = g(\alpha) + h(\alpha) \qquad \text{and} \qquad (gh)(\alpha) = g\big(h(\alpha)\big).$$

[N.B. $(gh)(\alpha)$ is not equal to the product $g(\alpha)h(\alpha)$.] In this exercise you will verify that $\mathrm{End}(A)$ is a ring. It is called the *ring of
endomorphisms of A*.
(a) Prove that $(g + h)$ and (gh) are in $\mathrm{End}(A)$.
(b) Verify that with this addition and multiplication, $\mathrm{End}(A)$ is a
(not necessarily commutative) ring with multiplicative identity.
(c) The *automorphism group of A*, denoted $\mathrm{Aut}(A)$, is the group of
units in the ring $\mathrm{End}(A)$. Prove that the elements of $\mathrm{Aut}(A)$ are
isomorphisms from A to itself.

6.9. (a) If A is a cyclic group of order n, prove that

$$\mathrm{End}(A) \cong \frac{\mathbb{Z}}{n\mathbb{Z}} \qquad \text{and} \qquad \mathrm{Aut}(A) \cong \left(\frac{\mathbb{Z}}{n\mathbb{Z}}\right)^{*}.$$

(b) If A is a direct sum of r cyclic groups of order n, prove that

$$\mathrm{Aut}(A) \cong \mathrm{GL}_r\left(\frac{\mathbb{Z}}{n\mathbb{Z}}\right).$$

(c) If A is a direct sum of r cyclic groups of order n, describe the ring
$\mathrm{End}(A)$ in terms of a ring of matrices.

6.10. Let C be an elliptic curve given by a Weierstrass equation with coefficients in $\mathbb{Q}$. Prove that there is a one-to-one homomorphism

$$\text{Gal}\big(\mathbb{Q}(C[n])/\mathbb{Q}\big) \longrightarrow \text{Aut}\big(C[n]\big)$$

defined by the rule that $\sigma \in \text{Gal}\big(\mathbb{Q}(C[n])/\mathbb{Q}\big)$ goes to the map

$$C[n] \longrightarrow C[n], \qquad P \longmapsto \sigma(P).$$

Further, show that $\text{Aut}\big(C[n]\big) \cong \text{GL}_2(\mathbb{Z}/n\mathbb{Z})$, thereby recovering the representation $\rho_n : \text{Gal}\big(\mathbb{Q}(C[n])/\mathbb{Q}\big) \to \text{GL}_2(\mathbb{Z}/n\mathbb{Z})$ defined in Section 3.

6.11. Let F be a field, and let V be an F-vector space of dimension r. Let

$$\text{Aut}_F(V) = \left\{ \begin{matrix} \text{one-to-one and onto } F\text{-linear} \\ \text{transformations } V \to V \end{matrix} \right\}.$$

Prove that $\text{Aut}_F(V)$ is isomorphic to $\text{GL}_r(F)$, the group of invertible $r \times r$ matrices with coefficients in F.

6.12. Let C be the elliptic curve $y^2 = x^3 + x$. The points $P_1 = (i, 0)$ and $P_2 = (-i, 0)$ are generators for $C[2]$ (cf. Example 2 of Section 3). The Galois group $\text{Gal}\big(\mathbb{Q}(C[2])/\mathbb{Q}\big)$ consists of two elements, the identity σ_0 and complex conjugation σ_1. Find the matrix $\rho_2(\sigma_1) \in \text{GL}_2(\mathbb{Z}/2\mathbb{Z})$ if the representation ρ_2 is defined using the generators P_1 and P_2.

6.13. For each of the following elliptic curves, determine $\text{Gal}\big(\mathbb{Q}(C[2])/\mathbb{Q}\big)$, the Galois group of the field generated by the points of order two.

(a) $y^2 = x^3 - x$.

(b) $y^2 = x^3 - x - 2$.

(c) $y^2 = x^3 + x - 2$.

(d) $y^2 = x^3 - 3x + 1$.

6.14. For each of the elliptic curves in the previous exercise, choose a basis for $C[2]$ and write down the matrices $\rho_2(\sigma)$ for each element σ in $\text{Gal}\big(\mathbb{Q}(C[2])/\mathbb{Q}\big)$ (cf. Section 3, Examples 1–3).

6.15. Let $\mathbb{C}/L$ be an elliptic curve with a complex multiplication

$$f : \frac{\mathbb{C}}{L} \longrightarrow \frac{\mathbb{C}}{L}, \qquad f(z) = cz.$$

(a) Prove that there are integers $A, B \in \mathbb{Z}$ such that

$$c^2 + Ac + B = 0.$$

(*Hint.* Write $L = \mathbb{Z}\omega_1 + \mathbb{Z}\omega_2$ and use the fact that both $c\omega_1$ and $c\omega_2$ are in L.)

(b) Prove that the integers A, B in (a) satisfy $A^2 < 4B$.

(c) Prove that the field $\mathbb{Q}(c)$ is a degree 2 extension of $\mathbb{Q}$, and that $\mathbb{Q}(c)$ is not contained in $\mathbb{R}$ (i.e., $\mathbb{Q}(c)$ is a quadratic imaginary field).

6.16. (a) Let C be an elliptic curve. Define the *endomorphism ring of C* to be

$$\text{End}(C) = \{\text{endomorphisms } C \to C\}.$$

Note this is a little different from the endomorphism ring of C considered as an abelian group because we are not taking all homomorphisms from C to itself, but only those which are defined by rational functions. In other words, $\text{End}(C)$ is the set of *algebraic* endomorphisms of C. Prove that the addition and multiplication rules

$$(\phi_1 + \phi_2)(P) = \phi_1(P) + \phi_2(P) \quad \text{and} \quad (\phi_1\phi_2)(P) = \phi_1(\phi_2(P))$$

make $\text{End}(C)$ into a ring.

(b) Let $L \subset \mathbb{C}$ be a lattice. Define a set of complex numbers R_L by

$$R_L = \{c \in \mathbb{C} : cL \subset L\}.$$

Prove that R_L is a ring.

(c) Let $C(\mathbb{C}) = \mathbb{C}/L$ be an elliptic curve. For each $\phi \in \text{End}(C)$, we showed in Section 4 that ϕ corresponds to a map

$$f : \frac{\mathbb{C}}{L} \longrightarrow \frac{\mathbb{C}}{L}, \qquad f(z) = c_\phi z,$$

where $c_\phi \in \mathbb{C}$ satisfies $c_\phi L \subset L$. Prove that the map

$$\text{End}(C) \longrightarrow R_L, \qquad \phi \longmapsto c_\phi,$$

is a one-to-one homomorphism of rings.

(d) * Prove that the homomorphism in (c) is an isomorphism.

6.17. Let C be the elliptic curve

$$C : y^2 = x^3 + x, \qquad \text{and let} \quad K_n = \mathbb{Q}(i)(C[n])$$

be the field considered in Section 5. We proved that K_n is Galois over $\mathbb{Q}$.

(a) Let $\tau : \mathbb{C} \to \mathbb{C}$ be complex conjugation, which we consider as an element of $\text{Gal}(K_n/\mathbb{Q})$ (i.e., Start by fixing an inclusion $K_n \subset \mathbb{C}$). Prove that every element $\sigma \in \text{Gal}(K_n/\mathbb{Q})$ can be written uniquely in the form $\sigma = st$ with $s \in \text{Gal}(K_n/\mathbb{Q}(i))$ and $t \in \{e, \tau\}$.

(b) Prove that for all $s \in \text{Gal}(K_n/\mathbb{Q}(i))$,

$$s\tau = \tau s^{-1}.$$

(c) Prove that $\text{Gal}(K_n/\mathbb{Q})$ is abelian if and only if every element $s \in \text{Gal}(K_n/\mathbb{Q}(i))$ satisfies $s^2 = e$.

6.18. Let C be the elliptic curve $C : y^2 = x^3 + x$, and let

$$\beta = \sqrt[4]{\frac{8\sqrt{3} - 12}{9}}.$$

(See the Example in Section 2.)

(a) Prove that the minimal polynomial of β over $\mathbb{Q}$ is

$$27x^8 + 72x^4 - 16 = 0.$$

(b) Prove that $\mathbb{Q}\big(C[3]\big) = \mathbb{Q}(\beta, i)$.

(c) Compute the Galois group of $\mathbb{Q}(\beta, i)$ over $\mathbb{Q}(i)$. In particular, verify that it is abelian.

6.19. Let C be the elliptic curve

$$C : y^2 = x^3 + 1.$$

For each integer $n \geq 1$, let

$$K_n = \mathbb{Q}\left(\sqrt{-3}\right)\left(C[n]\right)$$

be the extension field of $\mathbb{Q}\left(\sqrt{-3}\right)$ generated by the coordinates of the points of order n on C. Note that C has complex multiplication (cf. Section 4, Example 2).

(a) Prove that K_n is a Galois extension of $\mathbb{Q}$.

(b) Prove that

$$\text{Gal}\left(K_n / \mathbb{Q}\left(\sqrt{-3}\right)\right)$$

is abelian.

6.20. Let C be the elliptic curve

$$C : y^2 = x^3 + 4x^2 + 2x.$$

(a) Prove that the formula

$$\phi(P) = \begin{cases} \left(\dfrac{-y^2}{2x^2}, \dfrac{-y(x^2 - 2)}{2\sqrt{-2}\,x^2}\right), & \text{if } P = (x, y) \neq (0, 0), \\ \mathcal{O}, & \text{if } P = (0, 0) \text{ or } P = \mathcal{O}, \end{cases}$$

gives an endomorphism $\phi : C \to C$.

(b) Prove that C has complex multiplication. (*Hint.* Look at the kernel of ϕ.)

(c) Let

$$K_n = \mathbb{Q}\left(\sqrt{-2}\right)\left(C[n]\right)$$

be the extension field of $\mathbb{Q}\left(\sqrt{-2}\right)$ generated by the coordinates of the points of order n on C. Prove that K_n is a Galois extension of $\mathbb{Q}$.

(d) Prove that

$$\text{Gal}\left(K_n / \mathbb{Q}\left(\sqrt{-2}\right)\right)$$

is abelian.

6.21. Let C be the elliptic curve $C : y^2 = x^3 + x$, let L be the lattice $\mathbb{Z} + \mathbb{Z}i$,
and let $g_3 = 140 \sum\limits_{\omega \in L,\, \omega \neq 0} \dfrac{1}{\omega^6}$ be the quantity defined in Chapter II,
Section 2.

(a) Prove that $g_3 = 0$.

(b) Prove that there is a complex number γ so that the map

$$\frac{\mathbb{C}}{L} \longrightarrow C(\mathbb{C}), \qquad z \longmapsto \big(4\gamma^2 \wp(z),\, 4\gamma^3 \wp'(z)\big),$$

is an isomorphism. Here $\wp$ is the Weierstrass $\wp$ function described in
Chapter II, Section 2.

(c) Show that the complex multiplication map

$$C(\mathbb{C}) \longrightarrow C(\mathbb{C}), \qquad (x, y) \longmapsto (-x, -iy)$$

corresponds to the map

$$\frac{\mathbb{C}}{L} \longrightarrow \frac{\mathbb{C}}{L}, \qquad z \longmapsto iz.$$

In other words, verify the formulas

$$\wp(iz) = -\wp(z) \qquad \text{and} \qquad \wp'(iz) = -i\wp'(z).$$

APPENDIX A

Projective Geometry

In this appendix we have tried to summarize the basic properties of the projective plane and projective curves that are used elsewhere in this book. For further reading about projective algebraic geometry, the reader might profitably consult Brieskorn-Knörrer [1], Fulton [1], or Reid [1]. More high-powered accounts of modern algebraic geometry are given in Hartshorne [1] and Griffiths and Harris [1].

1. Homogeneous Coordinates and the Projective Plane

There are many ways to approach the construction of the projective plane. We will describe two constructions, one algebraic and one geometric, since each in its own way provides enlightenment.

We begin with a famous question from number theory, namely the solution of the equation

$$x^N + y^N = 1 \qquad \text{(Fermat Equation \#1)}$$

in rational numbers x and y. Suppose that we have found a solution, say $x = a/c$, $y = b/d$, where we write the fractions in lowest terms and with positive denominators. Substituting and clearing denominators gives the equation

$$a^N d^N + b^N c^N = c^N d^N.$$

It follows that $c^N | a^N d^N$. But $\gcd(a, c) = 1$ by assumption, so we conclude that $c^N | d^N$, and hence $c | d$. Similarly $d^N | b^N c^N$ and $\gcd(b, d) = 1$, which implies that $d | c$. Therefore $c = \pm d$, and since we've assumed that c and d are positive, we find that $d = c$. Thus any solution to the Fermat Equation \#1 in rational numbers has the form $(a/c, b/c)$, and thus gives a solution in integers (a, b, c) to the homogeneous equation

$$X^N + Y^N = Z^N. \qquad \text{(Fermat Equation \#2)}$$

Conversely, any integer solution (a, b, c) to the second Fermat Equation with $c \neq 0$ will give a rational solution $(a/c, b/c)$ to the first. However, different integer solutions (a, b, c) may lead to the same rational solution. For example, if (a, b, c) is an integer solution to Fermat's Equation #2, then for any integer t the triple (ta, tb, tc) will also be a solution, and clearly (a, b, c) and (ta, tb, tc) give the same rational solution to Fermat's Equation #1. The moral is that in solving Fermat's Equation #2, we should really treat triples (a, b, c) and (ta, tb, tc) as being the same solution, at least for non-zero t. This leads to the notion of *homogeneous coordinates* which we will describe in more detail later.

There is one more observation we wish to make before leaving this example, namely the "problem" that Fermat's Equation #2 may have some integer solutions that do not correspond to rational solutions of Fermat's Equation #1. First, the point $(0, 0, 0)$ is always a solution of the second equation, but this solution is so trivial that we will just discard it. Second, and potentially more serious, is the fact that if N is odd, then Fermat's Equation #2 has the solutions $(1, -1, 0)$ and $(-1, 1, 0)$ which do not give solutions to the first Fermat Equation. To see what is happening, suppose that we take a sequence of solutions (a_i, b_i, c_i), $i = 1, 2, 3, \ldots$, such that $(a_i, b_i, c_i) \to (1, -1, 0)$ as $i \to \infty$. Of course, we cannot do this with integer solutions, so now we'll let the a_i, b_i, c_i's be real numbers. The corresponding solutions to the first Fermat Equation are $(a_i/c_i, b_i/c_i)$, and we see that these solutions approach (∞, ∞) as $(a_i, b_i, c_i) \to (1, -1, 0)$. In other words, the extra solutions $(1, -1, 0)$ and $(-1, 1, 0)$ to Fermat's Equation #2 somehow correspond to solutions of the first Fermat Equation which lie "at infinity." As we will see, the theory of solutions of polynomial equations becomes neater and clearer if we treat these extra points "at infinity" just like we treat all of the other points.

We are now ready for our first definition of the projective plane, which is essentially an algebraic definition. We define the *projective plane* to be the set of triples $[a, b, c]$, with a, b, c not all zero, such that two triples $[a, b, c]$ and $[a', b', c']$ are considered to be the same point if there is a non-zero t such that $a = ta'$, $b = tb'$, $c = tc'$. The numbers a, b, c are called *homogeneous coordinates* for the point $[a, b, c]$. We will denote the projective plane by $\mathbb{P}^2$. In other words, we define an equivalence relation $\sim$ on the set of triples $[a, b, c]$ by the rule

$$[a, b, c] \sim [a', b', c'] \quad \text{if there is a non-zero } t \text{ so that } a = ta', \ b = tb', \ c = tc'.$$

Then $\mathbb{P}^2$ consists of the set of equivalence classes of triples $[a, b, c]$ except that we exclude the triple $[0, 0, 0]$:

$$\mathbb{P}^2 = \frac{\left\{ [a, b, c] : a, b, c \text{ are not all zero} \right\}}{\sim}.$$

More generally, for any integer $n \geq 1$ we define *projective n-space* to be the set of equivalence classes of homogeneous $n + 1$-tuples,

$$\mathbb{P}^n = \frac{\left\{ [a_0, a_1, \ldots, a_n] : a_0, a_1, \ldots, a_n \text{ not all zero} \right\}}{\sim},$$

where $[a_0, a_1, \ldots, a_n] \sim [a_0', a_1', \ldots, a_n']$ if there is a non-zero t so that $a_0 = ta_0', \ldots, a_n = ta_n'$.

We eventually want to do geometry in the projective plane, so we need to define some geometric objects. In the next section we will study quite general curves, but for the moment we will be content to describe the lines in $\mathbb{P}^2$. We define a *line* *in* $\mathbb{P}^2$ to be the set of points $[a, b, c] \in \mathbb{P}^2$ whose coordinates satisfy an equation of the form

$$\alpha X + \beta Y + \gamma Z = 0$$

for some constants α, β, γ not all zero. Note that if $[a, b, c]$ satisfies such an equation, then so does $[ta, tb, tc]$ for any t, so to check if a point of $\mathbb{P}^2$ is on a given line, one can use any homogeneous coordinates for the point.

In order to motivate our second description of the projective plane, we consider a geometric question. It is well-known that two points in the usual (x, y) plane determine a unique line, namely the line which goes through them. Similarly, two lines in the plane determine a unique point, namely the point where they intersect, unless the two lines happen to be parallel. From both an aesthetic and a practical viewpoint, it would be nice to provide these poor parallel lines with an intersection point of their own. Since the plane itself doesn't contain the requisite points, we will add on the extra points by fiat. How many extra points do we need? For example, would it suffice to use one extra point P and decree that any two parallel lines intersect at the point P? The answer is no, and here's why.

Let L_1 and L_2 be parallel lines, and let P be the extra point where they are to intersect. Similarly, let L_1' and L_2' be parallel lines which intersect at the extra point P'. (See Figure A.1.) Suppose that L_1 and L_1' are not parallel. Then L_1 and L_1' intersect at some ordinary point, say $L_1 \cap L_1' = \{Q\}$. But two lines are allowed to have only one point in common, so it follows that the points $P \in L_1$ and $P' \in L_1'$ must be distinct. So we really need to add an extra point for each distinct direction in the ordinary plane, and then decree that a line L consists of its usual points together with the extra point determined by its direction.

This leads to our second definition of the projective plane, this time in purely geometric terms. For simplicity, we will denote the usual *Euclidean* (also called *affine*) *plane* by

$$\mathbb{A}^2 = \big\{(x, y) : x \text{ and } y \text{ any numbers}\big\}.$$

Then we define the projective plane to be

$$\mathbb{P}^2 = \mathbb{A}^2 \cup \big\{\text{the set of directions in } \mathbb{A}^2\big\},$$

where *direction* is a non-oriented notion. Two lines have the same direction if and only if they are parallel. Logically we could define a direction in this sense to be an equivalence class of parallel lines, that is, a direction is a collection of all lines parallel to a given line. The extra points in $\mathbb{P}^2$

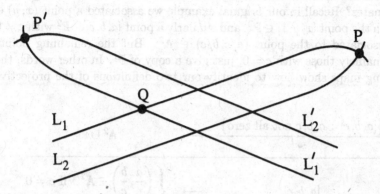

Parallel Lines With Intersection Points "At Infinity"

Figure A.1

associated to directions, that is the points in $\mathbb{P}^2$ that are not in $\mathbb{A}^2$, are often called *points at infinity*.

As indicated above, a line in $\mathbb{P}^2$ then consists of a line in $\mathbb{A}^2$ together with the point at infinity specified by its direction. The intersection of two parallel lines is the point at infinity corresponding to their common direction. Finally, the set of all points at infinity is itself considered to be a line L_∞, and the intersection of any other line L with L_∞ is the point at infinity corresponding to the direction of L. With these conventions, it is easy to see that there is a unique line going through any two distinct points of $\mathbb{P}^2$, and further any two distinct lines in $\mathbb{P}^2$ intersect in exactly one point. So the projective plane in this geometric incarnation eliminates the need to make a distinction between parallel and non-parallel lines. In fact, $\mathbb{P}^2$ has no parallel lines at all!

We now have two definitions of the projective plane, so it behooves us to show that they are equivalent. First we need a more analytic description of the set of directions in $\mathbb{A}^2$. One way to describe these directions is by the set of lines in $\mathbb{A}^2$ going through the origin, since every line in $\mathbb{A}^2$ is parallel to a unique line through the origin. Now the lines through the origin are given by equations

$$Ay = Bx$$

with A and B not both zero. However, it is possible for two pairs to give the same line. More precisely, the pairs (A, B) and (A', B') will give the same line if and only if there is a non-zero t such that $A = tA'$ and $B = tB'$. Thus the set of directions in $\mathbb{A}^2$ is naturally described by the points $[A, B]$ of the projective line $\mathbb{P}^1$. This allows us to write our second description of $\mathbb{P}^2$ in the form

$$\mathbb{P}^2 = \mathbb{A}^2 \cup \mathbb{P}^1;$$

a point $[A, B] \in \mathbb{P}^1 \subset \mathbb{P}^2$ corresponds to the direction of the line $Ay = Bx$.

How is this related to the definition of $\mathbb{P}^2$ in terms of homogeneous coordinates? Recall in our original example we associated a point $(x, y) \in \mathbb{A}^2$ with the point $[x, y, 1] \in \mathbb{P}^2$, and similarly a point $[a, b, c] \in \mathbb{P}^2$ with $c \neq 0$ was associated to the point $(a/c, b/c) \in \mathbb{A}^2$. But the remaining points in $\mathbb{P}^2$, namely those with $c = 0$, just give a copy of $\mathbb{P}^1$. In other words, the following maps show how to identify our two definitions of the projective plane:

$$\frac{\{[a, b, c] \ : \ a, b, c \text{ not all zero}\}}{\sim} \quad \longleftrightarrow \quad \mathbb{A}^2 \cup \mathbb{P}^1$$

$$[a, b, c] \quad \longrightarrow \quad \begin{cases} \left(\dfrac{a}{c}, \dfrac{b}{c}\right) \in \mathbb{A}^2 & \text{if } c \neq 0 \\[2mm] [a, b] \in \mathbb{P}^1 & \text{if } c = 0 \end{cases}$$

$$[x, y, 1] \quad \longleftarrow \quad (x, y) \in \mathbb{A}^2$$
$$[A, B, 0] \quad \longleftarrow \quad [A, B] \in \mathbb{P}^1$$

It's easy to check that these two maps are inverses. For example, if $c \neq 0$ then

$$[a, b, c] \longmapsto \left(\frac{a}{c}, \frac{b}{c}\right) \longmapsto \left[\frac{a}{c}, \frac{b}{c}, 1\right] = [a, b, c].$$

We'll leave the remaining verifications to you.

Each of our definitions of the projective plane came with a description of what constitutes a line, so we should also check that the lines match up properly. For example, a line L in $\mathbb{P}^2$ using homogeneous coordinates is the set of solutions $[a, b, c]$ to an equation

$$\alpha X + \beta Y + \gamma Z = 0.$$

Suppose first that α and β are not both zero. Then any point $[a, b, c] \in L$ with $c \neq 0$ is sent to the point

$$\left(\frac{a}{c}, \frac{b}{c}\right) \quad \text{on the line} \quad \alpha x + \beta y + \gamma = 0 \quad \text{in } \mathbb{A}^2.$$

And the point $[-\beta, \alpha, 0] \in L$ is sent to the point $[-\beta, \alpha] \in \mathbb{P}^1$, which corresponds to the direction of the line $-\beta y = \alpha x$. This is exactly right, since the line $-\beta y = \alpha x$ is precisely the line going through the origin that is parallel to the line $\alpha x + \beta y + \gamma = 0$. This takes care of all of the lines except for the line $Z = 0$ in $\mathbb{P}^2$. But the line $Z = 0$ is sent to the line in $\mathbb{A}^2 \cup \mathbb{P}^1$ consisting of all of the points at infinity. So the lines in our two descriptions of $\mathbb{P}^2$ are consistent.

2. Curves in the Projective Plane

An *algebraic curve* in the affine plane $\mathbb{A}^2$ is defined to be the set of solutions to a polynomial equation in two variables,

$$f(x, y) = 0.$$

For example, the equation $x^2 + y^2 - 1 = 0$ is a circle in $\mathbb{A}^2$, and $2x - 3y^2 + 1 = 0$ is a parabola.

In order to define curves in the projective plane $\mathbb{P}^2$, we will need to use polynomials in three variables, since points in $\mathbb{P}^2$ are represented by homogeneous triples. But there is the further difficulty that each point in $\mathbb{P}^2$ can be represented by many different homogeneous triples. It thus makes sense to look only at polynomials $F(X, Y, Z)$ with the property that $F(a, b, c) = 0$ implies that $F(ta, tb, tc) = 0$ for all t. These turn out to be the homogeneous polynomials, and we use them to define curves in $\mathbb{P}^2$.

More formally, a polynomial $F(X, Y, Z)$ is called a *homogeneous polynomial of degree d* if it satisfies the identity

$$F(tX, tY, tZ) = t^d F(X, Y, Z).$$

This identity is equivalent to the statement that F is a linear combination of monomials $X^i Y^j Z^k$ with $i + j + k = d$. We define a *projective curve C* in the projective plane $\mathbb{P}^2$ to be the set of solutions to a polynomial equation

$$C : F(X, Y, Z) = 0,$$

where F is a non-constant homogeneous polynomial. We will also call C an *algebraic curve* or sometimes just a *curve* if it is clear we are working in $\mathbb{P}^2$. The *degree of the curve C* is the degree of the polynomial F. For example,

$$C_1 : X^2 + Y^2 - Z^2 = 0 \quad \text{and} \quad C_2 : Y^2 Z - X^3 - X Z^2 = 0$$

are projective curves, where C_1 has degree 2 and C_2 has degree 3.

In order to check if a point $P \in \mathbb{P}^2$ is on the curve C, we can take any homogeneous coordinates $[a, b, c]$ for P and check if $F(a, b, c)$ is zero. This is true because any other homogeneous coordinates for P look like $[ta, tb, tc]$ for some non-zero t. Thus, $F(a, b, c)$ and $F(ta, tb, tc) = t^d F(a, b, c)$ are either both zero or both non-zero.

This tells us what a projective curve is when we use the definition of $\mathbb{P}^2$ by homogeneous coordinates. It will be very illuminating to relate this to the description of $\mathbb{P}^2$ as $\mathbb{A}^2 \cup \mathbb{P}^1$, where $\mathbb{A}^2$ is the usual affine plane and the points at infinity (i.e. the points in $\mathbb{P}^1$) correspond to the directions in $\mathbb{A}^2$. Let $C \subset \mathbb{P}^2$ be a curve given by a homogeneous polynomial of degree d,

$$C : F(X, Y, Z) = 0.$$

If $P = [a, b, c] \in C$ is a point of C with $c \neq 0$, then according to the identification $\mathbb{P}^2 \leftrightarrow \mathbb{A}^2 \cup \mathbb{P}^1$ described in Section 1, the point $P \in C \subset \mathbb{P}^2$ corresponds to the point

$$\left(\frac{a}{c}, \frac{b}{c} \right) \in \mathbb{A}^2 \subset \mathbb{A}^2 \cup \mathbb{P}^1.$$

On the other hand, combining $F(a, b, c) = 0$ with the fact that F is homogeneous of degree d shows that

$$0 = \frac{1}{c^d} F(a, b, c) = F\left(\frac{a}{c}, \frac{b}{c}, 1 \right).$$

In other words, if we define a new (non-homogeneous) polynomial $f(x, y)$ by

$$f(x, y) = F(x, y, 1),$$

then we get a map

$$\{ [a, b, c] \in C : c \neq 0 \} \longrightarrow \{ (x, y) \in \mathbb{A}^2 : f(x, y) = 0 \}.$$

$$[a, b, c] \longmapsto \left(\frac{a}{c}, \frac{b}{c} \right)$$

And it is easy to see that this map is one-to-one and onto, since if $(r, s) \in \mathbb{A}^2$ satisfies the equation $f(x, y) = 0$, then clearly $[r, s, 1] \in C$. We call the curve $f(x, y) = 0$ in $\mathbb{A}^2$ the *affine part* of the projective curve C.

It remains to look at the points $[a, b, c]$ on C with $c = 0$ and describe them geometrically in terms of the affine part of C. The points $[a, b, 0]$ on C satisfy the equation $F(X, Y, 0) = 0$, and they are sent to the points at infinity $[a, b] \in \mathbb{P}^1$ in $\mathbb{A}^2 \cup \mathbb{P}^1$. We claim that these points, which recall are really directions in $\mathbb{A}^2$, correspond to the limiting tangent directions of the affine curve $f(x, y) = 0$ as we move along the affine curve out to infinity. In other words, and this is really the intuition to keep in mind, an affine curve $f(x, y) = 0$ is somehow "missing" some points which lie out at infinity, and the points that are missing are the (limiting) directions as one moves along the curve out towards infinity.

Rather than giving a general proof, we will illustrate with two examples. First, consider the line

$$L : \alpha X + \beta Y + \gamma Z = 0,$$

say with $\alpha \neq 0$. The affine part of L is the line $L_0 : \alpha x + \beta y + 1 = 0$ in $\mathbb{A}^2$. The points at infinity on L correspond to the points with $Z = 0$. There is only one such point, namely $[-\beta, \alpha, 0] \in L$, which corresponds to the point at infinity $[-\beta, \alpha] \in \mathbb{P}^1$, which in turn corresponds to the direction $-\beta y = \alpha x$ in $\mathbb{A}^2$. This direction is exactly the direction of the line L_0. Thus L consists of the affine line L_0 together with the single point at infinity corresponding to the direction of L_0.

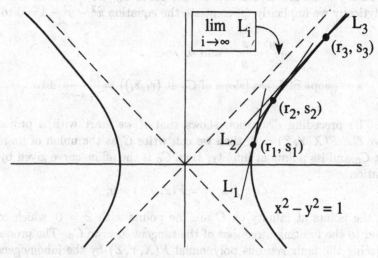

$$\lim_{i \to \infty} L_i$$

L_3
(r_3, s_3)

(r_2, s_2)

L_2 (r_1, s_1)

L_1

$x^2 - y^2 = 1$

Points At Infinity Are Limits Of Tangent Directions

Figure A.2

Next we look at the projective curve

$$C : X^2 - Y^2 - Z^2 = 0.$$

There are two points on C with $Z = 0$, namely $[1, 1, 0]$ and $[1, -1, 0]$. These two points correspond respectively to the points at infinity $[1, 1], [1, -1] \in \mathbb{P}^1$, or equivalently to the directions $y = x$ and $y = -x$ in $\mathbb{A}^2$. The affine part of C is the hyperbola

$$C_0 : x^2 - y^2 - 1 = 0.$$

Suppose we take a sequence of points $(r_1, s_1), (r_2, s_2), \ldots$ on C_0 such that these points tend towards infinity, say $|s_2| \to \infty$. If we rewrite $r_i^2 - s_i^2 - 1 = 0$ as

$$\left(\frac{r_i}{s_i} - 1 \right) \left(\frac{r_i}{s_i} + 1 \right) = \frac{1}{s_i^2},$$

then the right-hand side goes to 0 as $i \to \infty$, so we see that either

$$\lim_{i \to \infty} \frac{r_i}{s_i} = 1 \quad \text{or} \quad \lim_{i \to \infty} \frac{r_i}{s_i} = -1,$$

depending on which branch of the hyperbola we travel on. (See Figure A.2.)

Let L_i be the tangent line to C_0 at the point (r_i, s_i). We claim that as $i \to \infty$, the direction of the tangent line L_i approaches the direction of one of the lines $y = \pm x$. This is nothing more than the assertion that the lines $y = \pm x$ are asymptotes for the curve C_0. To check this assertion analytically we implicitly differentiate the equation $x^2 - y^2 - 1 = 0$ to get

$$\frac{dy}{dx} = \frac{x}{y}, \quad \text{and so}$$

$$(\text{slope of } L_i) = (\text{slope of } C_0 \text{ at } (r_i, s_i)) = \frac{r_i}{s_i} \xrightarrow[i \to \infty]{} \pm 1.$$

The preceding discussion shows that if we start with a projective curve $C : F(X, Y, Z) = 0$, then we can write C as the union of its affine part C_0 and its points at infinity. Here C_0 is the affine curve given by the equation

$$C_0 : f(x, y) = F(x, y, 1) = 0;$$

and the points at infinity on C are the points with $Z = 0$, which correspond to the limiting directions of the tangent lines to C_0. The process of replacing the homogeneous polynomial $F(X, Y, Z)$ by the inhomogeneous polynomial $f(x, y) = F(x, y, 1)$ is called *dehomogenization (with respect to the variable Z)*. We would now like to reverse this process.

Thus suppose we begin with an affine curve C_0 given by an equation $f(x, y) = 0$. We want to find a projective curve C whose affine part is C_0, or equivalently we want to find a homogeneous polynomial $F(X, Y, Z)$ so that $F(x, y, 1) = f(x, y)$. This is easy to do, although we want to be careful not to also include the line at infinity in our curve. If we write the polynomial $f(x, y)$ as $\sum a_{ij} x^i y^j$, then the *degree of f* is defined to be the largest value of $i + j$ for which the coefficient a_{ij} is not zero. For example,

$$\deg(x^2 + xy + x^2 y^2 + y^3) = 4 \quad \text{and} \quad \deg(y^2 - x^3 - ax^2 - bx - c) = 3.$$

Then we define the *homogenization* of the polynomial $f(x, y) = \sum a_{ij} x^i y^j$ to be

$$F(X, Y, Z) = \sum_{i,j} a_{ij} X^i Y^j Z^{d-i-j}, \quad \text{where } d = \deg(f).$$

It is clear from this definition that F is homogeneous of degree d and that $F(x, y, 1) = f(x, y)$. Further, our choice of d ensures that $F(X, Y, 0)$ is not identically zero, so the curve defined by $F(X, Y, Z) = 0$ does not contain the entire line at infinity. Thus by using homogenization and dehomogenization we obtain a one-to-one correspondence between affine curves and projective curves that do not contain the line at infinity.

We should also mention that there is nothing sacred about the variable Z. We could just as well dehomogenize a curve $F(X, Y, Z)$ with respect to one of the other variables, say Y, to get an affine curve $F(x, 1, z) = 0$ in the affine xz plane. It is sometimes convenient to do this if we are especially interested in one of the points at infinity on the projective curve C. In essence what we are doing is taking a different line, in this case the line $Y = 0$, and making it into the "line at infinity." An example should make this clearer. Suppose we want to study the curve

$$C : Y^2 Z - X^3 - Z^3 = 0 \quad \text{and the point} \quad P = [0, 1, 0] \in C.$$

If we dehomogenize with respect to Z, then the point P becomes a point at infinity on the affine curve $y^2 - x^3 - 1 = 0$. So instead we dehomogenize with respect to Y, which means setting $Y = 1$. We then get the affine curve

$$z - x^3 - z^3 = 0, \quad \text{and the point } P \text{ becomes the point } (x, z) = (0, 0).$$

In general, by taking different lines to be the line at infinity, we can break a projective curve C up into a lot of overlapping affine parts, and then these affine parts can be "glued" together to form the entire projective curve.

Up to now we have been working with polynomials without worrying overmuch about what the coefficients of our polynomials look like, and similarly we've talked about solutions of polynomial equations without specifying what sorts of solutions we mean. Classical algebraic geometry is concerned with describing the complex solutions to systems of polynomial equations, but in order to study number theory we will be more interested in finding solutions whose coordinates are in non-algebraically closed fields like $\mathbb{Q}$, or even in rings like $\mathbb{Z}$. That being the case, it makes sense to look at curves given by polynomial equations with rational or integer coefficients.

We call a curve C *rational* if it is the set of zeros of a polynomial having rational coefficients.[†] Note that the solutions of the equation $F(X, Y, Z) = 0$ and the equation $cF(X, Y, Z) = 0$ are the same for any non-zero c. This allows us to clear the denominators of the coefficients, so a rational curve is in fact the set of zeros of a polynomial with integer coefficients. All of the examples given above are rational curves, since their equations have integer coefficients.

Let C be a projective curve that is rational, say C is given by the equation $F(X, Y, Z) = 0$ for a homogeneous polynomial F having rational coefficients. The *set of rational points on C*, which we denote by $C(\mathbb{Q})$, is the set of points of C having rational coordinates:

$$C(\mathbb{Q}) = \{[a, b, c] \in \mathbb{P}^2 : F(a, b, c) = 0 \quad \text{and} \quad a, b, c \in \mathbb{Q}\}.$$

Note that if $P = [a, b, c]$ is in $C(\mathbb{Q})$, it is not necessary that a, b, c themselves be rational, since a point P has many different homogeneous coordinates. All one can say is that $[a, b, c] \in C$ is a rational point of C (i.e. is in $C(\mathbb{Q})$) if and only if there is a non-zero number t so that $ta, tb, tc \in \mathbb{Q}$.

Similarly, if C_0 is an affine curve that is rational, say $C : f(x, y) = 0$, then the set of rational points on C_0 is denoted $C_0(\mathbb{Q})$ and consists of

---

[†] We should warn the reader that this terminology is non-standard. In the usual terminology of algebraic geometry, a curve is called rational if it is isomorphic to the projective line $\mathbf{P}^1$, and a curve given by polynomials with rational coefficients is said to be defined over $\mathbb{Q}$.

all $(r, s) \in C$ with $r, s \in \mathbb{Q}$. It is easy to see that if C_0 is the affine piece of a projective curve C, then $C(\mathbb{Q})$ consists of $C_0(\mathbb{Q})$ together with those points at infinity which happen to be rational. Some of the most famous questions in number theory involve the set of rational points $C(\mathbb{Q})$ on certain curves C. For example, the Nth Fermat curve C_N is the projective curve

$$C_N : X^N + Y^N = Z^N,$$

and Fermat's last theorem asserts that $C_N(\mathbb{Q})$ consists of only those points with one of X, Y, or Z equal to zero.

The theory of Diophantine equations also deals with integer solutions of polynomial equations. Let C_0 be an affine curve that is rational, say given by an equation $f(x, y) = 0$. We define the *set of integer points of C_0*, which we denote $C_0(\mathbb{Z})$, to be the set of points of C_0 having integer coordinates:

$$C_0(\mathbb{Z}) = \{(r, s) \in \mathbb{A}^2 : f(r, s) = 0 \quad \text{and} \quad r, s \in \mathbb{Z}\}.$$

Why do we only talk about integer points on affine curves and not on projective curves? The answer is that for a projective curve, the notion of integer point and rational point coincide. Here we might say that a point $[a, b, c] \in \mathbb{P}^2$ is an integer point if its coordinates are integers. But if $P \in \mathbb{P}^2$ is any point which is given by homogeneous coordinates $P = [a, b, c]$ that are rational, then we can find an integer t to clear the denominators of a, b, c, and so $P = [ta, tb, tc]$ also has homogenous coordinates which are integers. So for a projective curve C we would have $C(\mathbb{Q}) = C(\mathbb{Z})$.

It is also possible to look at polynomial equations and their solutions in rings and fields other than $\mathbb{Z}$ or $\mathbb{Q}$ or $\mathbb{R}$ or $\mathbb{C}$. For example, one might look at polynomials with coefficients in the finite field $\mathbb{F}_p$ with p elements and ask for solutions whose coordinates are also in the field $\mathbb{F}_p$. You may worry about your geometric intutions in situations like this. How can one visualize points and curves and directions in $\mathbb{A}^2$ when the points of $\mathbb{A}^2$ are pairs (x, y) with $x, y \in \mathbb{F}_p$? There are two answers to this question. The first and most reassuring is that you can continue to think of the usual Euclidean plane (i.e., $\mathbb{R}^2$) and most of your geometric intuitions concerning points and curves will still be true when you switch to coordinates in $\mathbb{F}_p$. The second and more practical answer is that the affine and projective planes and affine and projective curves are defined algebraically in terms of ordered pairs (r, s) or homogenous triples $[a, b, c]$ without any reference to geometry. So in proving things one can work algebraically using coordinates, without worrying at all about geometric intuitions. We might summarize this general philosophy as

> *Think Geometrically, Prove Algebraically.*

One of the fundamental questions answered by the differential calculus is that of finding the tangent line to a curve. If $C : f(x, y) = 0$ is an affine curve, then implicit differentiation gives the relation $\dfrac{\partial f}{\partial x} + \dfrac{\partial f}{\partial y}\dfrac{dy}{dx} = 0$. So if $P = (r, s)$ is a point on C, the tangent line to C at P is given by the equation

$$\frac{\partial f}{\partial x}(r, s)(x - r) + \frac{\partial f}{\partial y}(r, s)(y - s) = 0.$$

This is the answer as provided by elementary calculus. But we clearly have a problem if both of the partial derivatives are 0. For example, this happens for each of the curves

$$C_1 : y^2 = x^3 + x^2 \qquad \text{and} \qquad C_2 : y^2 = x^3$$

at the point $P = (0, 0)$. If we sketch these curves, we see that they look a bit strange at P. (See Figures 1.13 and 1.14 near the end of Section 3 of Chapter I.) The curve C_1 crosses over itself at P, so it has two distinct tangent directions there. The curve C_2, on the other hand, has a cusp at P, which means it comes to a sharp point at P. We will say that P is a *singular point* of the curve $C : f(x, y) = 0$ if

$$\frac{\partial f}{\partial x}(P) = \frac{\partial f}{\partial y}(P) = 0.$$

We call P a *non-singular point* if it is not singular, and we say that C is a *non-singular curve* (or *smooth curve*) if every point of C is non-singular. If P is a non-singular point of C, then we define the *tangent line to C at P* to be the line described above.

For a projective curve $C : F(X, Y, Z) = 0$ described by a homogeneous polynomial we make similar definitions. More precisely, if $P = [a, b, c]$ is a point on C with $c \neq 0$, then we go to the affine part of C and check whether or not the point

$$P_0 = \left(\frac{a}{c}, \frac{b}{c}\right) \qquad \text{is singular on the affine curve} \quad C_0 : F(x, y, 1) = 0.$$

And if $c = 0$, then we can dehomogenize in some other way. For example, if $a \neq 0$, we check whether or not the point

$$P_0 = \left(\frac{b}{a}, \frac{c}{a}\right) \qquad \text{is singular on the affine curve} \quad C_0 : F(1, y, z) = 0.$$

We say that C is non-singular or smooth if all of its points, including the points at infinity, are non-singular. If P is a non-singular point of C, we define the tangent line to C at P by dehomogenizing, finding the tangent line to the affine part of C at P, and then homogenizing the equation of that tangent line to get a line in $\mathbb{P}^2$. (An alternative method to check for singularity and find tangent lines on projective curves is described in the exercises.)

When one is faced with a complicated equation, it is natural to try to make a change of variables in order to simplify it. Probably the first significant example of this that you have seen is the process of completing the square to solve a quadratic equation. Thus to solve $Ax^2 + Bx + C = 0$ we multiply through by $4A$ and rewrite the equation as

$$(2Ax + B)^2 + 4AC - B^2 = 0.$$

This suggests the substitution $x' = 2Ax + B$, and then we can solve $x'^2 + 4AC - B^2 = 0$ for $x' = \pm\sqrt{B^2 - 4AC}$. The crucial final step uses the fact that our substitution is invertible, so we can solve for x in terms of x' to obtain the usual quadratic formula

$$x = \frac{-B + x'}{2A} = \frac{-B \pm \sqrt{B^2 - 4AC}}{2A}.$$

More generally, suppose we are given a projective curve C of degree d, say defined by an equation $C : F(X, Y, Z) = 0$. In order to change coordinates on $\mathbb{P}^2$ we make a substitution

$$\begin{aligned}
X &= m_{11}X' + m_{12}Y' + m_{13}Z', \\
Y &= m_{21}X' + m_{22}Y' + m_{23}Z', \\
Z &= m_{31}X' + m_{32}Y' + m_{33}Z'.
\end{aligned} \tag{$*$}$$

Then we get a new curve C' given by the equation $F'(X', Y', Z') = 0$, where F' is the polynomial

$$\begin{aligned}
F'(X', Y', Z') = F(m_{11}X' + m_{12}Y' + m_{13}Z', \\
m_{21}X' + m_{22}Y' + m_{23}Z', m_{31}X' + m_{32}Y' + m_{33}Z').
\end{aligned}$$

The change of coordinates ($*$) gives a map from C' to C; that is, given a point $[a', b', c'] \in C'$, we substitute $X' = a'$, $Y' = b'$, and $Z' = c'$ into ($*$) to get a point $[a, b, c] \in C$. Further, this map $C' \to C$ will have an inverse provided that the matrix $M = (m_{ij})_{1 \le i, j \le 3}$ is invertible. More precisely, if $M^{-1} = N = (n_{ij})$, then the change of coordinates

$$\begin{aligned}
X' &= n_{11}X + n_{12}Y + n_{13}Z, \\
Y' &= n_{21}X + n_{22}Y + n_{23}Z, \\
Z' &= n_{31}X + n_{32}Y + n_{33}Z,
\end{aligned}$$

will map C to C'. We call a change of coordinates on $\mathbb{P}^2$ given by an invertible 3×3 matrix a *projective transformation*. Note that if the matrix has rational coefficients, then the corresponding projective transformation gives a one-to-one correspondence between $C(\mathbb{Q})$ and $C'(\mathbb{Q})$. So the number theoretic problem of finding the rational points on the curve C is equivalent to the problem of finding the rational points on C'.

3. Intersections of Projective Curves

Recall that our geometric construction of the projective plane was based on the desire that every pair of distinct lines should intersect in exactly one point. In this section we are going to discuss the intersection of curves of higher degree.

How many intersection points should two curves have? Let's begin with a thought experiment, and then we'll consider some examples and see to what extent our intuition is correct. Let C_1 be an affine curve of degree d_1 and let C_2 be an affine curve of degree d_2. Thus, C_1 and C_2 are given by polynomials

$$C_1 : f_1(x,y) = 0 \quad \text{with } \deg(f_1) = d_1, \text{ and}$$
$$C_2 : f_2(x,y) = 0 \quad \text{with } \deg(f_2) = d_2.$$

The points in the intersection $C_1 \cap C_2$ are the solutions to the simultaneous equations $f_1(x,y) = f_2(x,y) = 0$. Suppose now that we consider f_1 as a polynomial in the variable y whose coefficients are polynomials in x. Then $f_1(x,y) = 0$, being a polynomial equation of degree d_1 in y, should in principle have d_1 roots $y_1, y_2, \ldots, y_{d_1}$. Now we substitute each of these roots into the second equation $f_2(x,y) = 0$ to find d_1 equations for x, namely

$$f_2(x,y_1) = 0, \qquad f_2(x,y_2) = 0, \qquad \cdots \qquad f_2(x,y_{d_1}) = 0.$$

Each of these equations is a polynomial in x of degree d_2, so in principal each equation should yield d_2 values for x. Altogether we appear to get $d_1 d_2$ pairs (x,y) which satisfy $f_1(x,y) = f_2(x,y) = 0$, which seems to indicate we should have $\#(C_1 \cap C_2) = d_1 d_2$. For example, a curve of degree 2 and a curve of degree 4 should intersect in 8 points, as is illustrated in Figure A.3. This assertion, that curves of degree d_1 and d_2 intersect in $d_1 d_2$ point, is indeed true provided that it is interpreted properly. However, matters are considerably more complicated than they appear at first glance, as will be clear from the following examples. [Can you find all of the ways in which our plausibility argument fails to be a valid proof? For example, the "roots" $y_1, \ldots, y_{d_1}$ really depend on x, so we should write $f_2(x, y_i(x)) = 0$, and then it is not at all clear how many roots we should expect.]

Curves of degree one are lines, and curves of degree two are called *conics*. We already know that two lines in $\mathbb{P}^2$ intersect in a unique point, so the next simplest case is the intersection of a line and a conic. Our discussion above leads us to expect two intersection points, so we look at some examples to see what really happens. The (affine) line and conic

$$C_1 : x + y + 1 = 0 \quad \text{and} \quad C_2 : x^2 + y^2 = 1$$

intersect in the two points $(-1,0)$ and $(0,-1)$, as is easily seen by substituting $y = -x - 1$ into the equation for C_2 and solving the resulting quadratic equation for x. (See Figure A.4(a).) Similarly,

$$C_1 : x + y = 0 \quad \text{and} \quad C_2 : x^2 + y^2 = 1$$

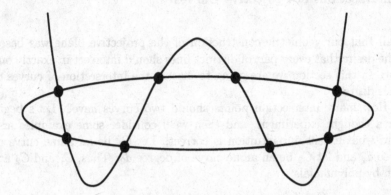

Curves Of Degree Two And Degree Four Intersect In Eight Points

Figure A.3

intersect in the two points $\left(\frac{1}{2}\sqrt{2}, -\frac{1}{2}\sqrt{2}\right)$ and $\left(-\frac{1}{2}\sqrt{2}, \frac{1}{2}\sqrt{2}\right)$. Note that we have to allow real coordinates for the intersection points, even though C_1 and C_2 are rational curves. (See Figure A.4(b).)

What about the intersection of the line and conic

$$C_1 : x + y + 2 = 0 \quad \text{and} \quad C_2 : x^2 + y^2 = 1?$$

They do not intersect at all in the usual Euclidean plane $\mathbb{R}^2$, but if we allow complex numbers then we again find two intersection points

$$\left(-1 + \frac{\sqrt{2}}{2}i, -1 - \frac{\sqrt{2}}{2}i\right) \quad \text{and} \quad \left(-1 - \frac{\sqrt{2}}{2}i, -1 + \frac{\sqrt{2}}{2}i\right).$$

(See Figure A.4(c).) Of course, it is reasonable to allow complex coordinates, since even for polynomials in one variable we need to use complex numbers to ensure that a polynomial of degree d actually has d roots counting multiplicities.

Next we look at

$$C_1 : x + 1 = 0 \quad \text{and} \quad C_2 : x^2 - y = 0.$$

These curves appear to intersect in the single point $(-1, 1)$, but appearances can be deceiving. (See Figure A.4(d).) Remember even for two lines, we may need to also look at the points at infinity in $\mathbb{P}^2$. In our case the line C_1 is in the vertical direction, and the tangent lines to the parabola C_2 approach the vertical direction, so geometrically C_1 and C_2 should have a common point at infinity corresponding to the vertical direction. Following our maxim from Section 2, we now check this algebraically. First we homogenize the equations for C_1 and C_2 to get the corresponding projective curves

$$\bar{C}_1 : X + Z = 0 \quad \text{and} \quad \bar{C}_2 : X^2 - YZ = 0.$$

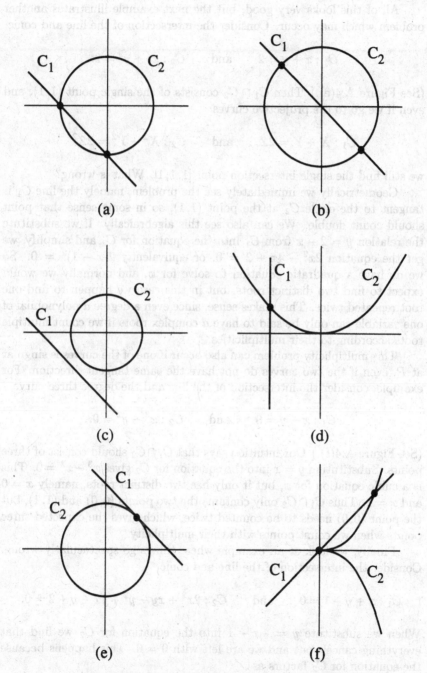

Some Of The Ways In Which Curves Can Intersect

Figure A.4

Then $\bar{C}_1 \cap \bar{C}_2$ consists of the two points $[-1, 1, 1]$ and $[0, 1, 0]$, as can be seen by substituting $X = -Z$ into the equation for $\bar{C}_2$. So we get the expected two points provided we work with projective curves.

All of this looks very good, but the next example illustrates another problem which may occur. Consider the intersection of the line and conic

$$C_1 : x + y = 2 \quad \text{and} \quad C_2 : x^2 + y^2 = 2.$$

(See Figure A.4(e).) Then $C_1 \cap C_2$ consists of the single point $(1, 1)$, and even if we go to the projective curves

$$\bar{C}_1 : X + Y = 2Z \quad \text{and} \quad \bar{C}_2 : X^2 + Y^2 = 2Z^2$$

we still find the single intersection point $[1, 1, 1]$. What is wrong?

Geometrically we immediately see the problem, namely the line C_1 is tangent to the circle C_2 at the point $(1, 1)$, so in some sense that point should count double. We can also see this algebraically. If we substitute the relation $y = 2 - x$ from C_1 into the equation for C_2 and simplify, we get the equation $2x^2 - 4x + 2 = 0$, or equivalently $2(x - 1)^2 = 0$. So we do have a quadratic equation to solve for x, and normally we would expect to find two distinct roots, but in this case we happen to find one root repeated twice. This makes sense, since even a degree d polynomial of one variable can only be said to have d complex roots if we count multiple roots according to their multiplicities.

This multiplicity problem can also occur if one of the curves is singular at P, even if the two curves do not have the same tangent direction. For example, consider the intersection of the line and the degree three curve

$$C_1 : x - y = 0 \quad \text{and} \quad C_2 : x^3 - y^2 = 0.$$

(See Figure A.4(f).) Our intuition says that $C_1 \cap C_2$ should consist of three points. Substituting $y = x$ into the equation for C_2 gives $x^3 - x^2 = 0$. This is a cubic equation for x, but it only has two distinct roots, namely $x = 0$ and $x = 1$. Thus $C_1 \cap C_2$ only contains the two points $(0, 0)$ and $(1, 1)$, but the point $(0, 0)$ needs to be counted twice, which gives the expected three points when we count points with their multiplicity.

Finally, we look at an example where things go spectacularly wrong. Consider the intersection of the line and conic

$$C_1 : x + y + 1 = 0 \quad \text{and} \quad C_2 : 2x^2 + xy - y^2 + 4x + y + 2 = 0.$$

When we substitute $y = -x - 1$ into the equation for C_2 we find that everything cancels out and we are left with $0 = 0$. This happens because the equation for C_2 factors as

$$2x^2 + xy - y^2 + 4x + y + 2 = (x + y + 1)(2x - y + 2),$$

so every point on C_1 lies on C_2. Notice that C_2 is the union of two curves, namely C_1 and the line $2x - y + 2 = 0$.

In general, if C is a curve given by an equation $C : f(x, y) = 0$, then we factor f into a product of irreducible polynomials

$$f(x, y) = p_1(x, y)p_2(x, y) \cdots p_n(x, y).$$

Note that $\mathbb{C}[x, y]$ is a unique factorization domain, so every polynomial has an essentially unique factorization into such a product. Then the *irreducible components of the curve* C are the curves

$$p_1(x, y) = 0, \qquad p_2(x, y) = 0, \qquad \cdots \qquad p_n(x, y) = 0.$$

We say that C is *irreducible* if it has only one irreducible component, or equivalently if $f(x, y)$ is an irreducible polynomial. Next, if C_1 and C_2 are two curves, we say that C_1 and C_2 *have no common components* if their irreducible components are distinct. It is not hard to prove that $C_1 \cap C_2$ consists of a finite set of points if and only if C_1 and C_2 have no common components. Finally, if we work instead with projective curves C, C_1, C_2, then we make the same definitions using factorizations into products of irreducible homogeneous polynomials in $\mathbb{C}[X, Y, Z]$.

We now consider the general case of projective curves C_1 and C_2, which we assume to have no common components. The intersection $C_1 \cap C_2$ is then a finite set of points with complex coordinates. To each point $P \in \mathbb{P}^2$ we assign a *multiplicity* or *intersection index* $I(C_1 \cap C_2, P)$. This is a non-negative integer reflecting the extent to which C_1 and C_2 are tangent to one another at P or are not smooth at P. We will give a formal definition in Section 4, but one can get a good feeling for the intersection index from the following properties:

(i) If $P \notin C_1 \cap C_2$, then $I(C_1 \cap C_2, P) = 0$.

(ii) If $P \in C_1 \cap C_2$, if P is a non-singular point of C_1 and C_2, and if C_1 and C_2 have different tangent directions at P, then $I(C_1 \cap C_2, P) = 1$. (One often says in this case that C_1 and C_2 intersect *transversally at P*.)

(iii) If $P \in C_1 \cap C_2$ and if C_1 and C_2 do not intersect transversally at P, then $I(C_1 \cap C_2, P) \geq 2$.

(For a proof of these properties, see the last part of Section 4.)

With these preliminaries, we are now ready to formally state the theorem which justifies the plausibility argument we gave at the beginning of this section.

Bezout's Theorem. *Let C_1 and C_2 be projective curves with no common components. Then*

$$\sum_{P \in C_1 \cap C_2} I(C_1 \cap C_2, P) = (\deg C_1)(\deg C_2),$$

where the sum is over all points of $C_1 \cap C_2$ having complex coordinates. In particular, if C_1 and C_2 are smooth curves with only transversal intersections, then $\#(C_1 \cap C_2) = (\deg C_1)(\deg C_2)$; and in all cases there is an inequality

$$\#(C_1 \cap C_2) \leq (\deg C_1)(\deg C_2).$$

PROOF. We will give the proof of Bezout's theorem in Section 4.   □

It would be hard to overestimate the importance of Bezout's theorem in the study of projective geometry. We should stress how amazing a theorem it is. The projective plane was constructed so as to ensure that any two lines (i.e., curves of degree 1) intersect in exactly one point, so one could say that the projective plane is formed by taking the affine plane and adding just enough points to make Bezout's theorem true for curves of degree 1. It then turns out that the projective plane has enough points to make Bezout's theorem true for all projective curves!

Sometimes Bezout's theorem is used to determine if two curves are the same, or at least have a common component. For example, if C_1 and C_2 are conics, and if C_1 and C_2 have five points in common, then Bezout's theorem tells us that they have a common component. Since the degree of a component can be no larger than the degree of the curve, it follows that either there is some line L contained in both C_1 and C_2, or else $C_1 = C_2$. Thus there is only one conic going through any given five points as long as no three of the points are collinear. This is analogous to the fact that there is a unique line going through two given points. More generally, one sees from Bezout's theorem if C_1 and C_2 are irreducible curves of degree d with $d^2 + 1$ points in common, then $C_1 = C_2$. Note, however, that for $d \geq 3$ there is in general no curve of degree d going through $d^2 + 1$ preassigned points. This is because the number $d^2 + 1$ of conditions to be met is greater than the number $(d+1)(d+1)/2$ unknown coefficients of a homogeneous polynomial of degree d.

We now want to consider a slightly more complicated situation. Suppose that C_1 and C_2 are two cubic curves, that is curves of degree 3, which intersect in 9 distinct points $P_1, \ldots, P_9$. Suppose further that D is another cubic curve which happens to go through the first 8 points $P_1, \ldots, P_8$. We claim that D also goes through the ninth point P_9. To see why this is true, we consider the collection of all cubic curves in $\mathbb{P}^2$, which we will denote by $\mathcal{C}^{(3)}$. An element $C \in \mathcal{C}^{(3)}$ is given by a homogeneous equation

$$C : aX^3 + bX^2Y + cXY^2 + dY^3 + eX^2Z + fXZ^2 +$$
$$gY^2Z + hYZ^2 + iZ^3 + jXYZ = 0,$$

so C is determined by the ten coefficients $a, b, \ldots, j$. Of course, if we multiply the equation for C by any non-zero constant, then we get the same curve, so really C is determined by the homogeneous 10-tuple $[a, b, \ldots, j]$.

Conversely, if two 10-tuples give the same curve, then they differ by multiplication by a constant. In other words, the set of cubic curves $\mathcal{C}^{(3)}$ is in a
very natural way isomorphic to the projective space $\mathbb{P}^9$.

Suppose we are given a point $P \in \mathbb{P}^2$ and ask for all cubic curves
that go through P. This describes a certain subset of $\mathcal{C}^{(3)} \cong \mathbb{P}^9$, and it
is easy to see what this subset is. If P has homogeneous coordinates $P =
[X_0, Y_0, Z_0]$, then substituting P into the equation for C shows that C will
contain P if and only if the 10-tuple $[a, b, \ldots, j]$ satisfies the homogeneous
linear equation

$$(X_0^3)a + (X_0^2 Y_0)b + (X_0 Y_0^2)c + (Y_0^3)d + (X_0^2 Z_0)e + (X_0 Z_0^2)f +$$
$$(Y_0^2 Z_0)g + (Y_0 Z_0^2)h + (Z_0^3)i + (X_0 Y_0 Z_0)j = 0.$$

[N.B. This is a linear equation in the 10 variables $a, b, \ldots, j$.] In other
words, for a given point $P \in \mathbb{P}^2$, the set of cubic curves $C \in \mathcal{C}^{(3)}$ which
contain P corresponds to the zeros of a homogeneous linear equation in $\mathbb{P}^9$.

Similarly, if we fix two points $P, Q \in \mathbb{P}^2$, then the set of cubic curves
$C \in \mathcal{C}^{(3)}$ containing both P and Q is given by the common solutions of two
linear equation in $\mathbb{P}^9$, where one linear equation is specified by P and the
other by Q. Continuing in this fashion, we find that for a collection of n
points $P_1, P_2, \ldots, P_n \in \mathbb{P}^2$ there is a one-to-one correpsondence between
the two sets

$$\{C \in \mathcal{C}^{(3)} : P_1, \ldots, P_n \in C\} \quad \text{and} \quad \left\{ \begin{array}{c} \text{simultaneous solutions of a} \\ \text{certain system of } n \text{ homo-} \\ \text{geneous linear equations in } \mathbb{P}^9 \end{array} \right\}.$$

For example, suppose we take $n = 9$. The solutions to a system of 9
homogeneous linear equations in 10 variables generally consists of the multiples of a single solution. In other words, if $\mathbf{v_0}$ is a non-zero solution,
then every solution will have the form $\lambda \mathbf{v_0}$ for some constant λ. Now
let $C_1 : F_1(X, Y, Z) = 0$ and $C_2 : F_2(X, Y, Z) = 0$ be cubic curves in $\mathbb{P}^2$,
each going through the given 9 points. The coefficients of F_1 and F_2 are
then 10-tuples which are solutions to the given system of linear equations,
so we conclude that $F_1 = \lambda F_2$, and hence $C_1 = C_2$. Thus, we find that, in
general, there is exactly one cubic curve in $\mathbb{P}^2$ that passes through 9 given
points. Note, however, that for special sets of nine points it is possible to
have a one parameter family of cubic curves going through them.

That is the situation of our original problem, to which we now return.
Namely, we take two cubic curves C_1 and C_2 in $\mathbb{P}^2$ that intersect in nine
distinct points $P_1, \ldots, P_9$. Let C_1 and C_2 be given by the equations

$$C_1 : F_1(X, Y, Z) = 0 \quad \text{and} \quad C_2 : F_2(X, Y, Z) = 0.$$

We consider the set of all cubic curves $C \in \mathcal{C}^{(3)}$ which pass through the
first 8 points $P_1, \ldots, P_8$. This set corresponds to the simultaneous solution
of 8 homogeneous linear equations in 10 variables. The set of solutions

of this system consists of all linear combinations of two linearly independent 10-tuples; in other words, if $\mathbf{v}_1$ and $\mathbf{v}_2$ are independent solutions, then every solution has the form $\lambda_1\mathbf{v}_1 + \lambda_2\mathbf{v}_2$ for some constants λ_1, λ_2. (In principle, the set of solutions might have dimension greater than two. We leave it as an exercise for you to check that because the eight points $P_1, \ldots, P_8$ are distinct, the corresponding linear equations will be independent.)

But we already know two cubic curves passing through $P_1, \ldots, P_8$, namely C_1 and C_2. The coefficients of their equations F_1 and F_2 will thus give two 10-tuples solving the system of 8 homogeneous linear equations, so they will span the complete solution set. This means that if D is any other cubic curve in $\mathbb{P}^2$ that contains the 8 points $P_1, \ldots, P_8$, then the equation for D has the form

$$D : \lambda_1 F_1(X, Y, Z) + \lambda_2 F_2(X, Y, Z) = 0 \qquad \text{for some constants } \lambda_1, \lambda_2.$$

But the ninth point P_9 is on both C_1 and C_2, so $F_1(P_9) = F_2(P_9) = 0$. It follows from the equation for D that D also contains the point P_9, which is exactly what we have been trying to demonstrate.

More generally, the following theorem is true.

Cayley-Bacharach Theorem. *Let C_1 and C_2 be curves in $\mathbb{P}^2$ without common components of respective degrees d_1 and d_2, and suppose that C_1 and C_2 intersect in $d_1 d_2$ distinct points. Let D be a curve in $\mathbb{P}^2$ of degree $d_1 + d_2 - 3$. If D passes through all but one of the points of $C_1 \cap C_2$, then D must pass through the remaining point also.*

It is not actually necessary that C_1 and C_2 intersect in distinct points. For example, if $P \in C_1 \cap C_2$ is a point of multiplicity 2, say because C_1 and C_2 have the same tangent direction at P, then one needs to require that D also have the same tangent direction at P. The most general result is somewhat difficult to state, so we will content ourselves with the following version:

Cubic Cayley-Bacharach Theorem. *Let C_1 and C_2 be cubic curves in $\mathbb{P}^2$ without common components, and assume that C_1 is smooth. Suppose that D is another cubic curve that contains 8 of the intersection points of $C_1 \cap C_2$ counting multiplicities. This means that if $C_1 \cap C_2 = \{P_1, \ldots, P_r\}$, then*

$$I(C_1 \cap D, P_i) \geq I(C_1 \cap C_2, P_i) \quad \text{for } 1 \leq i < r, \text{ and}$$
$$I(C_1 \cap D, P_r) \geq I(C_1 \cap C_2, P_r) - 1.$$

Then D goes through the ninth point of $C_1 \cap C_2$. In terms of multiplicities, this means that $I(C_1 \cap D, P_r) = I(C_1 \cap C_2, P_r)$.

We will conclude this section of the appendix by applying the Cayley-Bacharach theorem to prove a beautiful geometric result of Pascal. Let C be a smooth conic, for example a hyperbola, a parabola, or an ellipse. Choose

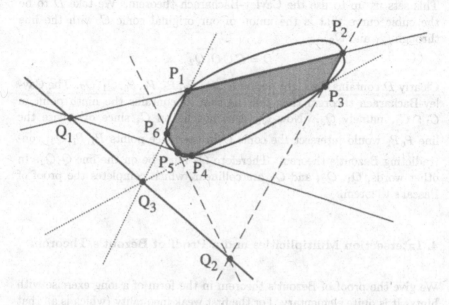

Pascal's Theorem

Figure A.5

any six points lying on the conic, say labeled consecutively as $P_1, P_2, \ldots, P_6$, and play connect-the-dots to draw a hexagon. Now take the lines through opposite sides of the hexagon and extend them to find the intersection points as illustrated in Figure A.5, say

$$\overleftrightarrow{P_1 P_2} \cap \overleftrightarrow{P_4 P_5} = \{Q_1\}, \quad \overleftrightarrow{P_2 P_3} \cap \overleftrightarrow{P_5 P_6} = \{Q_2\}, \quad \overleftrightarrow{P_3 P_4} \cap \overleftrightarrow{P_6 P_1} = \{Q_3\}.$$

Pascal's Theorem. *The three points Q_1, Q_2, Q_3 described above lie on a line.*

To prove Pascal's theorem, we consider the two cubic curves

$$C_1 = \overleftrightarrow{P_1 P_2} \cup \overleftrightarrow{P_3 P_4} \cup \overleftrightarrow{P_5 P_6} \quad \text{and} \quad C_2 = \overleftrightarrow{P_2 P_3} \cup \overleftrightarrow{P_4 P_5} \cup \overleftrightarrow{P_6 P_1}.$$

Why do we call C_1 and C_2 cubic curves? The answer is that if we choose an equation for the line $\overleftrightarrow{P_i P_j}$, say $\alpha_{ij} X + \beta_{ij} Y + \gamma_{ij} Z = 0$, then C_1 is given by the homogeneous cubic equation

$$C_1 : (\alpha_{12} X + \beta_{12} Y + \gamma_{12} Z)(\alpha_{34} X + \beta_{34} Y + \gamma_{34} Z)(\alpha_{56} X + \beta_{56} Y + \gamma_{56} Z) = 0,$$

and similarly for C_2.

Notice that all nine of the points

$$P_1, P_2, P_3, P_4, P_5, P_6, Q_1, Q_2, Q_3 \quad \text{are on both } C_1 \text{ and } C_2.$$

This sets us up to use the Cayley-Bacharach theorem. We take D to be the cubic curve that is the union of our original conic C with the line through Q_1 and Q_2,

$$D = C \cup \overleftrightarrow{Q_1 Q_2}.$$

Clearly D contains the eight points $P_1, P_2, P_3, P_4, P_5, P_6, Q_1, Q_2$. The Cayley-Bacharach theorem then tells us that D contains the ninth point in $C_1 \cap C_2$, namely Q_3. Now Q_3 does not lie on C, since otherwise the line $\overleftrightarrow{P_6 P_1}$ would intersect the conic C in the three points P_6, P_1, Q_3, contradicting Bezout's theorem. Therefore Q_3 must be on the line $\overleftrightarrow{Q_1 Q_2}$. In other words, Q_1, Q_2, and Q_3 are collinear, which completes the proof of Pascal's theorem.

4. Intersection Multiplicities and a Proof of Bezout's Theorem

We give the proof of Bezout's theorem in the form of a long exercise with hints. It is quite elementary. For the first weak inequality (which is all that is needed in many important applications of the theorem) we use only linear algebra and the notion of the dimension of a vector space. After that we need the concepts of commutative ring, ideal, and quotient ring, and the fact that unique factorization holds in polynomial rings, but that is about all.

Let C_1 and C_2 be curves in $\mathbb{P}^2$ of respective degrees n_1 and n_2, without common components. Until the last step of the proof we assume that the line at infinity is not a component of either curve, and we work with affine coordinates x, y. Let

$$f_1(x, y) = 0 \qquad \text{and} \qquad f_2(x, y) = 0$$

be the equations for the two curves in the affine plane $\mathbb{A}^2$. The assumptions we have made mean that the polynomials f_1 and f_2 have no common factor and are of degree n_1 and n_2 respectively.

The proof is pure algebra (though the geometric ideas behind it should be apparent) and works over any algebraically closed ground field k. The reader is welcome to take $k = \mathbb{C}$, but k could also be an algebraic closure of the finite field $\mathbb{F}_p$, for example.

Let $R = k[x, y]$ be a polynomial ring in two variables and let $(f_1, f_2) = Rf_1 + Rf_2$ be the ideal in R generated by the polynomials f_1 and f_2. The steps in the proof of Bezout's theorem are as follows:

(1) We prove the following two inequalities which, on eliminating the middle term, show that the number of intersection points of C_1 and C_2 in $\mathbb{A}^2$ is at most $n_1 n_2$:

$$\#(C_1 \cap C_2 \cap \mathbb{A}^2) \overset{(A)}{\leq} \dim\left(\frac{R}{(f_1, f_2)}\right) \overset{(B)}{\leq} n_1 n_2.$$

Note: In this section, dim means the dimension as a k-vector space.

(2) We show that (B) is an equality if C_1 and C_2 do not meet at infinity.

(3) We strengthen (A) to get

$$\sum_{P \in C_1 \cap C_2 \cap \mathbb{A}^2} I(C_1 \cap C_2, P) \overset{(A^+)}{\leq} \dim \left(\frac{R}{(f_1, f_2)} \right),$$

where $I(C_1 \cap C_2, P)$ is a suitably defined *intersection multiplicity* of C_1 and C_2 at P.

(4) We show that (A^+) is in fact an equality.

The fact that k is algebraically closed is not needed for the inequalities (1) and (3), but is essential for the equalities (2) and (4). Taken together, (2) and (4) give Bezout's theorem in case C_1 and C_2 do not meet at infinity. To get it in general there is one more step.

(5) We show that the definition of intersection multiplicity does not change when we make a projective transformation, and that there is a line L in $\mathbb{P}^2$ not meeting any intersection point. Changing coordinates so that the line L is the line at infinity, we then get Bezout in general.

To round out the argument we include one more segment:

(6) We prove some basic properties satisfied by the intersection multiplicity $I(C_1 \cap C_2, P)$ and show that it depends only on the initial part of the Taylor expansions of f_1 and f_2 at P.

Now we sketch the proof as a series of exercises with hints, breaking each of the segments (1)–(5) into smaller steps.

(1.1) Let $P_1, P_2, \ldots, P_m$ be m different points in the (x, y) plane. Show that for each i there is a polynomial $h_i = h_i(x, y)$ such that $h_i(P_i) = 1$ and $h_i(P_j) = 0$ for $j \neq i$. (*Idea.* Construct h_i as a product of linear polynomials, using the fact that for each $j \neq i$ there is a line through P_j not meeting P_i.)

(1.2) Suppose that the m points P_i from (1.1) lie in $C_1 \cap C_2$. Prove that the polynomials h_i are linearly independent modulo (f_1, f_2), and consequently that

$$m \leq \dim \left(\frac{R}{(f_1, f_2)} \right).$$

This proves inequality (A). (*Idea.* Consider a possible dependence

$$c_1 h_1 + c_2 h_2 + \cdots + c_m h_m = g_1 f_1 + g_2 f_2 \in (f_1, f_2)$$

with $c_i \in k$. Substitute P_i into the equation to show that every $c_i = 0$.)

This takes care of inequality (A). To prove (B) we define for each integer $d \geq 0$,

$$\phi(d) = \frac{1}{2}(d + 1)(d + 2) = \frac{1}{2}d^2 + \frac{3}{2}d + 1,$$
$$R_d = (\text{vector space of polynomials } f(x, y) \text{ of degree} \leq d),$$
$$W_d = R_{d - n_1} f_1 + R_{d - n_2} f_2.$$

Thus W_d is the k-vector space of polynomials of the form

$$f = g_1 f_1 + g_2 f_2 \quad \text{with} \quad \deg g_i \leq d - n_i \quad \text{for} \quad i = 1, 2.$$

Notice that $W_d = 0$ if $d < \max\{n_1, n_2\}$, and in any case $V_d \subset (f_1, f_2)$.
(1.3) Show that $\dim R_d = \phi(d)$. (*Idea.* One way to see this is to note that

$$\phi(d) - \phi(d-1) = (\text{number of monomial } x^i y^j \text{ of degree } d) = d + 1$$

and use induction on d.)
(1.4) For $d \geq n_1 + n_2$, show that

$$R_{d-n_1} f_1 \cap R_{d-n_2} f_2 = R_{d-n_1-n_2} f_1 f_2.$$

(Here we use the hypothesis that f_1 and f_2 have no common factor.)
(1.5) Prove that for $d \geq n_1 + n_2$,

$$\dim R_d - \dim W_d = \phi(d) - \phi(d-n_1) - \phi(d-n_2) + \phi(d-n_1-n_2) = n_1 n_2.$$

(*Idea.* If f is a non-zero polynomial, then $g \mapsto gf$ defines an isomorphism $R_{d-j} \xrightarrow{\sim} R_{d-j}f$; hence $\dim R_{d-j}f = \phi(d-j)$. Now use the lemma from linear algebra which says that

$$\dim(U + V) = \dim(U) + \dim(V) - \dim(U \cap V)$$

for subspaces U, V of a finite dimensional vector space.)
(1.6) Prove inequality (B) by showing that if g_j, $1 \leq j \leq n_1 n_2 + 1$, are elements of R, then they are linearly dependent modulo (f_1, f_2). (*Idea.* Take d so large that the g_j are in R_d and so that (1.5) holds. Then use (1.5) to show that there is a non-trivial linear combination $g = \sum c_j g_j$ such that $g \in W_d \subset (f_1, f_2)$.)

This finishes segment (1). For segment (2) we begin by recalling how one computes the intersections of an affine curve $f(x, y) = 0$ with the line at infinity.
(2.1) For each non-zero polynomial $f = f(x, y)$, let f^* denote the homogeneous part of f of highest degree. In other words, if

$$f = \sum_{i,j} c_{ij} x^i y^j \quad \text{has degree } n, \quad \text{then} \quad f^* = \sum_{i+j=n} c_{ij} x^i y^j.$$

Because k is algebraically closed, we can factor f^* into linear factors,

$$f^*(x, y) = \prod_{i=1}^{n} (a_i x + b_i y), \quad a_i, b_i \in k, \quad n = \deg f = \deg f^*.$$

Show that the points at infinity on the curve $f(x, y) = 0$ are the points with homogeneous coordinates

$$[X, Y, Z] = [b_i, -a_i, 0].$$

(*Idea.* Put $x = X/Z$, $y = Y/Z$, etc.)

An example should make this clearer. Consider the polynomials

$$f(x,y) = x^4 - x^2y^2 + 3x^3 + xy^2 + 2y^3 + 2y^2 + 8x + 3,$$
$$f^*(x,y) = x^4 - x^2y^2 = x^2(x+y)(x-y),$$

each of which has degree 4. The quartic curve $f(x,y) = 0$ thus meets the line at infinity in the points $(0,1,0)$, $(1,-1,0)$, $(1,1,0)$. The fact that x^2 divides $f^*(x,y)$ means that the curve is tangent to the line at infinity at the point $(0,1,0)$.

The remaining steps in segment 2 are as follows:

(2.2) If C_1 and C_2 do not meet at infinity, show that f_1^* and f_2^* have no common factor.

(2.3) If f_1^* and f_2^* have no common factor, show that $(f_1, f_2) \cap R_d = W_d$ for all $d \geq n_1 + n_2$.

(2.4) If $(f_1, f_2) \cap R_d = W_d$ and $d \geq n_1 + n_2$, show that

$$\dim\left(\frac{R}{(f_1, f_2)}\right) \geq n_1 n_2.$$

(*Idea.* (2.2) is an easy consequence of (2.1). To do (2.3) we suppose that $f \in (f_1, f_2) \cap R_d$ is written in the form $f = g_1 f_1 + g_2 f_2$ with g_1 and g_2 of smallest possible degree. If $\deg g_1 > d - n_1$, then looking at the terms of highest degree shows that $g_1^* f_1^* + g_2^* f_2^* = 0$. Then use the fact that f_1^* and f_2^* are relatively prime to show that there is an h such that

$$\deg(g_1 + hf_2) < \deg(g_1) \quad \text{and} \quad \deg(g_2 - hf_1) < \deg(g_2).$$

Deduce that $\deg g_i \leq d - n_i$, and hence that $f \in W_d$. For (2.4) note that by (1.5) there are $n_1 n_2$ element in R_d which are linearly independent modulo W_d, and that if $(f_1, f_2) \cap R_d = W_d$, then they are linearly independent as elements of R modulo (f_1, f_2). Hence, $\dim R/(f_1, f_2) \geq n_1 n_2$.)

To define intersection multiplicity we introduce the important notion of the *local ring* $\mathcal{O}_P$ of a point $P \in \mathbf{A}^2$. Let $K = k(x,y)$ be the fraction field of $R = k[x,y]$, that is, K is the field of rational functions of x and y. For a point $P = (a,b)$ in the (x,y) plane and a rational function $\phi = f(x,y)/g(x,y) \in K$, we say that ϕ is *defined at* P if $g(a,b) \neq 0$, and then we put

$$\phi(P) = \frac{f(a,b)}{g(a,b)} = \frac{f(P)}{g(P)}.$$

For a given point P we define the *local ring of P* to be the set of all $\phi \in K$ which are defined at P. We leave the following basic properties of $\mathcal{O}_P$ as exercises. First, $\mathcal{O}_P$ is a subring of K, and the map $\phi \mapsto \phi(P)$ is a homomorphism of $\mathcal{O}_P$ onto k which is the identity on k. Let

$$M_P = \{\phi \in \mathcal{O}_P : \phi(P) = 0\}$$

be the kernel of that homomorphism. Then $\mathcal{O}_P$ is equal to a direct sum $\mathcal{O}_P = k + M_P$, and $\mathcal{O}_P/M_P \cong k$. An element $\phi \in \mathcal{O}_P$ has a multiplicative inverse in $\mathcal{O}_P$ if and only if $\phi \notin M_P$. Every ideal of $\mathcal{O}_P$ other than $\mathcal{O}_P$ itself is contained in M_P, and so M_P is the unique maximal ideal of $\mathcal{O}_P$. (A ring having a unique maximal ideal is called a *local ring*. We used another local ring $R_p \subset \mathbb{Q}$ in Chapter II, Section 4. See also Exercise 2.7.)

Now let $(f_1, f_2)_P = \mathcal{O}_P f_1 + \mathcal{O}_P f_2$ denote the ideal in $\mathcal{O}_P$ generated by f_1 and f_2. Our definition of the *intersection multiplicity* (also called the *intersection index*) of C_1 and C_2 at P is

$$I(C_1 \cap C_2, P) = \dim \left(\frac{\mathcal{O}_P}{(f_1, f_2)_P} \right).$$

We are now ready to do segment (3), which means taking inequality (A) and strengthening it to inequality (A^+).
(3.1) Show that

$$\dim \left(\frac{\mathcal{O}_P}{(f_1, f_2)_P} \right) \leq \dim \left(\frac{R}{(f_1, f_2)} \right).$$

Deduce from inequality (B) that the intersection multiplicity $I(C_1 \cap C_2, P)$ is finite. (*Idea.* Note that any finite set of elements in $\mathcal{O}_P$ can be written over a common denominator. Show that if $g_1/h, g_2/h, \ldots, g_r/h$ are elements of $\mathcal{O}_P$ which are linearly independent modulo $(f_1, f_2)_P$, then $g_1, g_2, \ldots g_r$ are elements of R which are independent modulo (f_1, f_2).)
(3.2) Show that $\mathcal{O}_P = R + (f_1, f_2)_P$. (*Idea.* By (3.1) we may suppose that the elements g_i/h span $\mathcal{O}_P$ modulo $(f_1, f_2)_P$, and because $h^{-1} \in \mathcal{O}_P$, it follows that the polynomials g_i span $\mathcal{O}_P$ modulo $(f_1, f_2)_P$.)
(3.3) Show that if $P \notin C_1 \cap C_2$, then $I(C_1 \cap C_2, P) = 0$. Show that if $P \in C_1 \cap C_2$, then

$$(f_1, f_2)_P \subset M_P \quad \text{and} \quad I(C_1 \cap C_2, P) = 1 + \dim \left(\frac{M_P}{(f_1, f_2)_P} \right).$$

Conclude that if $P \in C_1 \cap C_2$, then $I(C_1 \cap C_2, P) \geq 1$, with equality if and only if $(f_1, f_2)_P = M_P$.
(3.4) Suppose that $P \in C_1 \cap C_2$. Let r satisfy $r \geq \dim(\mathcal{O}_P/(f_1, f_2)_P)$. Show that $M_P^r \subset (f_1, f_2)_P$. (*Idea.* We are to prove that, given any collection of r elements $t_1, t_2, \ldots, t_r$ in M_P, their product $t_1 t_2 \cdots t_r$ is in $(f_1, f_2)_P$. Define a sequence of ideals J_i in $\mathcal{O}_P$ by

$$J_i = t_1 t_2 \cdots t_i \mathcal{O}_P + (f_1, f_2)_P \quad \text{for } 1 \leq i \leq r, \quad \text{and} \quad J_{r+1} = (f_1, f_2)_P.$$

Then

$$M_P \supset J_1 \supset J_2 \supset \cdots \supset J_r \supset J_{r+1} = (f_1, f_2)_P.$$

Since $r > \dim\big(M_P/(f_1, f_2)_P\big)$, it follows that $J_i = J_{i+1}$ for some i with $1 \leq i \leq r$. If $i = r$, then $t_1 t_2 \cdots t_r \in (f_1, f_2)_P$ and we are done. If $i < r$, then we have

$$t_1 t_2 \cdots t_i = t_1 t_2 \cdots t_{i+1} \phi + \psi \qquad \text{with } \phi \in \mathcal{O}_P \text{ and } \psi \in (f_1, f_2)_P,$$

so $t_1 t_2 \cdots t_i (1 - t_{i+1}\phi) = \psi \in \mathcal{O}_P$. But $(1 - t_{i+1}\phi)(P) = 1$, so we have that $(1 - t_{i+1}\phi)^{-1} \in \mathcal{O}_P$. Hence $t_1 t_2 \ldots t_r = \psi t_{i+1} \cdots t_r (1 - t_{i+1}\phi)^{-1} \in \mathcal{O}_P$ as claimed.)

(3.5) Let $P \in C_1 \cap C_2 \cap \mathbf{A}^2$, and let $\phi \in \mathcal{O}_P$. Show that there exists a polynomial $g \in R$ such that

$$g \equiv \phi \,(\mathrm{mod}\,(f_1, f_2)_P) \quad \text{and}$$
$$g \equiv 0 \,(\mathrm{mod}\,(f_1, f_2)_Q) \quad \text{for all } Q \neq P, \, Q \in C_1 \cap C_2 \cap \mathbf{A}^2.$$

(*Idea.* The inequalities (A) and (B) already proved show that ony a finite number of points are involved here (at most $n_1 n_2$ in fact). Hence, by (1.1) there is a polynomial $h = h(x, y) \in R$ such that $h(P) = 1$ and $h(Q) = 0$ for $Q \neq P$, $Q \in C_1 \cap C_2 \cap \mathbf{A}^2$. This means $h^{-1} \in \mathcal{O}_P$ and $h \in M_Q$ for each of the other points Q. For integers $r \geq 1$ we have $h^{-r} \in \mathcal{O}_P$, and if r is sufficiently large, then, by (3.4), we will have $h^r \in (f_1, f_2)_Q$ for the other points Q. By (3.2) there is a polynomial $f \in R$ such that $f \equiv \phi h^{-r} \,(\mathrm{mod}\,(f_1, f_2)_P)$. Then $g = f h^r$ solves the problem.)

(3.6) Show that the natural map

$$R \longrightarrow \prod_{P \in C_1 \cap C_2 \cap \mathbf{A}^2} \frac{\mathcal{O}_P}{(f_1, f_2)_P} \qquad (*)$$

given by

$$f \longmapsto \big(\cdots, f \bmod (f_1, f_2)_P, \cdots\big)_{P \in C_1 \cap C_2 \cap \mathbf{A}^2}$$

is surjective, and conclude that the inequality (A^+) holds. (*Idea.* Let J be the kernel of the map. Then $(f_1, f_2) \subset J$, so $\dim R/(f_1, f_2) \geq \dim(R/J)$. The surjectivity of the map follows easily from (3.5) and implies that

$$\dim \frac{R}{J} = (\text{dimension of the target space})$$

$$= \sum_P \dim \frac{\mathcal{O}_P}{(f_1, f_2)_P} = \sum_P I(C_1 \cap C_2, P).\,)$$

To prove that (A^+) is an equality is now seen to be the same as showing that the kernel J of the map $(*)$ is equal to (f_1, f_2). So we must show that $J \subset (f_1, f_2)$, the other inclusion being obvious. Let $f \in J$. Our strategy for showing $f \in (f_1, f_2)$ is to consider the set

$$L = \{g \in R : gf \in (f_1, f_2)\}$$

and to prove that $1 \in L$.

(4.1) Show that L is an ideal in R and that $(f_1, f_2) \subset L \subset R$.

(4.2) Show that L has the following property:

For every $P \in \mathbf{A}^2$ there is a polynomial $g \in L$ such that $g(P) \neq 0$. $(**)$

In fact, property $(**)$ alone implies that $1 \in L$ by the famous "Nullstellensatz" of Hilbert. But we don't need the Nullstellensatz in full generality, because we have an additional piece of information about L, namely that $(f_1, f_2) \subset L$, and hence $\dim(R/L)$ is finite. Using this, and assuming that $1 \notin L$ in order to prove a contradiction, verify the following assertion.

(4.3) There is an $a \in k$ such that $1 \notin L + R(x - a)$. (*Idea*. The powers of x cannot all be linearly indpendent modulo L, so there are constants $c_i \in k$ and an integer n such that $x^n + c_1 x^{n-1} + \cdots + c_n \in L$. Since k is algebraically closed, we can write this as $(x - a_1)(x - a_2) \cdots (x - a_n) \in L$ with suitable $a_i \in k$. Show that if $1 \in L + R(x - a_i)$ for all $i = 1, \ldots, n$, then we get a contradiction to the assumption that $1 \notin L$.)

(4.4) There is a $b \in k$ such that $1 \notin L + R(x - a) + R(y - b)$. (*Idea*. Replace L by $L + R(x - a)$ and x by y and repeat the argument of (4.3).)

(4.5) Let $P = (a, b)$ and show that $g(P) = 0$ for all $g \in L$. This contradicts (4.2) and shows that $1 \in L$. (*Idea*. Write

$$g(x, y) = g\big(a + (x - a), b + (y - b)\big) = g(a, b) + g_1(x, y)(x - a) + g_2(x, y)(y - b)$$

and conclude that $g(a, b) \in L$.)

Our next job is to describe $K, \mathcal{O}_P, M_P, (f_1, f_2)_P$ in terms of homogeneous coordinates, so that they make sense also for points P at infinity. This will allow us to check that they are invariant under an arbitrary projective coordinate change in $\mathbb{P}^2$. To see what to do we put as usual $x = X/Z$, $y = Y/Z$, and we view $R = k[x, y] = k[X/Z, Y/Z]$ as a subring of the field $k(X, Y, Z)$ of rational functions of X, Y, Z. Then $K = k(x, y)$ becomes identified with the set of all rational functions $\Phi = F/G$ of X, Y, Z which are *homogeneous of degree* 0 in the sense that F and G are homogeneous polynomials of the same degree. Indeed, for $\phi \in K$ we have

$$\phi(x, y) = \frac{f(x, y)}{g(x, y)} = \frac{Z^n f(X/Z, Y/Z)}{Z^n g(X/Z, Y/Z)} = \frac{F(X, Y, Z)}{G(X, Y, Z)} = \Phi(X, Y, Z), \text{ say,}$$

where F and G are homogeneous of the same degree $n = \max\{\deg f, \deg g\}$. On the other hand, if $\Phi = F/G$ is a quotient of forms of the same degree, then $\Phi(tX, tY, tZ) = \Phi(X, Y, Z)$, and

$$\Phi(X, Y, Z) = \Phi(x, y, 1) = \frac{F(x, y, 1)}{G(x, y, 1)} \in K.$$

If $P = [A, B, C]$ is a point in $\mathbb{P}^2$ and $\Phi = F/G \in K$, we say that Φ is *defined at* P if $G(A, B, C) \neq 0$, i.e., if P is not on the curve $G(X, Y, Z) = 0$.

If Φ is defined at P, we put $\Phi(P) = F(A, B, C)/G(A, B, C)$, this ratio being independent of the choice of homogeneous coordinate triple for P. Clearly we should put

$$\mathcal{O}_P = \{\Phi \in K : \Phi \text{ is defined at } P\} \quad \text{and} \quad M_P = \{\Phi \in \mathcal{O}_P : \Phi(P) = 0\}.$$

We leave it to the conscientious reader to check the following assertion.
(5.1) If $P = (a, b) = [a, b, 1] \in \mathbb{A}^2$, then these definitions of $\mathcal{O}_P$, of $\Phi(P)$ for $\Phi \in \mathcal{O}_P$, and of M_P coincide with our earlier definitions.

Now let $C_1 : F_1 = 0$ and $C_2 : F_2 = 0$ be two curves in $\mathbb{P}^2$ without any common components. Let $f_1(x, y) = F_1(x, y, 1)$ and $f_2(x, y) = F_2(x, y, 1)$ be the polynomials defining their affine parts. Define

$$(F_1, F_2)_P = \left\{ \frac{F}{G} \in \mathcal{O}_P : F \text{ is of the form } F = H_1 F_1 + H_2 F_2 \right\}.$$

(Do you see why we cannot just say that $(F_1, F_2)_P$ is the ideal in $\mathcal{O}_P$ generated by F_1 and F_2?)
(5.2) Check that if $P \in \mathbb{A}^2$, then $(F_1, F_2)_P = (f_1, f_2)_P$ is the ideal in $\mathcal{O}_P$ generated by f_1 and f_2.

Of course, we now define the intersection multiplicity of C_1 and C_2 at every point $P \in \mathbb{P}^2$ by

$$I(C_1 \cap C_2, P) = \dim \frac{\mathcal{O}_P}{(F_1, F_2)_P}.$$

We know from (5.2) this coincides with our earlier definition for $P \in \mathbb{A}^2$.
(5.3) Check that the definitions of $\mathcal{O}_P$ and $(F_1, F_2)_P$, and hence also of the intersection multiplicity $I(C_1 \cap C_2, P)$, are independent of our choice of homogeneous coordinates in $\mathbb{P}^2$, i.e., they are are invariant under a linear change of the coordinates X, Y, Z.

To finally complete our proof of Bezout's theorem, we must show that there is a line L in $\mathbb{P}^2$ which does not meet $C_1 \cap C_2$. Then we can take a new coordinate system in which L is the line at infinity, and thereby reduce to the case already proved. To show that L exists, we use the following:
(5.4) Prove that, given any finite set S of points in $\mathbb{P}^2$, there is a line L not meeting S. (*Idea.* Use that an algebraically closed field k is not finite.)

Finally, the next result allows us to apply (5.4).
(5.5) Prove that $C_1 \cap C_2$ is finite. (*Idea.* Use the fact that for every line L which is not a component of either C_1 or C_2, we know (by putting L at infinity and using part (1) of this proof) that $C_1 \cap C_2$ contains a finite number of points not on L.)

That completes our proof of Bezout's Theorem in all its gory detail. To study more closely the properties of the intersection multiplicity $I(C_1 \cap C_2, P)$ at one point P, we can without loss of generality choose coordinates so that $P = (0, 0) = [0, 0, 1]$ is the origin in the affine plane, and we can work with affine coordinates x, y. Let $R = k[x, y]$ as before, and let

$$M = \{f = f(x, y) \in R : f(P) = f(0, 0) = 0\}.$$

(6.1) Prove that $M = (x, y) = Rx + Ry$ and $M_P = \mathcal{O}_P x + \mathcal{O}_P y$.

It follows that for each $n \geq 1$, M^n is the ideal in R generated by the monomials $x^n, x^{n-1}y, \ldots, xy^{n-1}, y^n$. Hence every polynomial $f \in R$ can be written uniquely as polynomial of degree at most n plus a remainder polynomial $r \in M^{n+1}$:

$$f(x, y) = c_{00} + c_{10}x + c_{01}y + \cdots + c_{ij}x^i y^j + \cdots$$
$$+ c_{n0}x^n + c_{n-1,1}x^{n-1}y + \cdots + c_{0n}y^n + r. \tag{*}$$

(6.2) Prove that every $\phi = f/g \in \mathcal{O}_P$ can be written uniquely in the form (*) with $c_{ij} \in k$ and $r \in M_P^{n+1}$. In other words, the inclusion $R \subset \mathcal{O}_P$ induces an isomorphism $R/M^{n+1} \cong \mathcal{O}_P/M_P^{n+1}$ for every $n \geq 0$. (*Idea.* We must show that $\mathcal{O}_P = R + M_P^{n+1}$ and that $R \cap M_P^{n+1} = M^{n+1}$. For the first, show that every $\phi \in \mathcal{O}_P$ can be written in the form $\phi = f/(1 - h)$ with $f \in R$ and $h \in M$. Hence

$$\phi = \frac{f}{1 - h} = f \cdot (1 + h + \cdots + h^n) + \frac{fh^{n+1}}{1 - h} \in R + M_P^{n+1}.$$

The second reduces to showing that if $gf \in M^n$ and $g(P) \neq 0$, then $f \in M^n$. This can be done by considering the terms of *lowest degree* in g and f and gf.)

Now we can already compute some intersection indices to see if our definition gives answers which are geometrically reasonable. As a matter of notation we introduce the symbol

$$I(f_1, f_2) = \dim\left(\frac{\mathcal{O}_P}{(f_1, f_2)_P}\right)$$

for the intersection multiplicity of two curves $f_1 = 0$ and $f_2 = 0$ at the origin.

(6.3) Check that the curve $y = x^n$ and the x axis intersect with multiplicity n at the origin, i.e., show that $I(y - x^n, y) = n$. (*Idea.* Note first that the ideals $(y - x^n, y)$ and (x^n, y) are equal, and that this ideal contains M^n. Then, using what we know (6.2) about $\mathcal{O}_P/M_P^n$, show that $1, x, \ldots, x^{n-1}$ is a basis for the vector space $\mathcal{O}_P/(x^n, y)\mathcal{O}_P$.)

(6.4) (Nakayama's Lemma) Suppose J is an ideal of $\mathcal{O}_P$ contained in a finitely generated ideal $\Phi = (\phi_1, \phi_2, \ldots, \phi_m)\mathcal{O}_P$. Suppose some elements of J generate Φ modulo $M_P\Phi$, i.e., $\Phi = J + M_P\Phi$. Then $J = \Phi$. (*Idea.* The case $\Phi = (\phi_1, \phi_2)\mathcal{O}_P$ is all we need. To prove that case, write

$$\phi_1 = j_1 + \alpha\phi_1 + \beta\phi_2, \qquad \phi_2 = j_2 + \gamma\phi_1 + \delta\phi_2,$$

with $j_1, j_2 \in J$ and $\alpha, \beta, \gamma, \delta \in M_P$. Then use the fact that the determinant of the matrix $\begin{pmatrix} 1-\alpha & \beta \\ \gamma & 1-\delta \end{pmatrix}$ is non-zero in order to express the ϕ's in terms of the j's.)

(6.5) Suppose that

$$f_1 = ax + by + \text{(higher terms)}, \qquad f_2 = cx + dy + \text{(higher terms)},$$

where "higher terms" means elements of M^2. Show that the following are equivalent.

 (i) The curves $f_1 = 0$ and $f_2 = 0$ meet transversally at the origin, i.e., are smooth with distinct tangent directions there.
 (ii) The determinant $ad - bc$ is not equal to zero.
 (iii) $(f_1, f_2)_P = M_P$, i.e., $I(f_1, f_2) = 1$.

(*Idea.* (i) $\iff$ (ii) follows directly from the definitions. One way to do (ii) $\Rightarrow$ (iii) is to use (6.4) with $\phi_1 = x$, $\phi_2 = y$, and $J = (f_1, f_2)_P$. To do (iii) $\Rightarrow$ (ii) note that if $ad - bc = 0$, then

$$\dim\left(\frac{(f_1, f_2)_P + M_P^2}{M_P^2}\right) \leq 1,$$

whereas, by (6.2), $\dim(M_P/M_P^2) = 2$.
(6.6) Let $f(x, y) \in R$. Show that $I(f(x, y), y) = m$, where x^m is the highest power of x dividing $f(x, 0)$. (*Idea.* Use the fact that the ideal $(f(x, y), y)$ is the same as the ideal $(f(x, 0), y)$. Then argue as in (6.3).)
(6.7) Let $C : F(X, Y, Z) = 0$ be a curve in $\mathbb{P}^2$ that does not contain the line $L_\infty : Z = 0$. Show that for each point $Q = [a, b, 0] \in L_\infty$, we have $I(C \cap L_\infty, Q) = m$, where $(bX - aY)^m$ is the highest power of $bX - aY$ dividing $F(X, Y, 0)$. (*Idea.* Make a suitable coordinate change to reduce to (6.6).)

5. Reduction Modulo p

Let $\mathbb{P}^2(\mathbb{Q})$ denote the set of rational points in $\mathbb{P}^2$. We say that a homogeneous coordinate triple $[A, B, C]$ is *normalized* if A, B, C are integers with no common factors. Each point $P \in \mathbb{P}^2(\mathbb{Q})$ has a normalized coordinate triple which is unique up to sign. To obtain it we start with any triple of rational coordinates, multiply through by a common denominator, and then divide the resulting triple of integers by their greatest common divisor. For exampe

$$\left[\frac{4}{5}, -\frac{2}{3}, 2\right] = [12, -10, 30] = [6, -5, 15].$$

The other normalized coordinate triple for this point is $[-6, 5, -15]$.

Let p be a fixed prime number and for each integer $m \in \mathbb{Z}$, let $\tilde{m} \in \mathbb{F}_p = \mathbb{Z}/p\mathbb{Z}$ denote its residue modulo p. If $[l, m, n]$ is a normalized coordinate triple for a point $P \in \mathbb{P}^2(\mathbb{Q})$, then the triple $[\tilde{l}, \tilde{m}, \tilde{n}]$ defines a point $\tilde{P}$ in $\mathbb{P}^2(\mathbb{F}_p)$ because at least one of the three numbers l, m, and n is not divisible by p. Since P determines the triple $[l, m, n]$ up to sign, the point $\tilde{P}$ depends only on P, not on the choice of coordinates for P. Thus, $P \mapsto \tilde{P}$ gives a well-defined map

$$\mathbb{P}^2(\mathbb{Q}) \longrightarrow \mathbb{P}^2(\mathbb{F}_p),$$

called for obvious reasons the *reduction mod p map*. Note that reduction mod p does not map $\mathbf{A}^2(\mathbb{Q})$ into $\mathbf{A}^2(\mathbb{F}_p)$. For example,

$$P = \left(\frac{1}{p}, 0\right) = \left[\frac{1}{p}, 0, 1\right] = [1, 0, p] \longmapsto [1, 0, 0] \notin \mathbf{A}^2(\mathbb{F}_p).$$

In fact, if $P = (a, b) = [a, b, 1] \in \mathbf{A}^2(\mathbb{Q})$, then its reduction $\tilde{P}$ is in $\mathbf{A}^2(\mathbb{F}_p)$ if and only if the rational numbers a and b are "p-integral," i.e., have denominators prime to p.

Let $C : F(X, Y, Z) = 0$ be a rational curve in $\mathbb{P}^2$. By rational we mean as usual that the coefficients of F are rational numbers. Clearing the denominators of the coefficients and then dividing by the greatest common divisor of their numerators, we can suppose that the coefficients of F are integers with greatest common divisor equal to one. Call such an F *normalized*. Then $\tilde{F}$, the polynomial we obtain by reducing the coefficients of F modulo p, is non-zero and defines a curve $\tilde{C}$ in characteristic p. If $[l, m, n]$ is a normalized coordinate triple and if $F(l, m, n) = 0$, then $\tilde{F}(\tilde{l}, \tilde{m}, \tilde{n}) = 0$, because $x \to \tilde{x}$ is a homomorphism. In other words, if P is a rational point on C, then $\tilde{P}$ is a point on $\tilde{C}$; reduction mod p takes $C(\mathbb{Q})$ and maps it into $\tilde{C}(\mathbb{F}_p)$.

If C_1 and C_2 are two curves, it follows that

$$\left(\widetilde{C_1(\mathbb{Q}) \cap C_2(\mathbb{Q})}\right) \subset \tilde{C}_1(\mathbb{F}_p) \cap \tilde{C}_2(\mathbb{F}_p).$$

Is there some sense in which $(\widetilde{C_1 \cap C_2}) = \tilde{C}_1 \cap \tilde{C}_2$ if we count multiplicities? After all, the degrees of the reduced curves $\tilde{C}_i$ are the same as those of the C_i, so by Bezout's theorem the intersection before and after reduction has the same number of points if we count multiplicities. But Bezout's theorem requires that the ground field be algebraically closed, and we don't have the machinery to extend our reduction mod p map to that case. However, if we assume that all of the complex intersection points are rational, then everything is okay. We treat only the special case in which one of the curves is a line. This case suffices for the application to elliptic curves we are after, and it is easy to prove.

Proposition. *Suppose C is a rational curve and L is a rational line in $\mathbb{P}^2$. Suppose that all of the complex intersection points of C and L are rational. Let $C \cap L = \{P_1, P_2, \ldots, P_d\}$, where $d = \deg(C)$ and each point P_i is repeated in the list as many times as its multiplicity. Assume* that $\tilde{L}$ is not a component of $\tilde{C}$. Then $\tilde{C} \cap \tilde{L} = \{\tilde{P}_1, \tilde{P}_2, \ldots, \tilde{P}_d\}$ with the correct multiplicities.

PROOF. Suppose first that L is the line at infinity $Z = 0$. Let $F(X, Y, Z) = 0$ be a normalized equation for C. The assumption that $\tilde{L}$ is not a component of $\tilde{C}$ means that $\tilde{F}(X, Y, 0) \neq 0$, i.e., that some coefficient of $F(X, Y, 0)$ is not divisible by p. For each intersection point P_i, let $P_i = [l_i, m_i, 0]$ in normalized coordinates. Then

$$F(X, Y, 0) = c \prod_{i=1}^{d} (m_i X - l_i Y) \tag{$*$}$$

for some constant c. This is true because the intersection points of a curve $F = 0$ with the line $Z = 0$ correspond, with the correct multiplicities, to the linear factors of $F(X, Y, 0)$. Since each of the linear polynomials on the right of $(*)$ is normalized and some coefficient of F is not divisible by p, we see that c must be an integer not divisible by p. Therefore we can reduce $(*)$ modulo p to obtain

$$\tilde{F}(X, Y, 0) = \tilde{c} \prod_{i=1}^{d} (\tilde{m}_i X - \tilde{l}_i Y), \tag{$\tilde{*}$}$$

which shows that $\tilde{C} \cap \tilde{L} = \{\tilde{P}_1, \tilde{P}_2, \ldots, \tilde{P}_d\}$ as claimed.

What if the line L is not the line $Z = 0$? Then we just make a linear change of coordinates

$$\begin{pmatrix} X' \\ Y' \\ Z' \end{pmatrix} = \begin{pmatrix} n_{11} & n_{12} & n_{13} \\ n_{21} & n_{22} & n_{23} \\ n_{31} & n_{32} & n_{33} \end{pmatrix} \begin{pmatrix} X \\ Y \\ Z \end{pmatrix}$$

so that L is the line $Z' = 0$ in the new coordinate system.

Is that all there is to it? No, we must be careful to make sure our change of coordinates is compatible with reduction modulo p. This is not true for general linear changes with $n_{ij} \in \mathbb{Q}$. However, if we change using a matrix (n_{ij}) with integer entries and determinant 1, then the inverse matrix (m_{ij}) will have integer entries, and the reduced matrices $(\tilde{n}_{ij})$ and $(\tilde{m}_{ij})$ are inverses giving a corresponding coordinate change in characteristic p. And clearly, if we change coordinates with (n_{ij}) and reduce mod p, the result will be the same as if we first reduce mod p and then change coordinates with $(\tilde{n}_{ij})$.

Thus, to complete our proof we must show that for every rational line L in $\mathbb{P}^2$ there is an "integral" coordinate change such that in the new coordinates L is the line at infinity. To do this, we let

$$L : aX + bY + cZ = 0$$

be a normalized equation for the line L and use the following result.

Lemma.   *Let (a, b, c) be a triple of integers satisfying $\gcd(a, b, c) = 1$. Then there exists a 3×3 matrix with integer coefficients, determinant 1, and bottom line (a, b, c).*

PROOF. Let $d = \gcd(b, c)$, choose integers r and s such that $rc - sb = d$, and note for later use that r and s are necessarily relatively prime. Now $\gcd(a, d) = 1$, so we can choose t and u such that $td + ua = 1$. Finally, since $\gcd(r, s) = 1$, we can choose v and w such that $vs - wr = u$. Then the matrix

$$\begin{pmatrix} t & v & w \\ 0 & r & s \\ a & b & c \end{pmatrix}$$

has the desired properties.   □

Finally, we apply the proposition to show that the reduction mod p map respects the group law on a cubic curve.

Corollary.   *Let C be a non-singular rational cubic curve in $\mathbb{P}^2$ and let $\mathcal{O}$ be a rational point on C, which we take as the origin for the group law on C. Suppose that $\tilde{C}$ is non-singular and take $\tilde{\mathcal{O}}$ as the origin for the group law on $\tilde{C}$. Then the reduction mod p map $P \to \tilde{P}$ is a group homomorphism $C(\mathbb{Q}) \to \tilde{C}(\mathbb{F}_p)$.*

PROOF. Let $P, Q \in C(\mathbb{Q})$, and let $R = P + Q$. This means that there are lines L_1 and L_2 and a rational point $S \in C(\mathbb{Q})$ such that, in the notation of the proposition,

$$C \cap L_1 = \{P, Q, S\} \quad \text{and} \quad C \cap L_2 = \{S, \mathcal{O}, R\}.$$

Putting tildes on everything, as is allowed by the proposition, we conclude that $\tilde{P} + \tilde{Q} = \tilde{R}$.   □

EXERCISES

A.1. Let $\mathbb{P}^2$ be the set of homogeneous triples $[a, b, c]$ as usual, and recall that with this definition a line in $\mathbb{P}^2$ is defined to be the set of solutions of an equation of the form

$$\alpha X + \beta Y + \gamma Z = 0$$

for some numbers α, β, γ not all zero.
(a) Prove directly from this definition that any two distinct points in $\mathbb{P}^2$ are contained in a unique line.
(b) Similarly, prove that any two distinct lines in $\mathbb{P}^2$ intersect in a unique point.

A.2. Let K be a field, for example K might be the rational numbers or the real numbers or a finite field. Define a relation $\sim$ on $(n+1)$-tuples $[a_0, a_1, \ldots, a_n]$ of elements of K by the following rule:

$$[a_0, a_1, \ldots, a_n] \sim [a_0', a_1', \ldots, a_n'] \quad \text{if there is a non-zero } t \in K$$
$$\text{so that } a_0 = ta_0', \ a_1 = ta_1', \ldots, a_n = ta_n'.$$

(a) Prove that $\sim$ is an equivalence relation. That is, prove that for any $(n+1)$-tuples $\mathbf{a} = [a_0, a_1, \ldots, a_n]$, $\mathbf{b} = [b_0, b_1, \ldots, b_n]$, and $\mathbf{c} = [c_0, c_1, \ldots, c_n]$, the relation $\sim$ satisfies the following three conditions:

- (i) $\mathbf{a} \sim \mathbf{a}$ (Reflexive)
- (ii) $\mathbf{a} \sim \mathbf{b} \Longrightarrow \mathbf{b} \sim \mathbf{a}$ (Symmetric)
- (iii) $\mathbf{a} \sim \mathbf{b}$ and $\mathbf{b} \sim \mathbf{c} \Longrightarrow \mathbf{a} \sim \mathbf{c}$ (Transitive)

(b) Which of these properties (i), (ii), (iii) fails to be true if K is replaced by a ring R that is not a field? (There are several answers to this question, depending on what the ring R looks like.)

A.3. We saw in Section 1 that the directions in the affine plane $\mathbb{A}^2$ correspond to the points of the projective line $\mathbb{P}^1$. In other words, $\mathbb{P}^1$ can be described as the set of lines in $\mathbb{A}^2$ going through the origin.

(a) Prove similarly that $\mathbb{P}^2$ can be described as the set of lines in $\mathbb{A}^3$ going through the origin.

(b) Let $\Pi \subset \mathbb{A}^3$ be a plane in $\mathbb{A}^3$ that goes through the origin, and let S_Π be the collection of lines in $\mathbb{A}^3$ going through the origin and contained in Π. From (a), S_Π defines a subset L_Π of $\mathbb{P}^2$. Prove that L_Π is a line in $\mathbb{P}^2$, and conversely that every line in $\mathbb{P}^2$ can be constructed in this way.

(c) Generalize (a) by showing that $\mathbb{P}^n$ can be described as the set of lines in $\mathbb{A}^{n+1}$ going through the origin.

A.4. Let $F(X, Y, Z) \in \mathbb{C}[X, Y, Z]$ be a homogeneous polynomial of degree d.

(a) Prove that the three partial derivatives of F are homogeneous polynomials of degree $d-1$.

(b) Prove that

$$X\frac{\partial F}{\partial X} + Y\frac{\partial F}{\partial Y} + Z\frac{\partial F}{\partial Z} = d \cdot F(X, Y, Z).$$

(*Hint.* Differentiate $F(tX, tY, tZ) = t^d F(X, Y, Z)$ with respect to t.)

A.5. Let $C : F(X, Y, Z) = 0$ be a projective curve given by a homogeneous polynomial $F \in \mathbb{C}[X, Y, Z]$, and let $P \in \mathbb{P}^2$ be a point.

(a) Prove that P is a singular point of C if and only if

$$\frac{\partial F}{\partial X}(P) = \frac{\partial F}{\partial Y}(P) = \frac{\partial F}{\partial Z}(P) = 0.$$

(b) If P is a non-singular point of C, prove that the tangent line to C at P is given by the equation

$$\frac{\partial F}{\partial X}(P)X + \frac{\partial F}{\partial Y}(P)Y + \frac{\partial F}{\partial Z}(P)Z = 0.$$

A.6. Let C be the projective curve given by the equation
$$C : Y^2 Z - X^3 - Z^3 = 0.$$

(a) Show that C has only one point at infinity, namely the point $[0, 1, 0]$ corresponding to the vertical direction $x = 0$.

(b) Let $C_0 : y^2 - x^3 - 1 = 0$ be the affine part of C, and let (r_i, s_i) be a sequence of point on C_0 with $r_i \to \infty$. Let L_i be the tangent line to C_0 at the point (r_i, s_i). Prove that as $i \to \infty$, the slopes of the lines L_i approach infinity, i.e., they approach the slope of the line $x = 0$.

A.7. Let $f(x, y)$ be a polynomial.

(a) Expand $f(tx, ty)$ as a polynomial in t whose coefficients are polynomials in x and y. Prove that the degree of $f(tx, ty)$, considered as a polynomial in the variable t, is equal to the degree of the polynomial $f(x, y)$.

(b) Prove that the homogenization $F(X, Y, Z)$ of $f(x, y)$ is given by
$$F(X, Y, Z) = Z^d f\left(\frac{X}{Z}, \frac{Y}{Z}\right), \qquad \text{where } d = \deg(f).$$

A.8. For each of the given affine curves C_0, find a projective curve C whose affine part is C_0. Then find all of the points at infinity on the projective curve C.

(a) $C_0 : 3x - 7y + 5 = 0$.

(b) $C_0 : x^2 + xy - 2y^2 + x - 5y + 7 = 0$.

(c) $C_0 : x^3 + x^2 y - 3xy^2 - 3y^3 + 2x^2 - x + 5 = 0$.

A.9. For each of the following curves C and points P, either find the tangent line to C at P or else verify that C is singular at P.

(a) $C : y^2 = x^3 - x,$ $P = (1, 0)$.

(b) $C : X^2 + Y^2 = Z^2,$ $P = [3, 4, 5]$.

(c) $C : x^2 + y^4 + 2xy + 2x + 2y + 1 = 0,$ $P = (-1, 0)$.

(d) $C : X^3 + Y^3 + Z^3 = XYZ,$ $P = [1, -1, 0]$.

A.10. (a) Prove that a projective transformation of $\mathbb{P}^2$ sends lines to lines.

(b) More generally, prove that a projective transformation of $\mathbb{P}^2$ sends curves of degree d to curves of degree d.

A.11. Let P, P_1, P_2, P_3 be points in $\mathbb{P}^2$ and let L be a line in $\mathbb{P}^2$.

(a) If P_1, P_2, and P_3 do not lie on a line, prove that there is a projective transformation of $\mathbb{P}^2$ so that
$$P_1 \longmapsto [0, 0, 1], \qquad P_2 \longmapsto [0, 1, 0], \qquad P_3 \longmapsto [1, 0, 0].$$

(b) If no three of P_1, P_2, P_3 and P lie on a line, prove that there is a unique projective transformation as in (a) which also sends P to $[1, 1, 1]$.

(c) Prove that there is a projective transformation of $\mathbb{P}^2$ so that L is sent to the line $Z' = 0$.

(d) More generally, if P does not lie on L, prove that there is a projective transformation of $\mathbb{P}^2$ so that L is sent to the line $Z' = 0$ and P is sent to the point $[0, 0, 1]$.

A.12. For each of the pairs of curves C_1, C_2, find all of the points in the intersection $C_1 \cap C_2$. Be sure to include points with complex coordinates and points at infinity.

(a) $C_1 : x - y = 0,$ $C_2 : x^2 - y = 0.$
(b) $C_1 : x - y - 1 = 0,$ $C_2 : x^2 - y^2 + 2 = 0.$
(c) $C_1 : x - y - 1 = 0,$ $C_2 : x^2 - 2y^2 - 5 = 0.$
(d) $C_1 : x - 2 = 0,$ $C_2 : y^2 - x^3 + 2x = 0.$

A.13. For each of the pairs of curves C_1, C_2, compute the intersection index $I(C_1 \cap C_2, P)$ at the indicated point P. Also sketch the curves and the point in $\mathbb{R}^2$.

(a) $C_1 : x - y = 0,$ $C_2 : x^2 - y = 0,$ $P = (0,0).$
(b) $C_1 : y = 0,$ $C_2 : x^2 - y = 0,$ $P = (0,0).$
(c) $C_1 : x - y = 0,$ $C_2 : x^3 - y^2 = 0,$ $P = (0,0).$
(d) $C_1 : x^2 - y = 0,$ $C_2 : x^3 - y = 0,$ $P = (0,0).$
(e) $C_1 : x + y = 2,$ $C_2 : x^2 + y^2 = 2,$ $P = (1,1).$

A.14. Let $\mathcal{C}^{(d)}$ be the collection of curves of degree d in $\mathbb{P}^2$.

(a) Show that $\mathcal{C}^{(d)}$ is naturally isomorphic to the projective space $\mathbb{P}^N$ for a certain value of N, and find N explicitly in terms of d.

(b) In Section 3 we gave a plausibility arguement for why the Cayley-Bacharach theorem is true for curves of degree 3. Give a similar argument for general curves C_1, C_2, and D of degrees d_1, d_2, and $d_1 + d_2 - 3$ respectively.

A.15. Let $P \in \mathbb{A}^2$. In this exercise we ask you to verify various properties of $\mathcal{O}_P$, the local ring of P, as defined in Section 4.

(a) Prove that $\mathcal{O}_P$ is a subring of $K = k(x, y)$.

(b) Prove that the map $\phi \mapsto \phi(P)$ is a homomorphism of $\mathcal{O}_P$ onto k. Let M_P be the kernel of this homomorphism.

(c) Prove that $\mathcal{O}_P$ equals the direct sum $k + M_P$.

(d) Prove that $\phi \in \mathcal{O}_P$ is a unit if and only if $\phi \notin M_P$.

(e) Let $I \subset \mathcal{O}_P$ be an ideal of $\mathcal{O}_P$. Prove that either $I = \mathcal{O}_P$, or else $I \subset M_P$. Deduce that M_P is the unique maximal ideal of $\mathcal{O}_P$.

A.16. Let P_1, P_2, P_3, P_4, P_5 be five distinct point in $\mathbb{P}^2$.

(a) Show that there exists a conic C (i.e., a curve of degree two) passing through the five points.

(b) Show that C is unique if and only if no four of the five points lie on a line.

(c) Show that C is irreducible if and only if no three of the five points lie on a line.

A.17. In this exercise we guide you in proving the cubic Cayley-Bacharach theorem in the case that the eight points are distinct. Let $C_1 : F_1 = 0$ and $C_2 : F_2 = 0$ be cubic curves in $\mathbb{P}^2$ without common component which have eight distinct points $P_1, P_2, \ldots, P_8$ in common. Suppose that $C_3 : F_3 = 0$ is a third cubic curve passing through these same eight points. Prove that C_3 is on the "line of cubics" joining C_1

and C_2, i.e., prove that there are constants λ_1 and λ_2 such that

$$F_3 = \lambda_1 F_1 + \lambda_2 F_2.$$

In order to prove this result, assume that no such λ_1, λ_2 exist and derive a contradiction as follows:

(i) Show that F_1, F_2, and F_3 are linearly independent.

(ii) Let P' and P'' be any two points in $\mathbb{P}^2$ different from each other and different from the P_i. Show that there is a cubic curve C passing through all ten points $P_1, \ldots, P_8, P', P''$. (*Hint.* Show that there exist constants $\lambda_1, \lambda_2, \lambda_3$ such that $F = \lambda_1 F_1 + \lambda_2 F_2 + \lambda_3 F_3$ is not identically 0 and such that the curve $F = 0$ does the job.)

(iii) Show that no four of the eight points P_i are collinear, and no seven of them lie on a conic. (*Hint.* Use the fact that C_1 and C_2 have no common component.)

(iv) Use the previous exercise to observe that there is a unique conic Q going through any five of the eight points $P_1, \ldots, P_8$.

(v) Show that no three of the eight points P_i are collinear. (*Hint.* If three are on a line L, let Q be the unique conic going through the other five, choose P' on L and P'' not on Q. Then use (ii) to get a cubic C which has L as a component, so is of the form $C = L \cup Q'$ for some conic Q'. This contradicts the fact that Q is unique.)

(vi) To get the final contradiction, let Q be the conic through the five points $P_1, P_2, \ldots, P_5$. By (iii), at least one (in fact two) of the remaining three points is not on Q. Call it P_6, and let L be the line joining P_7 and P_8. Choose P' and P'' on L so that again the cubic C through the ten points has L as a component. Show that this gives a contradiction.

A.18. Show that if C_1 and C_2 are both singular at the point P, then their intersection index satisfies $I(C_1 \cap C_2, P) \geq 3$.

A.19. Consider the affine curve $C : y^4 - xy - x^3 = 0$. Show that at the origin $(x, y) = (0, 0)$, C meets the y axis four times, C meets the x axis three times, and C meets every other line through the origin twice.

A.20. Show that the separation of real conics into hyperbolas, parabolas, and ellipses is an affine business and has no meaning projectively, by giving an example of a quadratic homogeneous polynomial $F(X, Y, Z)$ with real coefficients such that

$$F(x, y, 1) = 0 \text{ is a hyperbola in the real } (x, y) \text{ plane},$$
$$F(x, 1, z) = 0 \text{ is a parabola in the real } (x, z) \text{ plane},$$
$$F(1, y, z) = 0 \text{ is an ellipse in the real } (y, z) \text{ plane}.$$

Bibliography

Artin [1]

    Artin, M., *Algebra,* Prentice Hall, Englewood Cliffs, N.J., 1991.

Baker [1]

    Baker, A., Contributions to the theory of Diophantine equations (II). The Diophantine equation $y^2 = x^3 + k$. *Philos. Trans. Roy. Soc. London* **263** (1967/68), 193–208.

Baker [2]

    ———, *Transcendental Number Theory,* Cambridge University Press, Cambridge, 1975.

Billing-Mahler [1]

    Billing, G., Mahler, K., On exceptional points on cubic curves. *J. London Math. Soc.* **15** (1940), 32–43.

Bremner-Cassels [1]

    Bremner, A.; Cassels, J.W.S., On the equation $Y^2 = X(X^2+p)$. *Math. Comp.* **42** (1984), 257–264.

Brieskorn-Knörrer [1]

    Brieskorn, E., Knörrer, H., *Plane Algebraic Curves,* transl. by J. Stillwell, Birkhäuser, Basel, 1986.

Chahal [1]

    Chahal, J.S., *Topics in Number Theory,* Plenum Press, New York-London, 1988.

Deligne [1]

    Deligne, P., La conjecture de Weil I. *Publ. Math. IHES* **43** (1974), 273–307.

Faltings [1]

    Faltings, G., Diophantine approximation on abelian varieties. *Annals of Math.* **133** (1991), 549–576.

Fueter [1]

    Fueter, R., Über kubische diophantische Gleichungen. *Comm. Math. Helv.* **2** (1930), 69–89.

Fulton [1]

    Fulton, W., *Algebraic Curves,* Benjamin, 1969.

Griffiths and Harris [1]

    Griffiths, P., Harris, J., *Principles of Algebraic Geometry,* John Wiley & Sons, New York, 1978.

Hartshorne [1]

    Hartshorne, R., *Algebraic Geometry,* Springer-Verlag, New York, 1977.

Hasse [1]

    Hasse, H., Beweis des Analogons der Riemannschen Vermutung für die Artinschen und F. K. Schmidtschen Kongruenzzeta-funktionen in gewissen elliptischen Fällen. *Nachr. Ges. Wiss. Göttingen, Math.-Phys. K.* (1933), 253–262.

Heath-Brown and Patterson [1]
    Heath-Brown, D.R., Patterson, S.J., The distribution of Kummer sums at prime arguments. *J. Reine Angew. Math.* **310** (1979), 111–130.

Herstein [1]
    Herstein, I.N., *Topics in Algebra*, Xerox College Publishing, Lexington, MA, 1975.

Husemöller [1]
    Husemöller, D., *Elliptic Curves*, Springer-Verlag, New York, 1987.

Jacobson [1]
    Jacobson, N., *Basic Algebra I,II*, W.H. Freeman & Co., New York, 1985.

Koblitz [1]
    Koblitz, N., *Introduction to Elliptic Curves and Modular Forms*, Springer-Verlag, New York, 1984.

Koblitz [2]
    ————, *A Course in Number Theory and Cryptography*, Springer-Verlag, New York, 1987.

Kummer [1]
    Kummer, E.E., De residuis cubicis disquisitiones nonnullae analyticae. *J. Reine Angew. Math.* **32** (1846), 341–359.

Lang [1]
    Lang, S., *Elliptic Functions*, Addison-Wesley, Reading, MA, 1973.

Lang [2]
    ————, *Elliptic Curves: Diophantine Analysis*, Springer-Verlag, Berlin, 1978.

Lang [3]
    ————, *Fundamentals of Diophantine Geometry*, Springer-Verlag, New York, 1983.

Lenstra [1]
    Lenstra, H.W., Factoring integers with elliptic curves. *Annals of Math.* **126** (1987), 649–673.

Luck-Moussa-Waldschmidt [1]
    Luck, J.M., Moussa, P., Waldschmidt, M., eds., *Number Theory and Physics, Proc. in Physics* **47**, Springer-Verlag, Berlin, 1990.

Lutz [1]
    Lutz, E., Sur l'equation $y^2 = x^3 - Ax - B$ dans les corps p-adic. *J. Reine Angew. Math.* **177** (1937), 237–247.

Mazur [1]
    Mazur, B., Modular curves and the Eisenstein ideal. *IHES Publ. Math.* **47** (1977), 33–186.

Mazur [2]
    ————, Rational isogenies of prime degree. *Invent. Math.* **44** (1978), 129–162.

Mestre [1]
      Mestre, J.-F., Formules explicites et minorations de conducteurs
      de variétiés algebriques. *Compos. Math.* **58** (1986), 209-232.
Mestre [2]
      ———, Private communication, January, 1992.
Mordell [1]
      Mordell, L.J., On the rational solutions of the indeterminate
      equations of the third and fourth degrees. *Proc. Camb. Philos.*
      Soc. **21** (1922), 179–192.
Nagell [1]
      Nagell, T., Solution de quelque problémes dans la théorie arith-
      métique des cubiques planes du premier genre. *Wid. Akad.*
      Skrifter Oslo I (1935), Nr. 1.
Pollard [1]
      Pollard, J.M., Theorems on factorization and primality testing.
      Proc. Camb. Philos. Soc. **76** (1974), 521–528.
Reid [1]
      Reid, M., *Undergraduate Algebraic Geometry*, London Math.
      Soc. Student Texts 12, Cambridge University Press, Cambridge,
      1988.
Robert [1]
      Robert, A., *Elliptic Curves*, Lecture Notes in Math. 326, Spring-
      er-Verlag, Berlin, 1973.
Schmidt [1]
      Schmidt, W., Simultaneous approximation to algebraic numbers
      by rationals. *Acta Math.* **125** (1970), 189–201.
Schmidt [2]
      ———, *Diophantine Approximation*, Lecture Notes in Math.
      785, Springer-Verlag, Berlin, 1980.
Selmer [1]
      Selmer, E., The diophantine equation $ax^3 + by^3 + cz^3 = 0$. *Acta*
      Math. **85** (1951), 203–362.
Serre [1]
      Serre, J.-P., *A Course in Arithmetic*, Springer-Verlag, New
      York, 1973.
Serre [2]
      ———, *Linear Representations of Finite Groups*, Springer-
      Verlag, New York, 1973.
Serre [3]
      ———, *Abelian ℓ-adic Representations*, Benjamin, New York,
      1968.
Serre [4]
      ———, Propriétés galoisiennes des points d'ordre fini des
      courbes elliptiques. *Invent. Math.* **15** (1972), 259–331.
Siegel [1]
      Siegel, C.L., The integer solutions of the equation $y^2 = ax^n +$
      $bx^{n-1} + \cdots + k$. *J. London Math. Soc.* **1** (1926), 66–68.

Siegel [2]

————, Über einige Anwendungen diophantischer Approximationen (1929). In *Collected Works,* Springer-Verlag, Berlin, 1966, 209–266.

Silverman [1]

Silverman, J.H., Integer points and the rank of Thue elliptic curves. *Invent. Math.* **66** (1982), 395–404.

Silverman [2]

————, *The Arithmetic of Elliptic Curves,* Springer-Verlag, New York, 1986.

Skolem [1]

Skolem, Th., *Diophantische Gleichungen,* Springer-Verlag, Berlin, 1938.

Tate [1]

Tate, J., The arithmetic of elliptic curves. *Invent. Math.* **23** (1974), 171–206.

Thue [1]

Thue, A., Über Annäherungswerte Algebraischer Zahlen. *J. Reine Angew. Math.* **135** (1909), 284–305.

Vojta [1]

Vojta, P., Siegel's theorem in the compact case. *Annals of Math.* **133** (1991), 509–548.

Walker [1]

Walker, R.J., *Algebraic Curves,* Dover, 1962.

Weil [1]

Weil, A., *Sur les Courbes Algébriques et les Variétés qui s'en Déduisent,* Hermann, Paris, 1948.

Weil [2]

————, Number of solutions of equations over finite fields. *Bull. Amer. Math. Soc.* **55** (1949), 497–508.

List of Notation

$\alpha_1, \alpha_2, \alpha_3$	cubic Gauss sums, 114
θ_p	angle related to number of points modulo p, 120
$\pi(X)$	number of primes less than X, 120
$z \to \tilde{z}$	reduction modulo p, 121
$\tilde{C}$	reduction of C modulo p, 121
$\tilde{P}$	reduction of a point modulo p, 121
Φ	points of finite order in $C(\mathbb{Q})$, 122
$\mathcal{H}$	upper half plane, 141
$\| \ \|$	size of a vector, 158
$F^{(k)}(X, Y)$	k^{th} X-derivative, divided by $k!$, 161
$F(X, Y)$	auxiliary polynomial, 163
$P(X) + Q(X)Y$	auxiliary polynomial, 163
$W(X)$	Wronskian polynomial, 168
$\tau(d)$	exponent for Diophantine approximation, 174
$[K : F]$	the degree of K over F, 180
$\text{Aut}(K)$	group of automorphisms $\sigma : K \to K$, 180
$\text{Gal}(K/\mathbb{Q})$	Galois group of $K/\mathbb{Q}$, 181
$\text{Aut}_F(K)$	group of automorphisms of K fixing F, 183
$\text{Gal}(K/F)$	Galois group of K/F, 183
$C(K)$	K rational points on C, 185
$\sigma(P)$	action of Galois group on a point, 186
λ_n	n^{th}-power homomorphism on $\mathbb{C}^*$, 188
λ_n	multiplication-by-n map on C, 188
$C[n]$	kernel of multiplication-by-n map, 188
$\mathbb{Q}(C[n])$	field of definition of $C[n]$ over $\mathbb{Q}$, 193
$F(C[n])$	field of definition of $C[n]$ over F, 193
$\text{GL}_r(R)$	general linear group, 195
ρ_n	Galois representation on points of order n, 195
t	cyclotomic representation, 196
i	$= \sqrt{-1}$, 197
ψ_n	division polynomial, 214
ϕ_n	multiplication-by-n polynomial, 214
ω_n	multiplication-by-n polynomial, 214
R_L	endomorphism ring of the lattice L, 217
$[a, b, c]$	homogeneous triple, a point in $\mathbb{P}^2$, 221
$\mathbb{P}^2$	the projective plane, 221
$\sim$	equivalence relation on triples $[a, b, c]$, 221
$\mathbb{P}^n$	projective n-space, 221
$[a_0, \ldots, a_n]$	homogeneous $n + 1$-tuple, a point in $\mathbb{P}^n$, 221
$\sim$	equivalence relation on $n + 1$-tuples, 221
$\mathbb{A}^2$	Euclidean (or affine) plane, 222
L_∞	line at infinity in $\mathbb{P}^2$, 223
$C(\mathbb{Q})$	set of rational points on a projective curve C, 229
$C_0(\mathbb{Q})$	set of rational points on an affine curve C_0, 230
$C_0(\mathbb{Z})$	set of integer points on an affine curve C_0, 230
$I(C_1 \cap C_2, P)$	intersection index of C_1 and C_2 at P, 237

Index

Undergraduate Texts in Mathematics

(continued from page ii)

Gamelin: Complex Analysis.

Gordon: Discrete Probability.

Hairer/Wanner: Analysis by Its History. *Readings in Mathematics.*

Halmos: Finite-Dimensional Vector Spaces. Second edition.

Halmos: Naive Set Theory.

Hämmerlin/Hoffmann: Numerical Mathematics. *Readings in Mathematics.*

Harris/Hirst/Mossinghoff: Combinatorics and Graph Theory.

Hartshorne: Geometry: Euclid and Beyond.

Hijab: Introduction to Calculus and Classical Analysis.

Hilton/Holton/Pedersen: Mathematical Reflections: In a Room with Many Mirrors.

Hilton/Holton/Pedersen: Mathematical Vistas: From a Room with Many Windows.

Iooss/Joseph: Elementary Stability and Bifurcation Theory. Second edition.

Isaac: The Pleasures of Probability. *Readings in Mathematics.*

James: Topological and Uniform Spaces.

Jänich: Linear Algebra.

Jänich: Topology.

Jänich: Vector Analysis.

Kemeny/Snell: Finite Markov Chains.

Kinsey: Topology of Surfaces.

Klambauer: Aspects of Calculus.

Lang: A First Course in Calculus. Fifth edition.

Lang: Calculus of Several Variables. Third edition.

Lang: Introduction to Linear Algebra. Second edition.

Lang: Linear Algebra. Third edition.

Lang: Short Calculus: The Original Edition of "A First Course in Calculus."

Lang: Undergraduate Algebra. Second edition.

Lang: Undergraduate Analysis.

Laubenbacher/Pengelley: Mathematical Expeditions.

Lax/Burstein/Lax: Calculus with Applications and Computing. Volume 1.

LeCuyer: College Mathematics with APL.

Lidl/Pilz: Applied Abstract Algebra. Second edition.

Logan: Applied Partial Differential Equations.

Lovász/Pelikán/Vesztergombi: Discrete Mathematics.

Macki-Strauss: Introduction to Optimal Control Theory.

Malitz: Introduction to Mathematical Logic.

Marsden/Weinstein: Calculus I, II, III. Second edition.

Martin: Counting: The Art of Enumerative Combinatorics.

Martin: The Foundations of Geometry and the Non-Euclidean Plane.

Martin: Geometric Constructions.

Martin: Transformation Geometry: An Introduction to Symmetry.

Millman/Parker: Geometry: A Metric Approach with Models. Second edition.

Moschovakis: Notes on Set Theory.

Owen: A First Course in the Mathematical Foundations of Thermodynamics.

Palka: An Introduction to Complex Function Theory.

Pedrick: A First Course in Analysis.

Peressini/Sullivan/Uhl: The Mathematics of Nonlinear Programming.

Prenowitz/Jantosciak: Join Geometries.

Priestley: Calculus: A Liberal Art. Second edition.